METHODS IN MOLECULAR BIOLOGY

Series Editor
John M. Walker
School of Life and Medical Sciences
University of Hertfordshire
Hatfield, Hertfordshire, AL10 9AB, UK

For further volumes:
http://www.springer.com/series/7651

Root Development

Methods and Protocols

Edited by

Daniela Ristova

Gregor Mendel Institute of Molecular Plant Biology
Vienna, Wien, Austria

Elke Barbez

Institute of Applied Genetics & Cell Biology
University of Natural Resources and Life Sciences
Vienna, Wien, Austria

Humana Press

Editors
Daniela Ristova
Gregor Mendel Institute of Molecular Plant Biology
Vienna, Wien, Austria

Elke Barbez
Institute of Applied Genetics & Cell Biology
University of Natural Resources and Life Sciences
Vienna, Wien, Austria

ISSN 1064-3745 ISSN 1940-6029 (electronic)
Methods in Molecular Biology
ISBN 978-1-4939-7746-8 ISBN 978-1-4939-7747-5 (eBook)
https://doi.org/10.1007/978-1-4939-7747-5

Library of Congress Control Number: 2018934877

Printed on acid-free paper

This Humana Press imprint is published by the registered company Springer Science+Business Media, LLC part of Springer Nature
The registered company address is: 233 Spring Street, New York, NY 10013, U.S.A.

Preface

The root system, the part of plants hidden in the soil, represents an essential organ for plant growth and productivity. Root development exhibits a high level of developmental plasticity, continuously translating environmental information into developmental decisions. Environmental changes have become an increased challenge for crop production. Therefore, deciphering root developmental mechanisms in detail will be crucial for understanding how plants cope with environmental changes and how to translate the fundamental knowledge into applications.

In this book, we aim to provide the scientific community with a selection of current protocols and experimental approaches relevant to root development, written by researchers in their specific expertise. This book covers methods from genetic screens and phenotypic analysis, cell biology methods, to systems biology tools and genome-wide approaches. The complexity of the protocols ranges from fundamental methods for quantification of different root developmental processes to complex methods that require sophisticated equipment. The first eight chapters are dedicated to genetic and phenotypic analyses, opening with phenotyping of diverse root traits under controlled and field conditions, as well as genetic/phenotypic screenings of different developmental processes including root meristem, lateral roots, root hairs, adventitious roots, and gravitropism. The second part of the book is dedicated to a variety of microscopy and cell biology techniques used in root biology, comprised with eight chapters: from analysis of calcium dynamics, immunolocalization, and histological profiling to long-term *in vivo* imaging and immunoprecipitation. In the third section, we aim to cover various systems biology tools and genome-wide approaches. Here we include six chapters from transcriptomics, ChIP-seq, micro-ChIP-seq, and proteomics to network analysis and genome-wide association studies.

This book provides diverse elegant methods, complemented with existing protocols, which are optimized for the current needs in plant root biology as well as for use in plant species other than *Arabidopsis thaliana*. The authors in this book guide the reader through their presented methods by providing step-by-step instructions as well as numerous tips and tricks that they learned from their experiences.

Root Development: Methods and Protocols is a useful tool for plant biologists specialized in root development, from beginners to experienced researchers in the field. We wish our readers an inspiring journey through the methods and expertise of their colleagues.

Vienna, Wien, Austria

Daniela Ristova
Elke Barbez

Contents

Contributors

PAOLO BAESSO • *Department of Life Sciences, Imperial College London, London, UK*
KHURRAM BASHIR • *Plant Genomic Network Research Team, RIKEN Center for Sustainable Resource Sciences, Yokohama, Kanagawa, Japan; CREST, JST, Saitama, Japan*
TOM BEECKMAN • *Department of Plant Biotechnology and Bioinformatics, Ghent University, Ghent, Belgium; VIB Center for Plant Systems Biology, Ghent, Belgium*
MALCOLM J. BENNETT • *Centre for Plant Integrative Biology, School of Biosciences, University of Nottingham, Loughborough, UK*
SANDRA BENSMIHEN • *Laboratoire des Interactions Plantes Micro-Organismes (LIPM), Université de Toulouse, INRA, CNRS, Castanet-Tolosan, France*
VARODOM CHAROENSAWAN • *The Sainsbury Laboratory, University of Cambridge, Cambridge, UK; Department of Biochemistry, Faculty of Science, Mahidol University, Bangkok, Thailand; Integrative Computational BioScience (ICBS) Center, Mahidol University, Nakhon Pathom, Thailand; Systems Biology of Diseases Research Unit, Faculty of Science, Mahidol University, Nakhon Pathom, Thailand*
YI-CHIA CHOU • *Institute of Plant and Microbial Biology, Academia Sinica, Taipei, Taiwan*
ORLANDO CONTRERAS-LÓPEZ • *Departamento de Genética Molecular y Microbiología, FONDAP Center for Genome Regulation, Millennium Institute for Integrative Systems and Synthetic Biology (MIISSB), Pontificia Universidad Católica de Chile, Santiago, Chile*
SANDRA CORTIJO • *The Sainsbury Laboratory, University of Cambridge, Cambridge, UK*
GLORIA M. CORUZZI • *NYU Center for Genomics and Systems Biology, New York, NY, USA*
ALEX COSTA • *Department of Biosciences, University of Milan, Milan, Italy*
LUCINDA DE-ARAUJO • *Department of Applied Genetics and Cell Biology (DAGZ), University of Natural Resources and Life Sciences (BOKU), Vienna, Austria*
BENJAMIN M. DELORY • *Ecosystem Functioning and Services, Institute of Ecology, Leuphana University, Lüneburg, Germany*
JOSEPH G. DUBROVSKY • *Departamento de Biología Molecular de Plantas, Instituto de Biotecnología, Universidad Nacional Autónoma de México (UNAM), Cuernavaca, Morelos, México*
RICHARD VAN DUIJNEN • *Ecosystem Functioning and Services, Institute of Ecology, Leuphana University, Lüneburg, Germany*
ELENA FERARU • *Department of Applied Genetics and Cell Biology (DAGZ), University of Natural Resources and Life Sciences (BOKU), Vienna, Austria*
ANDY FRENCH • *Centre for Plant Integrative Biology, School of Biosciences, University of Nottingham, Loughborough, UK*
JIŘÍ FRIML • *Institute of Science and Technology Austria (IST Austria), Klosterneuburg, Austria*
DANNY GEELEN • *In Vitro Biology and Horticulture, Department of Plant Production, Faculty of Bioscience Engineering, University of Ghent, Ghent, Belgium*
KRIS GEVAERT • *VIB-UGent Center for Medical Biotechnology, VIB, Ghent, Belgium; Department of Biochemistry, Ghent University, Ghent, Belgium*
MIRIAM L. GIFFORD • *School of Life Sciences, The University of Warwick, Coventry, UK*

Rodrigo A. Gutiérrez • *Departamento de Genética Molecular y Microbiología, FONDAP Center for Genome Regulation, Millennium Institute for Integrative Systems and Synthetic Biology (MIISSB), Pontificia Universidad Católica de Chile, Santiago, Chile*
Ian Hands-Portman • *School of Life Sciences, The University of Warwick, Coventry, UK*
Violaine Herrbach • *Laboratoire des Interactions Plantes Micro-Organismes (LIPM), Université de Toulouse, INRA, CNRS, Castanet-Tolosan, France*
Ellie Himschoot • *Department of Plant Biotechnology and Bioinformatics, Ghent University, Ghent, Belgium; VIB Center for Plant Systems Biology, Ghent, Belgium*
Rana Muhammad Fraz Hussain • *School of Life Sciences, The University of Warwick, Coventry, UK*
Kei Iida • *Graduate School of Medicine, Kyoto University, Kyoto, Japan*
Wann-Neng Jane • *Institute of Plant and Microbial Biology, Academia Sinica, Taipei, Taiwan*
Michael Karampelias • *Department of Molecular Genetics, Centre for Research in Agricultural Genomics (CRAG), CSIC-IRTA-UAB-UB, Barcelona, Spain*
Eunice Han Xian Khor • *Institute of Plant and Microbial Biology, Academia Sinica, Taipei, Taiwan*
Jürgen Kleine-Vehn • *Department of Applied Genetics and Cell Biology (DAGZ), University of Natural Resources and Life Sciences (BOKU), Vienna, Austria*
Barbara Korbei • *Department of Applied Genetics and Cell Biology (DAGZ), University of Natural Resources and Life Sciences (BOKU), Vienna, Austria*
Melanie Krebs • *Centre for Organismal Studies, Plant Developmental Biology, University of Heidelberg, Heidelberg, Germany*
Beatriz Lagunas • *School of Life Sciences, The University of Warwick, Coventry, UK*
Ying Li • *Department of Horticulture and Landscape Architecture, Purdue University, West Lafayette, IN, USA; Purdue Center for Plant Biology, Purdue University, West Lafayette, IN, USA*
Fabienne Maillet • *Laboratoire des Interactions Plantes Micro-Organismes (LIPM), Université de Toulouse, INRA, CNRS, Castanet-Tolosan, France*
Riccardo Di Mambro • *Department of Biology, University of Pisa, Pisa, Italy*
Akihiro Matsui • *Plant Genomic Network Research Team, RIKEN Center for Sustainable Resource Sciences, Yokohama, Kanagawa, Japan*
Girish Mokkapati • *Institute of Plant and Microbial Biology, Academia Sinica, Taipei, Taiwan*
Tomás C. Moyano • *Departamento de Genética Molecular y Microbiología, FONDAP Center for Genome Regulation, Millennium Institute for Integrative Systems and Synthetic Biology (MIISSB), Pontificia Universidad Católica de Chile, Santiago, Chile*
Lukas Muller • *Centre for Plant Integrative Biology, School of Biosciences, University of Nottingham, Loughborough, UK*
Selene Napsucialy-Mendivil • *Departamento de Biología Molecular de Plantas, Instituto de Biotecnología, Universidad Nacional Autónoma de México (UNAM), Cuernavaca, Morelos, México*
Natalia Nikonorova • *Department of Plant Biotechnology and Bioinformatics, Ghent University, Ghent, Belgium; VIB Center for Plant Systems Biology, Ghent, Belgium*
Davy Opdenacker • *Department of Plant Biotechnology and Bioinformatics, Ghent University, Ghent, Belgium; VIB Center for Plant Systems Biology, Ghent, Belgium*
Loïc Pagès • *INRA, Centre PACA, UR 1115 PSH, Avignon Cedex 9, France*

ALESSIA PARA • *Weinberg College of Arts and Sciences, Northwestern University, Evanston, IL, USA*
RICARDO S. RANDALL • *Department of Life Sciences, Imperial College London, London, UK*
SULTANA RASHEED • *Plant Genomic Network Research Team, RIKEN Center for Sustainable Resource Sciences, Yokohama, Kanagawa, Japan; Kihara Institute for Biological Research, Yokohama City University, Yokohama, Japan*
FRANÇOIS ROUDIER • *Laboratoire de Reproduction et Développement des Plantes – ENS Lyon, Lyon Cedex 07, France*
SABRINA SABATINI • *Department of Biology and Biotechnology "C. Darwin", Sapienza University of Rome, Rome, Italy*
JORGE E. SALAZAR-HENAO • *Institute of Plant and Microbial Biology, Academia Sinica, Taipei, Taiwan*
ELIZABETH SARKEL • *Department of Applied Genetics and Cell Biology (DAGZ), University of Natural Resources and Life Sciences (BOKU), Vienna, Austria*
MARIA SCHÖLLER • *Department of Applied Genetics and Cell Biology (DAGZ), University of Natural Resources and Life Sciences (BOKU), Vienna, Austria*
WOLFGANG SCHMIDT • *Institute of Plant and Microbial Biology, Academia Sinica, Taipei, Taiwan*
MOTOAKI SEKI • *Plant Genomic Network Research Team, RIKEN Center for Sustainable Resource Sciences, Yokohama, Kanagawa, Japan; CREST, JST, Saitama, Japan; Kihara Institute for Biological Research, Yokohama City University, Yokohama, Japan*
GIOVANNI SENA • *Department of Life Sciences, Imperial College London, London, UK*
ÜMIT SEREN • *Gregor Mendel Institute (GMI), Austrian Academy of Sciences, Vienna BioCenter (VBC), Vienna, Austria*
IVE DE SMET • *Department of Plant Biotechnology and Bioinformatics, Ghent University, Ghent, Belgium; VIB Center for Plant Systems Biology, Ghent, Belgium*
DANIELA C. SOTO • *Departamento de Genética Molecular y Microbiología, FONDAP Center for Genome Regulation, Millennium Institute for Integrative Systems and Synthetic Biology (MIISSB), Pontificia Universidad Católica de Chile, Santiago, Chile*
ELISABETH STES • *Department of Plant Biotechnology and Bioinformatics, Ghent University, Ghent, Belgium; VIB Center for Plant Systems Biology, Ghent, Belgium; VIB-UGent Center for Medical Biotechnology, VIB, Ghent, Belgium; Department of Biochemistry, Ghent University, Ghent, Belgium; VIB Headquarters, Ghent, Belgium*
RANJAN SWARUP • *Centre for Plant Integrative Biology, School of Biosciences, University of Nottingham, Loughborough, UK*
MAHO TANAKA • *Plant Genomic Network Research Team, RIKEN Center for Sustainable Resource Sciences, Yokohama, Kanagawa, Japan*
RICARDO TEJOS • *Facultad de Recursos Naturales Renovables, Universidad Arturo Prat, Iquique, Chile*
VICKY M. TEMPERTON • *Ecosystem Functioning and Services, Institute of Ecology, Leuphana University, Lüneburg, Germany*
HOANG KHAI TRINH • *In Vitro Biology and Horticulture, Department of Plant Production, Faculty of Bioscience Engineering, University of Ghent, Ghent, Belgium*
STEFFEN VANNESTE • *Department of Plant Biotechnology and Bioinformatics, Ghent University, Ghent, Belgium; VIB Center for Plant Systems Biology, Ghent, Belgium*
INGE VERSTRAETEN • *Institute of Science and Technology Austria (IST Austria), Klosterneuburg, Austria*

LAM DAI VU • *Department of Plant Biotechnology and Bioinformatics, Ghent University, Ghent, Belgium; VIB Center for Plant Systems Biology, Ghent, Belgium; VIB-UGent Center for Medical Biotechnology, VIB, Ghent, Belgium; Department of Biochemistry, Ghent University, Ghent, Belgium*
SASCHA WAIDMANN • *Department of Applied Genetics and Cell Biology (DAGZ), University of Natural Resources and Life Sciences (BOKU), Vienna, Austria*
LIAM WALKER • *School of Life Sciences, The University of Warwick, Coventry, UK*
EMANUELA W. A. WEIDLICH • *Ecosystem Functioning and Services, Institute of Ecology, Leuphana University, Lüneburg, Germany*
DARREN M. WELLS • *Centre for Plant Integrative Biology, School of Biosciences, University of Nottingham, Loughborough, UK*
PHILIP A. WIGGE • *The Sainsbury Laboratory, University of Cambridge, Cambridge, UK*
KATE WOOLLEY-ALLEN • *Department of Computer Science, The University of Warwick, Coventry, UK*
WEI XUAN • *State Key Laboratory of Crop Genetics and Germplasm Enhancement and MOA Key Laboratory of Plant Nutrition and Fertilization in Lower-Middle Reaches of the Yangtze River, Nanjing Agricultural University, Nanjing, People's Republic of China*
LARRY M. YORK • *Noble Research Institute, Ardmore, OK, USA*

Part I

Genetic and Phenotypic Analyses

Chapter 1

Measuring Plant Root Traits Under Controlled and Field Conditions: Step-by-Step Procedures

Benjamin M. Delory, Emanuela W. A. Weidlich, Richard van Duijnen, Loïc Pagès, and Vicky M. Temperton

Abstract

In this chapter, we present methods that we routinely use to measure plant root traits in the field and under controlled environmental conditions (using rhizoboxes). We describe procedures to (1) collect, wash, and store root samples, (2) acquire images of washed root samples, and (3) measure root traits using image analysis. In addition, we also describe sampling methods for studying belowground productivity, soil exploration, and root distribution in the first soil layers at the community level (soil coring and ingrowth core method). Because the use of rhizoboxes allows a nondestructive and dynamic measurement of traits hardly accessible in the field, a section of this chapter is devoted to the acquisition and analysis of images of roots growing in rhizoboxes.

Key words Root traits, Phenotyping, Rhizobox, Soil coring, Ingrowth core method, Root washing, Root staining, Image analysis

1 Introduction

In plant ecology, trait-based approaches are a powerful way to investigate how population and community changes impact ecosystem processes [1]. Among all the possible traits that can be measured on individual plants, those focusing on roots are of major importance. Belowground, the spatial configuration of root systems [2], the morphological features of individual roots [3], their capacity to acquire water and nutrients and release exudates into the rhizosphere [4], as well as their ability to interact with soil organisms [5] are important drivers of ecosystem processes such as plant productivity, carbon storage, nutrient cycling, and soil formation/stabilization [1, 6, 7]. A summary of traits commonly used in functional root research is provided in Fig. 1.

In comparison with traits measured on aboveground plant parts, root traits are particularly difficult to measure, mainly because of the belowground location of roots. In addition, the impossibility

Daniela Ristova and Elke Barbez (eds.), *Root Development: Methods and Protocols*, Methods in Molecular Biology, vol. 1761, https://doi.org/10.1007/978-1-4939-7747-5_1, © Springer Science+Business Media, LLC 2018

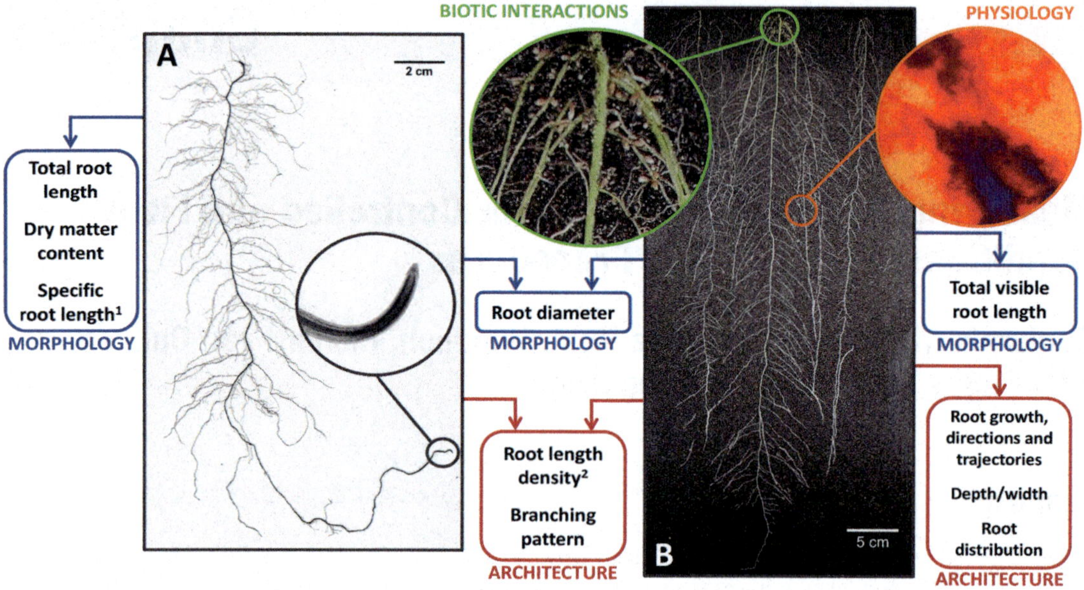

Fig. 1 Categories of root traits commonly measured on washed root samples (**a**) or on roots growing in a rhizobox (**b**). Biotic traits are illustrated by a picture of a *Trifolium pratense* root system showing a large amount of N_2-fixing nodules. Physiological traits are illustrated by a picture showing pH variations in the rhizosphere of tomato roots using planar optodes. [1]The specific root length is calculated as the ratio of total root length to root dry weight. [2]The root length density is calculated as the ratio of total root length to soil volume

to use simple root morphological markers to separate roots of different species in plant communities growing in the field or in pots makes the task even more challenging [1]. When plant scientists and ecologists wish to measure functional root traits in the field, they often extract soil cores and bring them to the laboratory to isolate the roots for further analyses [8]. Under controlled experimental conditions, however, a variety of root traits can be nondestructively measured using rhizoboxes [9–11]. Because of their transparent front window, rhizoboxes allow one to study root growth and physiology over time. Although rhizoboxes have a finite size constraining root development and hence duration of experiments as well as limiting the analysis of root system architectures in two spatial dimensions, their use is common in functional root research and allows for the nondestructive and dynamic measurement of biotic and physiological root traits that are not easily accessible in the field [12, 13]. An example might be the use of planar optodes for the quantitative 2D imaging of rhizosphere pH, O_2, and CO_2 dynamics [14, 15].

In this chapter, we present methods allowing the measurement of root traits in the field and under controlled environmental conditions (using rhizoboxes). In addition, we also describe sampling methods for studying belowground productivity, soil exploration, and root distribution in the first soil layers at the community level. Although minirhizotrons are interesting methods to provide root length density estimates and fine root dynamics information in situ (e.g., root turnover) [16], their use is not described in this chapter. When plant root traits have to be measured, we describe methods that we routinely use in our laboratory to (1) collect, wash, and store root samples, (2) acquire images of washed root samples, and (3) measure root traits using image analysis. Methods describing how images of roots growing in a rhizobox can be acquired and analyzed are also included in this chapter.

2 Materials

2.1 Collecting Roots from Individual Plants

1. A garden fork or a spade.
2. Containers for transporting and washing the monoliths collected in the field (e.g., buckets).
3. A canister filled with tap water.
4. Plastic trays.
5. A scalpel or sharp dissecting scissors.
6. Containers to store the roots belonging to the target plants (e.g., plastic bottles).
7. Permanent markers (ink pens).

2.2 Collecting Roots from Soil Cores

1. Sharpened soil corers/augers.
2. Hammers.
3. A sharp knife and a ruler (for dividing the core into several layers).
4. Plastic bags.
5. Plastic labels.
6. Permanent markers (ink pens).
7. Cold boxes.

2.3 Collecting Roots with the Ingrowth Core Method

1. Opaque plastic ground sheet + material to hold the ground sheets in place in the field.
2. Shovels and spades.
3. Containers to store the soil collected from the field.
4. Sieves (mesh size: 2–5 mm).
5. Soil augers/corers (diameter: 5 cm).
6. Hammers.
7. Plastic trays.
8. A calibrated precision scale.

9. PVC tubes (diameter: 4–5 cm, length: 45 cm) + caps.
10. Mesh bags (polyamide fiber, mesh size: 1 mm, diameter: 4 cm, length: 45 cm).
11. A tool to compress the soil inside the mesh bags (e.g., a wooden stick).
12. Plastic labels (different colors).
13. Pliers.
14. Knives or sharp scissors.
15. Plastic bags (to store the ingrowth cores).
16. Permanent markers (ink pens).
17. Cold boxes.

2.4 Root Washing

1. A sink equipped with a soil collector.
2. A nozzle linked to a tap with a hose.
3. Nontransparent plastic trays.
4. Sieves (mesh sizes: 1–2 mm and 200–500 μm).
5. Tweezers.
6. A sharp knife and a template (for dividing the soil inside a rhizobox into several layers). A template could be a plastic-coated sheet of paper cut at the dimension of the soil layer.
7. A scalpel or sharp dissecting scissors.
8. Clean containers to store the roots after washing (e.g., plastic bottles).
9. Transparent plastic trays.
10. A colored background to put below the transparent plastic tray.
11. A soft paintbrush.
12. A desk light.
13. A Petri dish.
14. A plastic or a glass funnel.
15. Some filters (*see* **Note 1**).
16. A glass bottle used to recover the water after filtration.
17. Permanent markers (ink pens).

2.5 Root Staining

1. Some glass beakers for staining the root samples (one per sample).
2. Permanent markers (ink pens).
3. Staining solution: 1.7 mM neutral red (3-amino-7-dimethylamino-2-methylphenazine hydrochloride). Add about 200 mL of distilled water to a 1 L volumetric flask or a 1 L graduated cylinder. Weigh 0.5 g of neutral red and transfer it to the flask/

cylinder. Add distilled water to a volume of 1 L. Mix the solution at room temperature. Transfer the solution to a glass bottle. Seal the bottle with a cap. Store at 4 °C.

4. Plastic trays.
5. A plastic or a glass funnel.
6. Some filters (*see* **Note 1**).
7. A glass bottle used to recover the neutral red solution after filtration (waste bottle).
8. Tweezers.
9. Containers to store stained root samples (e.g., plastic bottles).

2.6 Acquiring Images of Washed Root Samples

1. An optical flatbed scanner equipped with a dual lighting system (e.g., Epson Perfection V800 Photo).
2. A plastic tray.
3. A scalpel or sharp dissecting scissors.
4. A waterproof transparent plastic tray (*see* **Note 2**).
5. Plastic tweezers.
6. Transparent plastic sheets cut at the dimension of the transparent plastic trays.
7. A syringe without the needle.
8. A plastic or a glass funnel.
9. Some filters (*see* **Note 1**).
10. A glass bottle.

2.7 Measuring Root Dry Weight

1. A drying oven.
2. A desiccator.
3. A calibrated precision scale.
4. Tweezers.

2.8 Acquiring Images of Roots Growing in a Rhizobox

If direct images of roots are taken with a scanner or a digital camera, only item 1 is needed. If roots are drawn on transparent acetate sheets, only items 2–4 are needed.

1. An image acquisition system relying on a scanner or a digital camera (*see* **Note 3**).
2. Transparent acetate sheets of the same size as the transparent front window of the rhizoboxes (1 sheet per rhizobox).
3. Ink pens of different colors.
4. A common flatbed scanner (reflective lighting system).

3 Methods

In this section, we present methods for measuring morphological, architectural, and biotic root traits on plants growing in the field or under controlled environmental conditions (using rhizoboxes). The main steps of these different protocols are presented in Fig. 2. In addition, we also describe sampling methods allowing the analysis of root productivity and root distribution at the community level (e.g., soil coring, ingrowth core method).

Please make sure to follow all waste disposal regulations when disposing waste materials.

3.1 Collecting Roots from Individual Plants (Soil Monolith Method)

1. In the field, select the plant on which root traits are to be measured.
2. With a garden fork or a spade, demarcate a soil monolith around the target plant. The size of the monolith will mainly depend on the plant species studied (age, size, root distribution pattern, etc.). Typically, soil monoliths can be 30–40 cm large and 30–50 cm deep [3].
3. Dig out the soil monolith and put it in a container.
4. Fill the container containing the soil monolith with tap water.
5. Wash the monolith until the roots are visible.

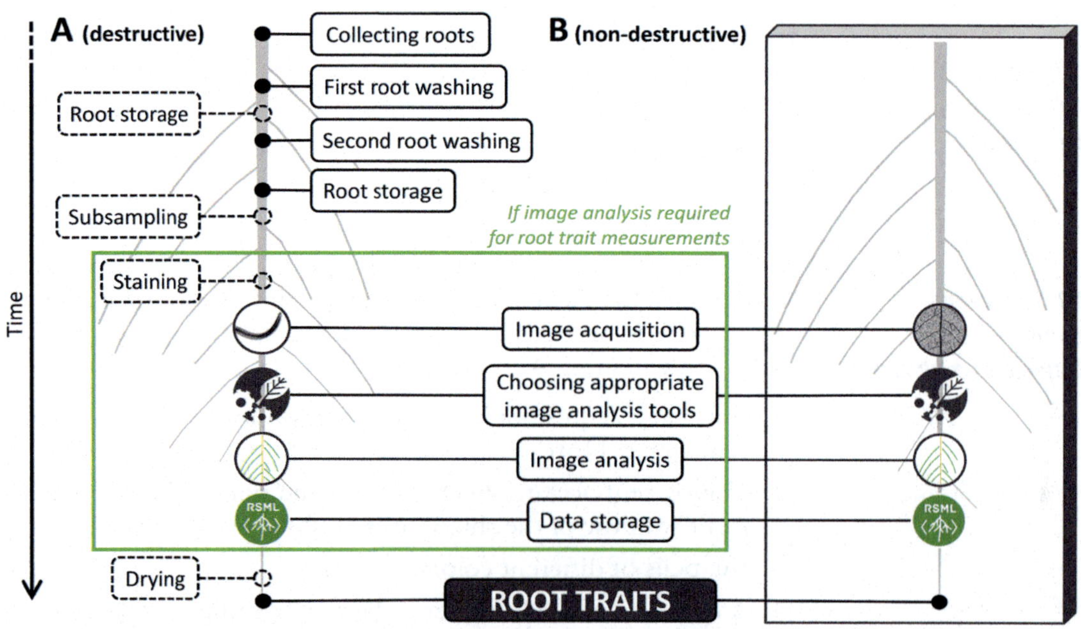

Fig. 2 Step-by-step procedures for measuring root traits on washed root samples (**a**) or on roots growing in a rhizobox (**b**). Boxes surrounded by a dashed line denote facultative steps (depending on the root traits that one wishes to measure). Logos of the Plant Image Analysis Database and the Root System Markup Language (RSML) were kindly provided by Guillaume Lobet

6. Trace the roots back to the shoot of the target plant. Remove the roots that do not belong to the target plant.
7. Put the roots of the target plant in a tray filled with tap water.
8. With a scalpel or sharp dissecting scissors, separate the roots from the shoot.
9. Store the roots inside a container labelled with the name of the sample and/or its identification number (*see* **Note 4**). Unwashed roots or roots that are not perfectly washed can be stored under humid and chilly conditions (e.g., 4 °C) for up to 3 days. For longer time periods, the samples should be stored in a freezer (−20 °C) (*see* **Note 5**).

3.2 Collecting Roots from Soil Cores

Soil coring is a technique commonly used in the field to study belowground productivity, root morphology, root distribution, and soil exploration by roots in the first soil layers at a given time point (*see* **Note 6**). With this technique, roots of several plant species are collected as fragments no longer connected to their parent. Because roots are heterogeneously distributed in the soil, a large number of samples are often required [16].

1. With a sharpened soil auger/corer, take a soil core at a spot where you wish to sample roots. Depending on your research question, you might want to collect roots only from the first soil layers or deeper (>0.5 m) into the soil. If deep and/or large cores are to be taken, more sophisticated mechanical devices will be required (e.g., soil corers mounted on vehicles) [16].
2. If you are interested in the vertical distribution of roots in the soil, divide your entire core into several depth layers (use a sharp knife and a ruler).
3. Store the samples in labelled plastic bags. If you divided the cores into several layers, store them in different plastic bags. Add an additional label annotated with the sample name and/or its identification number inside each bag (*see* **Notes 4** and 7).
4. In the field, store the samples in a cool box. In the lab, unwashed roots or roots that are not perfectly washed can be stored under humid and chilly conditions (e.g., 4 °C) for up to 3 days. For longer time periods, the samples should be stored in a freezer (−20 °C) (*see* **Note 5**).

3.3 Collecting Roots with the Ingrowth Core Method

In a field experiment, the ingrowth core (IGC) method allows the measurement of fine root productivity over time. This is done by installing the so-called ingrowth cores (i.e., mesh bags filled with sieved soil, IGCs) into pre-drilled holes. Roots of a plant community are allowed to grow into the IGCs for a given time period.

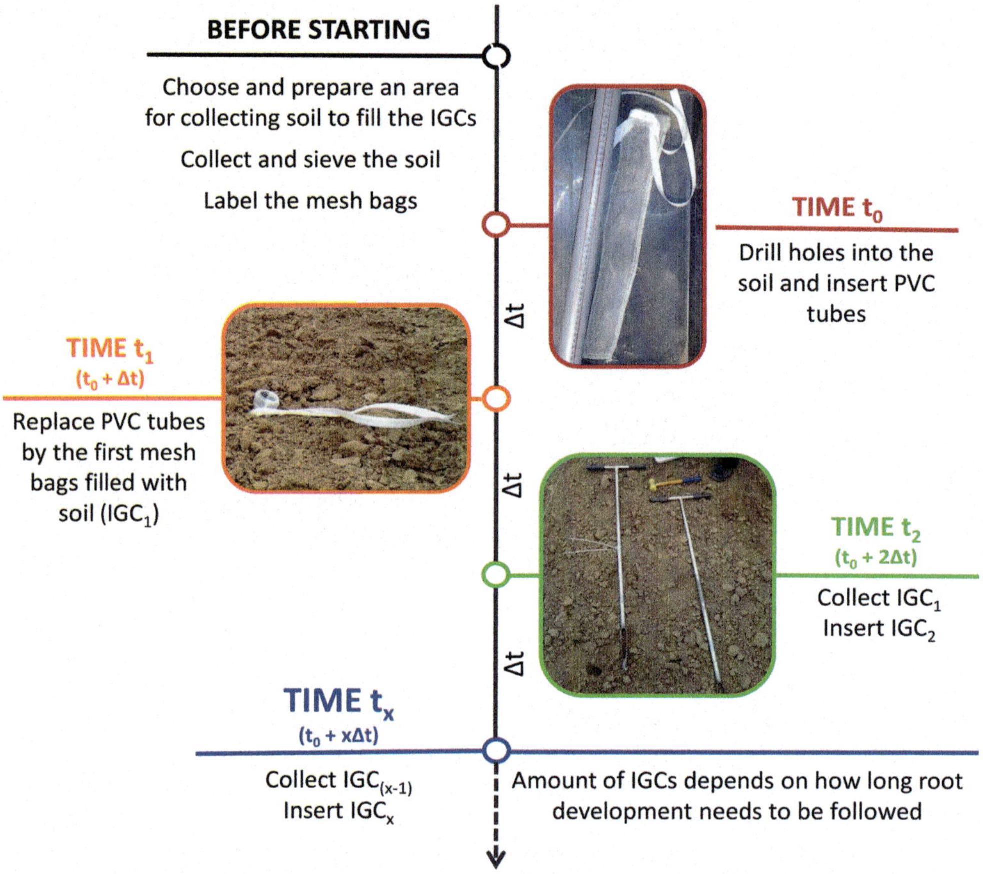

Fig. 3 Time line showing the main steps of the ingrowth core method used for measuring new root growth in the field. *See* Subheading 3.3 of this chapter (Collecting Roots with the Ingrowth Core Method)

Then, the IGCs are collected and fine root productivity is measured [17, 18]. In systems characterized by fast root proliferation (e.g., managed grasslands), the ingrowth time period is often short (e.g., 2 weeks), and the root decay inside the IGCs is considered negligible [19]. Therefore, the total amount of root material inside one IGC is considered to represent the total fine root productivity at a given location, for a given time period, and a fixed core volume [17]. At the plot level, fine root productivity can be followed continuously by activating new IGCs located in another part of the field plot at the end of each time period. A time line presenting the main steps of the IGC method is shown in Fig. 3. Hereafter, we present a protocol using an ingrowth time period of 2 weeks.

1. Choose an area adjacent to the main experiment with homogenous soil properties. Cover the soil surface of the chosen area with an opaque plastic sheet in order to suppress plant growth and obtain soil material free from living roots. After several months, collect soil from the topsoil layer (0–30 cm deep).

Store the soil in a container for transportation. Sieve the soil in order to remove any organic material (*see* **Note 8**).

2. Determine the dry bulk density of the soil at the location of your experiment. To do so, collect a number of soil cores from the study area. The volume of each core must be known. Bring the cores to the lab. Spread the soil samples in plastic trays, and let them dry at room temperature until constant mass is reached. Weigh the dry soil samples and determine the dry soil bulk density (ratio between the dry mass of a core and its volume).
3. Locate the spots where fine root productivity is to be tracked at the location of your experiment. Depending on the site of the area and the period of time you want to assess, you will need to sample a certain number of IGCs per plot. The exact locations need deciding before standing coring. Using a hand auger, drill holes into the ground at an angle of 45° to the soil surface. The holes should be deep enough to allow the insertion of the IGCs. Typically, 45 cm long IGCs allow the exploration of the first 30 cm of the soil. The number of drilled holes must be equal to the total number of IGCs that will be installed over the entire time frame of your experiment. Protect each hole by inserting one PVC tube. Cover the top of the PVC tubes with a plastic cap (*see* **Note 9**).
4. Calculate the amount of soil needed to fill one IGC (*see* **Note 10**). This value can be easily calculated by multiplying the volume of one IGC by the dry soil bulk density measured previously (point 2 of this subsection).
5. After 2 weeks, replace PVC tubes by IGCs at the locations where fine root productivity is to be measured for the first time period (2 weeks). To do so, take the PVC tube out of the soil, and insert it in an empty mesh bag. Put the mesh bag with the PVC tube inside back into the ground. Use the PVC tube as a funnel to fill the mesh bags with the appropriate amount of sieved soil (see **steps 1** and **4** of this subsection). During this process, gradually remove the PVC tube from the bag, so that only the mesh bag remains in the ground. A wooden stick can be used to press and compact the soil inside the mesh bag. Make sure that the soil compaction is similar between IGCs. When the mesh bag is filled with the required amount of sieved soil, close it with a knot and add a colored label on it (we advise you to use a different color for each time point) (*see* **Note 11**).
6. After 2 weeks, pull the IGCs out of the soil. Pliers might be useful at this step. The IGCs should be taken out carefully in order to avoid losing roots. If necessary, use a knife or sharp scissors to cut the roots growing outside the IGCs. Store the

IGCs in plastic bags, and keep them inside cold boxes until they are brought back to the lab. Be careful when labelling the samples collected from the field. For the IGCs, write the sampling name and/or its identification number in the bag's strap, as well as in an additional label connected to the IGC. Write the name of the sample on the plastic bag itself (*see* **Note 7**). Close the hole that remains in the soil after collecting an IGC.

7. In each plot, activate a new IGC as soon as one is removed from the soil. To do so, repeat **steps 5** and **6** of this subsection (*see* **Note 12**).
8. In the lab, unwashed roots or roots that are not perfectly washed can be stored under humid and chilly conditions (e.g., 4 °C) for up to 3 days. For longer time periods, the samples should be stored in a freezer (−20 °C). Mesh bags containing ingrowth cores can be stored in a freezer (−20 °C) as well (*see* **Note 5**).

3.4 Root Washing

The washing of roots is organized in two steps. The objective of the first washing is to remove most of the mineral particles and organic material that do not adhere to the roots (**steps 1–8** of this subsection). The operations described for the first root washing are best performed over a sink equipped with a soil collector underneath. For very sandy soils, this first washing is often enough to get clean roots. In most cases, a second washing will be required in order to separate the roots from small debris and other contaminants (**steps 9–16** of this subsection).

1. Fill a clean tray with a thin layer of tap water (this tray will be used to store all the roots collected during the washing of one sample). Hereafter, we will refer to this tray as "tray 1." If roots are from ingrowth cores, start with **step 2**, and skip **steps 3** and **4** of this subsection. For washing entire root systems growing in rhizoboxes, go to **step 3**, and skip **step 2** of this subsection. If roots are from soil cores taken from the field or from soil layers taken from a rhizobox, go to **step 4**, and skip **steps 2** and **3** of this subsection.
2. If soil samples are from mesh bags, start by opening the bags at the top. If roots are stuck in the mesh or in the lace of a bag, take these roots first, and store them in the tray 1. When a mesh bag is open, start washing the sample above a 1–2 mm sieve using running water. Place a tray or a collecting plate below the sieve, and regularly check for the presence of roots that might not have been retained by the sieve. Wash the ingrowth core progressively, from the top to the bottom of the core. Using tweezers, remove all visible roots from the sieve/tray/collecting plate, and place them in tray 1. Fine roots can be easily removed from the sieve by immersing the mesh in a thin layer of water (fine roots will float).

3. If roots were growing in a rhizobox, start by carefully removing the front window. If the soil does not have to be divided into layers to study the vertical root distribution, incline the rhizobox at a 30–40° angle to the horizontal, hold the plants at the top, and start washing the roots with running water. Move progressively from the bottom of the rhizobox to the top. When most of the soil is gone, place the roots in tray 1. If you wish to study the vertical root distribution, divide the soil into several layers soon after opening the rhizobox (use a sharp knife and a template). Then, go to **step 4** of this subsection, and follow the instructions for washing the different layers.
4. If roots and soil are intimately mixed but are not contained in a mesh bag or a rhizobox, directly place one sample inside a 1–2 mm sieve. Put a tray or a collecting plate below the sieve, and start washing the roots using running water. Manually remove big mineral and/or organic particles from the sieve. Wash the roots progressively, and regularly check for the presence of roots that might not have been retained by the sieve. Using tweezers, remove all visible roots from the sieve/tray/collecting plate, and place them in tray 1. Fine roots can be easily removed from the sieve by immersing the mesh in a thin layer of water (fine roots will float).
5. Filter the content of tray 1 through a 200–500 μm sieve. Rinse tray 1 with tap water and filter the suspension.
6. Place the sieve containing the roots in a tray, and incline the sieve at a 45° angle to the horizontal.
7. Under running water, finger-massage the roots, and remove the rest of the mineral and organic soil particles that do not strongly adhere to the roots.
8. Using tweezers, remove the roots from the sieve, and store them in an appropriate container. At this stage, roots can be stored under humid and chilly conditions (e.g., 4 °C) for up to 3 days. For longer time periods, the samples should be stored in a freezer (−20 °C) (*see* **Note 5**).
9. In order to remove debris and contaminants that strongly adhere to the roots, a second washing is often necessary. Start by putting one of your samples in a tray filled with a thin layer of distilled water (*see* **Note 13**). Hereafter, we will refer to this tray as "tray 2."
10. Deposit a transparent tray on your working table, and place a colored background behind it. Fine roots can be easily detected using a blue background. Fill the transparent tray with a thin layer of distilled water.
11. Transfer the whole or a portion of your sample into the transparent tray.

12. Using tweezers, manually separate the roots from the soil particles. A soft paintbrush might be helpful to remove small mineral particles adhering to the roots. Use a desk lamp to cast light on your sample. Store clean roots in a Petri dish filled with distilled water. Before starting to clean a new sample, replace the dirty water in the transparent tray by clean distilled water.
13. When you think that the totality of your sample is clean, check that you did not forget any fine roots by pouring the content of tray 2 into the transparent tray.
14. Recover the roots from the Petri dish using tweezers or by filtering the content of the Petri dish using a funnel and appropriate filters.
15. Store clean root samples inside appropriate containers. Write the name and/or the identification number of the samples on the containers.
16. Washed roots can be stored for short time periods at 4 °C in a 50% (v/v) ethanol solution. Roots can be stored in a freezer (−20 °C) for long time periods (*see* **Note 5**). Sometimes, roots have to be flash-frozen in liquid nitrogen and stored at −80 °C for further biochemical analyses.

3.5 Root Staining

Staining the roots with a dye is an efficient way to increase the contrast between roots and the background during scanning. Staining the roots is particularly useful if samples contain a large proportion of fine roots. However, this operation must be avoided if one wishes to do biochemical analyses on roots after scanning [20, 21].

1. Place the whole or a subsample of each clean root sample inside a glass beaker. Write the name and/or the identification number of the sample on its corresponding beaker.
2. Fill the beakers with the staining solution. Add an abundant volume of staining solution in order to fully immerse the roots. Store the beakers on a tray at room temperature (*see* **Note 14**).
3. After 24 h (*see* **Note 15**), recover the stained roots by filtration using a funnel and a filter. Collect the staining solution in a bottle.
4. Rinse the roots with an abundant volume of distilled water in order to remove excess dye.
5. Recover the roots from the filter using tweezers.
6. If you do not plan to acquire root images soon after staining, store the stained root samples in an appropriate and annotated container at 4 °C (short-term storage) or in a freezer (−20 °C, long-term storage) (*see* **Note 5**).

3.6 Acquiring Images of Washed Root Samples

1. Create a method for root image acquisition. Ideally, root images should be acquired as 8 bits per pixel gray level images using the transparency unit (TPU) of the optical scanner. Setting a resolution value appropriate for your sample is of great importance to perform accurate measurements using image analysis (*see* **Note 16**). If you do not know the root diameter distribution in your sample, setting the resolution to 400 dpi is a good starting point.
2. Fill the nontransparent plastic tray with a thin layer of distilled water, and spread your root sample in the water. The roots do not have to be well spread at this stage.
3. If structural information are not of interest, use a scalpel or sharp dissecting scissors to cut the roots into small segments (1–3 cm). It will reduce overlapping between roots, and it will make the spreading of the roots on the scanning area easier.
4. When acquiring images of washed root samples, two situations can be encountered depending on whether the spreading of roots inside the transparent tray and image acquisition are performed at the same location (*situation 1*) or at two different locations (*situation 2*). For *situation 1*, both the spreading of roots inside the transparent tray and image acquisition are performed on the scanner glass. When *situation 2* is encountered, it is important to find a way to keep the roots in place when the tray is moved to the scanning area. *Situation 2* is often met when several people work together to prepare the samples and acquire root images. For *situation 1*, start with **step 5** and skip **step 6** of this subsection. For *situation 2*, start with **step 6** and skip **step 5** of this subsection.
5. *Situation 1*. Position the transparent tray on the scanner glass, and fill it with a thin layer of distilled water. Depending on the size of your sample, transfer the whole or a portion of the roots (subsample) inside the transparent tray. Several images can be acquired for large root samples. Evenly spread the roots on the scanning area using plastic tweezers. Try to avoid overlapping between roots by not exceeding a root length density of 1 cm cm^{-2} on the scanning area.
6. *Situation 2*. Position the transparent tray on a table, and fill it with a thin layer of distilled water. Depending on the size of your sample, transfer the whole or a portion of the roots (subsample) into the transparent tray. Several images can be acquired for large root samples. Evenly spread the roots using plastic tweezers. The use of a colored background is very helpful for improving visibility of roots. Try to avoid overlapping between roots by not exceeding a root length density of 1 cm cm^{-2} on the scanning area. Once the roots are evenly spread, deposit a transparent plastic sheet on top of the water

layer, and remove the excess liquid from the tray using a syringe. Take the transparent tray and position it on the scanner glass.

7. Acquire an image (e.g., tiff, jpeg) and save it on your computer.
8. Remove the roots from the tray. Roots can be removed from the tray using plastic tweezers or by filtration using a funnel and appropriate filters. Recover the water in a glass bottle. Put the roots back in their container for storage. Store the roots in a freezer (−20 °C).
9. Proceed with the next sample (*see* **Note 17**).

3.7 Measuring Root Dry Weight

1. Dry the roots in an oven (60 °C) until constant mass is reached (e.g., 48 h, a longer time period will be required for thick and/or lignified roots).
2. Let the samples cool down in a desiccator at room temperature.
3. Weigh the samples using a calibrated precision scale.

3.8 Acquiring Images of Roots Growing in a Rhizobox

When roots are observed through the transparent front window of a rhizobox, two main strategies can be used to acquire root images and follow the development of roots over time.

1. The first strategy is to acquire direct images of roots with a scanner or a digital camera. This option is interesting if you do not want to lose information (root system architecture, root distribution, root diameters, presence/absence of nodules, presence/absence of root hairs, etc.). With a well-designed image acquisition system, root images can be quickly and easily acquired (<30 s/rhizobox). A description of our image acquisition system is provided in **Note 3**.
2. A second strategy is to manually draw the roots on a transparent acetate sheet using ink pens. A specific color can be used for each observation date, root type, or root order. This option is particularly interesting if you are only interested in following the root system architecture of plants over time. Be aware that some traits (e.g., root diameters) cannot be accurately measured with this technique. At the end of the experiment, the acetate sheets can be scanned using a common flatbed scanner. The resulting images can then be analyzed using image analysis software tools.

3.9 Root Image Analysis

1. Follow the method described in **Note 18** to select one or several root image analysis software solutions that meet your needs. Read **Note 19** if you are interested in counting the number of nodules along the roots of legumes using digital images of roots growing in a rhizobox.

2. Analyze your images following the instructions provided by the developers of the software tools (*see* **Note 20**).
3. Store and analyze your data (*see* **Note 21**).

4 Notes

1. Clean tea bags work fine for recovering the roots by filtration.
2. Transparent and waterproof plastic trays are easy to build yourself. For an A4 optical scanner glass, a 20 × 25 × 2 cm (width × depth × height) tray could be made out of transparent acrylic glass sheets (PMMA, polymethyl metacrylate). First, cut one 2–3 mm thick sheet into five pieces: one 20 × 25 cm piece for the bottom, two 2 × 25 cm pieces for the long edges, and two 2 × (20–2 × thickness) cm pieces for the short edges of the tray. Second, glue the pieces together using a PMMA glue (colorless, polymerization adhesive, UV/light curing) following manufacturer's instructions. Check if the tray is waterproof before using it on the scanner. PMMA trays can easily be scratched, so care must be taken during their manipulation.
3. Our image acquisition system is composed of a metallic frame holding the rhizobox vertically, a digital camera equipped with a 28 mm lens and connected to a computer, a camera holder, and two LED tubes (4300 K, 60 cm length). The camera, the camera holder, the LED tubes, and the frame holding the rhizobox are installed inside a closed box whose internal walls are entirely covered with dark sheets. The camera directly faces the rhizotron and is supported by a camera holder located 54 cm away from the transparent front window. The left and right sides of the transparent front window are illuminated by LED tubes positioned laterally (raking lighting). This setting provides a uniform lighting all over the height of a rhizobox and allows us to avoid reflection on the transparent front window during image acquisition. Images are acquired using the remote live view and shooting option of our camera before being saved on our computer (both jpeg and uncompressed raw images). We advise the reading of [22] for a presentation of inexpensive methods to acquire images of roots growing along a transparent front window.
4. In order to avoid mistakes when analyzing samples, it is sometimes easier to code sample names using identification numbers (or a combination of numbers/letters). This strategy might be particularly useful if you work with people that are not familiar with the names you give to your samples, or if your sample names are long and/or similar to each other. A file storing the sample names and their corresponding identification number

must be created and safely archived. In such file, we also advise you to write the level of each experimental factor for each sample. When data are analyzed using statistical software tools (e.g., R statistical software), the identification numbers of the samples can then be automatically replaced by the name of the samples and/or the factor levels using the information stored in the file created previously.

5. Because freezing compromises root cell integrity, it might influence root diameters. Therefore, root diameter measurements should be performed on fresh root samples (i.e., soon after their collection).
6. When working with soil cores coming from species-rich plant communities, studying root biomass distribution at the species level is one of biggest challenges because we cannot rely on simple root morphological markers to sort the roots of different species. To overcome this drawback, a quantitative DNA-based technique relying on the amplification of species-specific markers has been developed to calculate the relative species abundances in mixed root samples (see [23] for more information).
7. Always label samples twice and use permanent markers to identify the cores and the plastic bags used in the field.
8. In order to make the sieving easier, we recommend to air-dry the soil material at room temperature.
9. Instead of using PVC caps, the tubes can be closed using a plastic sheet hold in place with rubber bands.
10. Once you know the amount of soil you need to fill one mesh bag, we advise you to go to the field with plastic bags containing the exact amount of sieved soil needed for each mesh bag. A beaker marked with a line showing the desired volume of soil required for each mesh bag can also be used in the field.
11. Marking each activated IGC with a colored tape/strap makes their localization easier in the field, particularly when IGCs are surrounded by a dense vegetation.
12. If the goal is to follow fine root productivity for 8 weeks, each experimental plot should contain four IGCs. Every 2 weeks, one IGC is collected and a new one is activated. Only one IGC at a time is activated inside each plot. Therefore, the number of IGCs used depends on how long root development needs to be followed.
13. Adding sodium chloride (NaCl) and/or liquid soap to the water often helps separating the roots from the soil [3]. This technique must be avoided if other analyses have to be performed on roots (e.g., biochemical analyses).

14. An orbital shaker can be used to agitate the roots during staining.
15. See [20] for a discussion about the effect of the staining time period on root length measurements with WinRHIZO™.
16. Because objects appear as a set of square dots (pixels) on digital images, the resolution (usually expressed in dots per inch) is an important factor affecting the accuracy of length and diameter measurements based on image analysis [20]. Because the diameter of a single root varies from one species to another, as well as between different root types and root orders within individuals, care must be taken when one has to set the resolution of the images acquired with an optical scanner. For root diameter measurements, the minimum resolution (Res_{min}, dpi) can be estimated with Eq. 1, where D_{min} and P_{min} are the diameter of the finest root (μm) and the required minimum number of pixels transversally crossing the finest root on a given image, respectively. For accurate root diameter measurements, we suggest to have $P_{min} \geq 8$.

$$Res_{min} = 25400 \frac{P_{min}}{D_{min}} \tag{1}$$

17. If images of stained roots are acquired, we advise you to regularly replace the water layer inside the transparent tray by clean distilled water because the dye might leak out of the roots. This is particularly true when the root cell integrity is compromised by long-term storage in a freezer (−20 °C).
18. Depending on the research questions you want to address, one or a combination of root image analysis software packages will be necessary to extract the relevant biological structures from the captured images and compute static and dynamic features to describe your root systems locally (at the individual root level) and/or globally (at the whole root system scale). In order to choose the best software package(s) to analyze your images, you first have to identify your needs. To do so, we suggest you to ask yourself the following questions: [1] What traits/variables would you like to measure? [2] Do you have to deal with time series images? [3] What is the operating system of your computer (Windows, Mac, Linux)? [4] What automation level do you need (fully automated, semi-automated, manual)? [5] What are the properties of your images (format, size, resolution, color/grayscale, presence/absence of soil around the roots, etc.)? [6] Do you need a free software package or can you afford a commercial one? [7] With which file format would you like to save and store your data? [8] How portable should the data be? Once you have answered all these questions, we

advise you to use the Plant Image Analysis Database (http://www.plant-image-analysis.org/) to identify all the root image analysis software packages meeting your needs [24]. Finally, try the tools you selected with your own images, and find out the best one(s) to perform your analyses.

19. Because of the soil surrounding the roots, digital images of roots growing in a rhizobox are not easy to analyze. If a poor contrast exists between roots and background, images might have to be analyzed manually. In order to count the number of nodules visible through the transparent front window of a rhizobox, we analyze root images using Fiji [25]. Nodules are counted manually using the multipoint and ROI manager tools.
20. If you use WinRHIZO™ to analyze your images, we recommend the use of a global thresholding method to detect the roots. The threshold value can be chosen manually or automatically. In order to avoid missing roots, we advise you to adjust the threshold interactively for some parts of the image by using the "interactive modifications to gray level pixel classification" command. More information about the accuracy of the methods proposed in WinRHIZO™ can be found in [21].
21. The large diversity of software solutions available to analyze root images can be seen as an opportunity to use complementary tools to analyze a set of images. Unfortunately, it can be difficult to compare measurements acquired with different platforms because [1] each platform has its own representation of a given root system architecture and [2] the exported data do not share a common structure. To overcome these problems, Lobet and co-workers recently introduced the Root System Markup Language (RSML) as a convenient way to store, share, and increase the transferability of root architecture data [26]. At the time of writing, the RSML has been implemented in seven image analysis tools: SmartRoot [27], RootNav [28], RhizoScan [29], Root System Analyser [30], RooTrak [31], EZ-Rhizo [32], and GLO-RIA [33]. In addition, open-source packages have been developed in several languages (R, Excel, Python, Java, and C#) in order to analyze data saved with the RSML format [26]. Within the R statistical environment, the functions of the archiDART package allow the batch processing of RSML files containing 2D or 3D RSA data [34]. These functions are able to compute global RSA traits, root growth rates, root growth directions and trajectories, and lateral root distribution. It also contains plotting functions developed to visualize the dynamics of root system growth. This R package also allows the batch processing of files produced with Data Analysis of Root Tracings [35].

Acknowledgments

We would like to thank Dr. Thomas Niemeyer (Leuphana University, Germany) for its excellent technical support. We are also thankful to Dr. Guillaume Lobet for providing some images used in this book chapter. This research was funded by CNPq Brazil (Ciência Sem Fronteiras Program) with the PhD scholarship of Emanuela W.A. Weidlich, BMBF in Germany for the INPLAMINT PhD project of Richard van Duijnen within the BonaRes soils programme, as well as by the Chair of Ecosystem Functioning and Services, Leuphana University, Lüneburg, Germany.

References

1. Bardgett RD, Mommer L, De Vries FT (2014) Going underground: root traits as drivers of ecosystem processes. Trends Ecol Evol 29:692–699
2. Lynch JP (2013) Steep, cheap and deep: an ideotype to optimize water and N acquisition by maize root systems. Ann Bot 112:347–357
3. Pagès L (2016) Branching patterns of root systems: comparison of monocotyledonous and dicotyledonous species. Ann Bot 118:1337–1346
4. Delory BM, Delaplace P, Fauconnier M-L et al (2016) Root-emitted volatile organic compounds: can they mediate belowground plant-plant interactions? Plant Soil 402:1–26
5. Mommer L, Kirkegaard J, van Ruijven J (2016) Root–root interactions: towards a rhizosphere framework. Trends Plant Sci 21:209–217
6. Lynch J (1995) Root architecture and plant productivity. Plant Physiol 109:7–13
7. Faucon M-P, Houben D, Lambers H (2017) Plant functional traits: soil and ecosystem services. Trends Plant Sci 22(5):385–394
8. Pérez-Harguindeguy N, Díaz S, Garnier E et al (2013) New handbook for standardized measurement of plant functional traits worldwide. Aust J Bot 61:167–234
9. Nagel KA, Putz A, Gilmer F et al (2012) GROWSCREEN-Rhizo is a novel phenotyping robot enabling simultaneous measurements of root and shoot growth for plants grown in soil-filled rhizotrons. Funct Plant Biol 39:891–904
10. Faget M, Nagel KA, Walter A et al (2013) Root-root interactions: extending our perspective to be more inclusive of the range of theories in ecology and agriculture using in-vivo analyses. Ann Bot 112:253–266
11. Lobet G, Draye X (2013) Novel scanning procedure enabling the vectorization of entire rhizotron-grown root systems. Plant Methods 9:1–10
12. Oburger E, Schmidt H (2016) New methods to unravel rhizosphere processes. Trends Plant Sci 21:243–255
13. Faget M, Blossfeld S, von Gillhaussen P et al (2013) Disentangling who is who during rhizosphere acidification in root interactions: combining fluorescence with optode techniques. Front Plant Sci 4:1–8
14. Blossfeld S (2013) Light for the dark side of plant life: planar optodes visualizing rhizosphere processes. Plant Soil 369:29–32
15. Blossfeld S, Schreiber CM, Liebsch G et al (2013) Quantitative imaging of rhizosphere pH and CO_2 dynamics with planar optodes. Ann Bot 112:267–276
16. Maeght J, Rewald B, Pierret A et al (2013) How to study deep roots - and why it matters. Front Plant Sci 4:299
17. Steingrobe B, Schmid H, Claassen N (2000) The use of the ingrowth core method for measuring root production of arable crops - influence of soil conditions inside the ingrowth core on root growth. J Plant Nutr Soil Sci 163:617–622
18. Steingrobe B, Schmid H, Gutser R et al (2001) Root production and root mortality of winter wheat grown on sandy and loamy soils in different farming systems. Biol Fertil Soils 33:331–339
19. Chen S, Lin S, Reinsch T et al (2015) Comparison of ingrowth core and sequential soil core methods for estimating belowground net primary production in grass-clover swards. Grass Forage Sci 71:515–528
20. Bouma TJ, Nielsen KL, Koutstaal B (2000) Sample preparation and scanning protocol for computerised analysis of root length and diameter. Plant Soil 218:185–196

21. Delory BM, Weidlich EWA, Meder L et al (2017) Accuracy and bias of methods used for root length measurements in functional root research. Methods Ecol Evol 8 (11):1594–1606
22. Mohamed A, Monnier Y, Mao Z et al (2017) An evaluation of inexpensive methods for root image acquisition when using rhizotrons. Plant Methods 13:11
23. Mommer L, Wagemaker CAM, De Kroon H et al (2008) Unravelling below-ground plant distributions: a real-time polymerase chain reaction method for quantifying species proportions in mixed root samples. Mol Ecol Resour 8:947–953
24. Lobet G, Draye X, Périlleux C (2013) An online database for plant image analysis software tools. Plant Methods 9:1–7
25. Schindelin J, Arganda-carreras I, Frise E et al (2012) Fiji - an open source platform for biological image analysis. Nat Methods 9:676–682
26. Lobet G, Pound MP, Diener J et al (2015) Root system markup language: toward an unified root architecture description language. Plant Physiol 167:617–627
27. Lobet G, Pagès L, Draye X (2011) A novel image-analysis toolbox enabling quantitative analysis of root system architecture. Plant Physiol 157:29–39
28. Pound MP, French AP, Atkinson JA et al (2013) RootNav: navigating images of complex root architectures. Plant Physiol 162:1802–1814
29. Diener J, Nacry P, Périn C et al (2013) An automated image-processing pipeline for high-throughput analysis of root architecture in OpenAlea. In: 7th International conference on functional-structural plant models, Saariselkä, Finland, pp. 85–87
30. Leitner D, Felderer B, Vontobel P et al (2014) Recovering root system traits using image analysis exemplified by two-dimensional neutron radiography images of lupine. Plant Physiol 164:24–35
31. Mairhofer S, Zappala S, Tracy SR et al (2012) RooTrak: automated recovery of three-dimensional plant root architecture in soil from X-ray microcomputed tomography images using visual tracking. Plant Physiol 158:561–569
32. Armengaud P, Zambaux K, Hills A et al (2009) EZ-Rhizo: integrated software for the fast and accurate measurement of root system architecture. Plant J 57:945–956
33. Rellán-Álvarez R, Lobet G, Lindner H et al (2015) GLO-roots: an imaging platform enabling multidimensional characterization of soil-grown root systems. elife 4:1–26
34. Delory BM, Baudson C, Brostaux Y et al (2016) archiDART: an R package for the automated computation of plant root architectural traits. Plant Soil 398:351–365
35. Le Bot J, Serra V, Fabre J et al (2010) DART: a software to analyse root system architecture and development from captured images. Plant Soil 326:261–273

Chapter 2

Phenotyping Crop Root Crowns: General Guidance and Specific Protocols for Maize, Wheat, and Soybean

Larry M. York

Abstract

Root crown phenotyping allows measuring many parameters that describe the top portion of crop root systems and can be used for the basis of genome-wide association studies, QTL analysis, and physiology studies. Root crown phenotyping is a relatively simple process, with the major steps being excavation, transport, soil removal, and measurements. On the other hand, all steps require substantial manual labor and occur outside in challenging environments, and many factors influence the ultimate success of applying this methodology to research questions. Here, general guidance is given for how to optimize root crown phenotyping for unique research questions, and specific protocols are given for acquiring images of root crowns of three crop species: maize, wheat, and soybean.

Key words Corn, Crop, Phenomics, Nutrient use efficiency, Root system architecture, Soya, Trait, Water use efficiency

1 Introduction

1.1 Background

Roots are the interface of the plant with soil, and therefore understanding root system architecture, or the three dimensional configuration of roots over time [1], has become a major research focus in the new millennium. At the same time, plant phenotyping, or measuring physical properties of many plants, has been enabled by technological advances in sensors, automation, and data analysis. Root system phenotyping is used to link phenes, or fundamental units of the phenome [2], to genetic loci [3, 4] and to link phene to function [5–7]. Root phenotyping can be categorized based on sampling method, sensing method, plant growth stage, and throughput. Over the past few years, root crown phenotyping has become popularly known as "shovelomics," which was popularized in [8].

Root crown phenotyping in the field can be accomplished in many ways, and the specific methodology depends on the focal species, field environment, available resources, and research goals.

Daniela Ristova and Elke Barbez (eds.), *Root Development: Methods and Protocols*, Methods in Molecular Biology, vol. 1761, https://doi.org/10.1007/978-1-4939-7747-5_2, © Springer Science+Business Media, LLC 2018

Therefore, providing a single method has limited applicability. Instead, special considerations and a general overview of field root crown phenotyping will be presented, so that researchers may tailor and optimize their own phenotyping platform. Following this overview, example protocols are included from the author's research on maize, wheat, and soybean.

1.2 Imaging Platform

The types of platforms being used for root crown imaging can be divided into two categories: horizontal or vertical root crown methods. In horizontal root crown methods, the root crown is placed on a flat surface, usually matte black, and the camera is suspended from above facing down toward the root crown [3, 5, 9–12]. In vertical root crown methods, the root crown is attached from below or above such that the stem axis is vertical, and the camera is placed horizontally, and the background is not touching the root crown [13–16]. Horizontal methods have the benefit of being simpler and possibly faster for imaging, as no time is used affixing the root crown, and suspending root crowns vertically may cause some roots to bend. However, affixing the root crown vertically with the background not touching the crown allows the root crown to be more in focus than the background, which can aid ease of segmentation of the root crown. While a simple tripod with a head capable of being placed upside down over a root crown suffices, imaging boxes or tents with controlled lighting are the most promising method for automated image analysis.

1.3 Desired Throughput

Phenotyping exists along a continuum from extensive to intensive due to an inherent tradeoff between throughput and detailed measurements [17]. While imaging and image analysis help to alleviate this tradeoff, decisions still must be made to balance the number of samples that can be processed versus the number of measurements that can be taken for each sample. These decisions are important, because greater throughput with less detailed measurements can lead to missing important features with biological implications. At the same time, statistical power for genetic or functional correlations is lost when throughput is sacrificed for intensive measurements. For example, intensive phenotyping of maize root crowns that examined root phenes in every whorl of nodal roots indicated that the oldest and middle-aged nodal roots have more relations with shoot mass and nitrogen content [5]. However, the outer nodal roots are the youngest and what are typically measured in high-throughput protocols of maize root crown phenotyping [8, 10], so biological relevance is potentially being lost. Therefore, decisions about the intensity and extensity of measurements should be carefully made based on prior knowledge. In the protocols below, intensive phenotyping is described as an extra group of steps for each species. In all cases, increasing the number of workers will increase throughput.

2 Materials

2.1 Excavation

1. Round point (digging) shovel.
2. Wheelbarrow.
3. Slip-on horticultural tags (maize).
4. Plastic bags (100 micron thickness, 25 cm wide, 50 cm long, wheat, and soybean).

2.2 Washing

1. Two hoses with jet nozzles.
2. Plastic bins (large enough to hold water and several root crowns).
3. Dish soap (optional, add for clayey soils).
4. Set up a soak area with bins next to a wash area with appropriate soil disposal, possibly into the field.

2.3 Imaging

1. PVC or aluminum cubic frame 50 cm in all dimensions covered in black cloth except front.
2. Digital camera suspended from top of frame facing down.
3. Black background (*see* **Note 1**).
4. Markers for scale, round filter paper (*see* **Note 2**).
5. LED ring light (optional, creates more consistent images for automated analysis).
6. Pruning shears (for intensive maize method). Remote shutter for camera.

3 Methods

3.1 The Basics of Root Crown Phenotyping

Root crown phenotyping is a relatively simple process, with the major steps being:

1. Excavation—removing a root crown from soil.
2. Transport—moving excavated root crowns to washing or measurement station.
3. Soil removal—shaking a root crown, soaking in water, and rinsing.
4. Measurements
   (a) Manual—using a “scoreboard” or other measurement tools.
   (b) Imaging—using a camera.
   (c) Image analysis—using existing software or develop new software.

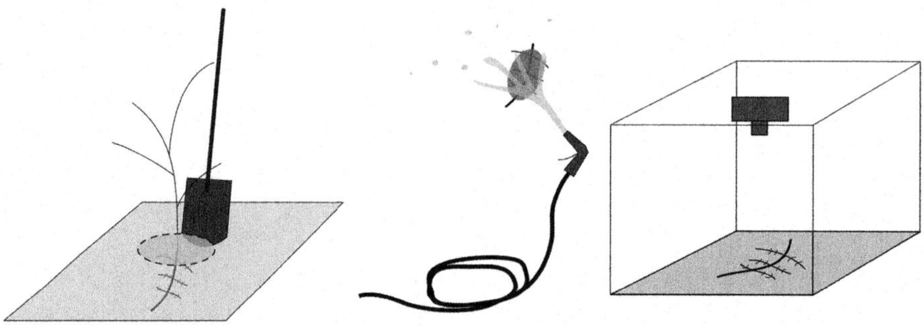

Fig. 1 Root crown phenotyping is divided into three main steps: excavation, washing, and imaging. Transporting of root crowns after excavating to be washed is not shown. Excavation is generally done manually with a shovel, then root crowns are soaked and then sprayed by a water hose, and finally root crowns are imaged against a background for further image analysis (Used with permission: York, Larry (2017): Root crown phenotyping (AKA shovelomics) cartoon. figshare. https:/doi.org/10.6084/m9.figshare.5149699.v3 [18])

A circle is made around the focal plant stem by placing the shovel blade straight down, fully into the soil at a distance from the stem so that no roots are cut (Fig. 1). At the last insertion, the root crown may be pried up by using the shovel as a lever, which may need repeated on two sides to remove the crown. When excavating, the goal is to remove the crown such that deeper roots are lost but the horizontal spread of the root is retained. Manually excavating with a shovel is tedious, and other approaches include the use of a potato digger or a tree spade.

Transporting soil-laden root crowns from the field can be a significant challenge. The larger the root crown, the more soil it will hold and the heavier it will be. At the same time, the identity of the root crown must be maintained. This can be done by using a tag around the stem of the plant or by placing the root crown in a labeled plastic bag. Use of a wheelbarrow or cart will aid in the removal of root crowns from the field.

Soil removal is a substantial yet little-considered bottleneck in root crown phenotyping. The amount of soil embedded within a root crown is a function of the overall extent of the crown, the density of roots (length $\times$ soil volume^{-1}), and the soil texture. Larger root crowns and greater root length density both contribute to more soil being contained in the crown. In sandier soils, soil may easily fall from the root crown after excavation, especially when dry. On the other hand, clayey soil tightly binds roots and is not easily removed. Soaking root crowns for at least 1 h in water greatly contributes to the loosening of the soil. Additionally, addition of soap aids the dispersion of clay particles. Following soaking, a water hose and nozzle are used to spray away soil as gently as possible in order to not break finer roots.

Measurement of the cleaned root crown represents the end point for the field component of root crown phenotyping. Manual measurements can be scores, counts, or measurements using rulers, calipers, and protractors. Common measurements include angles and diameters of various root classes, numbers of axial roots, lateral root density emerging from axial roots, and lateral root length. A printable scoreboard is available from the Lynch and Brown labs at Penn State (http://goo.gl/kAVXj) as a guide for collecting many of these measurements from maize root crowns and easily extendable to other crops.

Imaging the root crowns for subsequent image analysis at a later time can increase throughput in the field while also decreasing labor needs by decoupling root crown excavation and washing from measurements. Images that maximize the contrast between roots and background are essential for successful automated image analysis. Imaging solutions can range from cheap to more elaborate and expensive. Placing a root crown on a black, matte background with a camera suspended above is sufficient for nonautomated image analysis using *ObjectJ* in *ImageJ* to lead users through the measurements [5, 12] (Fig. 2). Nonautomated image analysis

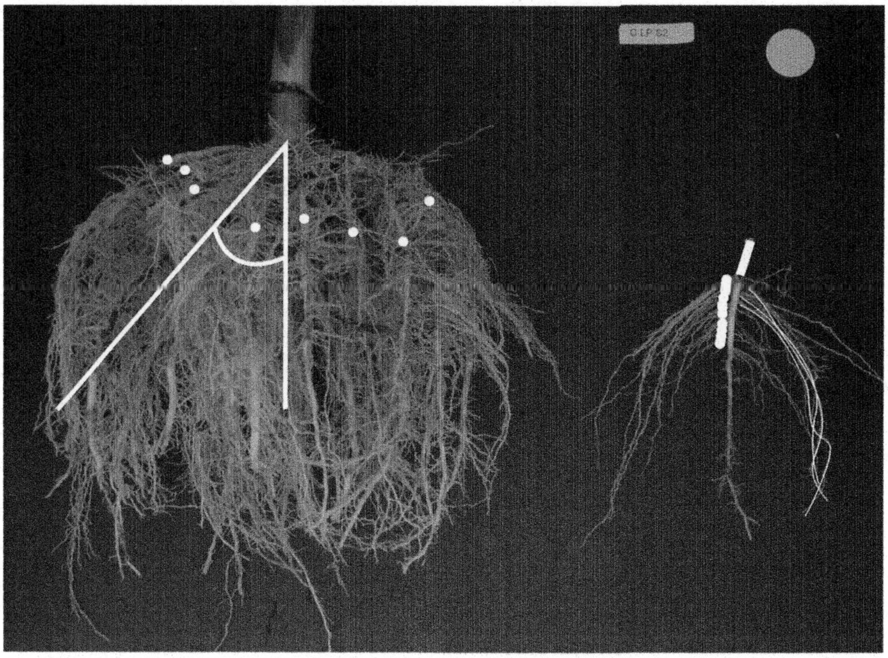

Fig. 2 A maize root crown from an intensive phenotyping platform [5] overlaid with several manual measurements taken using the *ObjectJ* plugin for *ImageJ*, including from the main root crown: nodal root number and nodal root angle. From the excised nodal root, the following measurements are made: distance to branching, nodal root diameter, lateral root branching density, and lateral root length. These are intuitive descriptors with biological significance, while features extracted using automated image analysis tend to be less intuitive and mathematical in nature. However, all measurements are useful and often correlated

measurements can include all the measurements listed for manual measurements above but may take substantial amounts of time to complete. However, while automated measurements require less labor, they may be harder to interpret with relation to biological significance.

Automated image analysis of root crowns was first described by TE Grift et al. [14] and first used for high-throughput phenotyping of crop root crowns in [10] and then T Colombi et al. [16]. Digital imaging of root traits (DIRT) is designed to work for both monocot and dicot crop root systems and is available as a web platform through CyVerse (http://dirt.iplantcollaborative.org). Currently, major root crown image analysis platforms take a single perspective of a root crown and flatten that image to a black and white image through the process of segmentation. During this step, a great deal of information is lost, especially with more complicated root systems where many roots occlude underlying roots. Segmentation allows many general purpose image analysis methods to be used, along with the generation of novel, root-specific measurements. To increase the success rate of segmentation, controlled illumination is needed for the imaging platform to create consistent backgrounds that contrast with the root crown. Basic measurements allowed by automated image analysis include the projected area, convex hull, number and size distribution of holes in image, average number of roots in each line of image, and greatest root crown width and height. Most of these relate to the size and complexity of the root crown, yet information about the classes of roots, such as axial or lateral, are lacking. Therefore, manual measurements may be described as biological, while automated features are more mathematical in nature. Future techniques will undoubtedly attempt to unravel the 3D structure of the root crown and identify more intuitive local measurements based on the types of roots in the crown.

3.2 Maize

3.2.1 Excavation

1. Tag focal plant with labeled slip-on tag at base of shoot.
2. Insert shovel fully a blade width away on four sides of focal plant; pry out.
3. Place entire plant with soil in wheel barrow; fill with plants to capacity.
4. Transport samples to washing station.

3.2.2 Washing

1. Place samples in bin with water to soak; have a surplus number soaking at all times.
2. Remove shoot(s) with lopper and oven-dry for shoot mass or discard.
3. Move crowns back and forth vigorously in water to remove as much soil as possible.

4. Spray crown with as low-pressure jet as possible to remove remaining soil.
5. If needed, return to bin for soaking before returning to **step 3**, or go to next step.
6. Image directly or store submerged in water in cold room for later imaging within 5 days.

3.2.3 Imaging

1. Select a root crown without excess water, yet don't let roots dry out (*see* **Note 3**).
2. Place a scale (3 cm circle or otherwise) and a plant tag on the background (*see* **Note 4**).
3. Excise a representative nodal root from the current node.
4. Place root crown on left of background and excised root on right.
5. Use remote shutter to take photograph.

3.2.4 Imaging (Intensive)

6. Cut away all nodal roots in previously imaged node using pruning shears (*see* **Note 5**).
7. Return to Imaging **step 3** above.
8. Repeat until all nodes of roots have been imaged.

3.3 Wheat

3.3.1 Excavation

1. Insert shovel a half blade width away along a row on both sides of focal plants as to dig up a line of four to six plants.
2. Place entire group of plants with soil into labeled plastic bag.
3. Place bag of plants and soil into wheel barrow; fill with plants to capacity.
4. Transport samples to washing station.

3.3.2 Washing

1. Place samples in bin with water to soak; have a surplus number soaking at all times.
2. Move crowns back and forth vigorously in water to remove as much soil as possible.
3. Select one or more representative root crowns from the group to image.
4. Spray crown with as low-pressure jet as possible to remove remaining soil.
5. Remove shoot(s) and oven-dry for shoot mass or discard.
6. Image directly or store submerged in water in cold room for later imaging.

3.3.3 Imaging

1. Select a root crown without excess water, yet don't let roots dry out.

2. Place a scale (3 cm circle or otherwise) and a plant tag on the background. Printed labels allow optical character recognition during image analysis, or use bar codes.
3. Place root crown on background.
4. Use remote shutter to take photograph.

3.3.4 Imaging (Intensive)

5. Separate root crown into main shoot and tillers. The main shoot may be recognized by the inclusion of the seminal root system.
6. Image the main shoot and all tillers as described above, separately.

3.4 Soybean

3.4.1 Excavation

1. Insert shovel a blade width away along a row on both sides of focal plants as to dig up several plants.
2. Place entire group of plants with soil into labeled plastic bag.
3. Place bag of plants and soil into wheel barrow; fill with plants to capacity.
4. Transport samples to washing station.

3.4.2 Washing

1. Place samples in bin with water to soak; have a surplus number soaking at all times.
2. Move crowns back and forth vigorously in water to remove as much soil as possible.
3. Select one or more representative root crowns from the group to image.
4. Spray crown with as low-pressure jet as possible to remove remaining soil.
5. Remove shoot(s) with pruning shears and oven-dry for shoot mass or discard.
6. Image directly, or store submerged in water in cold room for later imaging.

3.4.3 Imaging

1. Select a root crown without excess water, yet don't let roots dry out.
2. Place a scale (3 cm circle or otherwise) and a plant tag on the background. Printed labels allow optical character recognition during image analysis, or use bar codes.
3. Place root crown on background.
4. Use remote shutter to take photograph.

3.5 Image Analysis for All Species

For image analysis, readers are referred to automated software available from Bucksch et al. [10] and Colombi et al. [16] or nonautomated ImageJ plugins available from LM York and JP Lynch [5]. Review specifics of software imaging protocols, and make updates to imaging described above as necessary.

4 Notes

1. The choice of background for root imaging from above has three main criteria: contrast with roots, low reflectivity, and ability to be easily cleaned. Black chalkboard paint on metal is one example. Red backgrounds have also been successively used. Cutting board-style plastic has also been used. Most recently, backlit approaches are being explored for the upcoming RhizoVision-Crown platform from LM York.
2. For the DIRT platform, a circular scale is used in each image. However, a scale is unnecessary if the camera stays in a fixed position and the physical size to pixel count scale is known.
3. If storing roots submerged in water (recommended), place a few outside the container, and replace periodically such that they surface becomes dry without drying out the interior of roots.
4. Printed labels allow optical character recognition during image analysis in DIRT, or use bar codes.
5. Root nodes in maize can be difficult to differentiate. However, roots from the same node emerge in a similar plane and have similar diameters, besides each node being higher up on the shoot.

References

1. Lynch JP (1995) Root architecture and plant productivity. Plant Physiol 109:7–13
2. York LM, Nord EA, Lynch JP (2013) Integration of root phenes for soil resource acquisition. Front Plant Sci 4:1–15
3. Burridge JD, Schneider HM, Huynh BL, Roberts PA, Bucksch A, Lynch JP (2016) Genome-wide association mapping and agronomic impact of cowpea root architecture. Theor Appl Genet 130(2):419–431
4. Atkinson JA, Wingen LU, Griffiths M, Pound MP, Gaju O, Foulkes MJ, Gouis JL, Griffiths S, Bennett MJ, King J, Wells DM (2015) Phenotyping pipeline reveals major seedling root growth QTL in hexaploid wheat. J Exp Bot 66:2283–2292
5. York LM, Lynch JP (2015) Intensive field phenotyping of maize (Zea Mays L.) root crowns identifies phenes and phene integration associated with plant growth and nitrogen acquisition. J Exp Bot 66:5493–5505
6. Saengwilai P, Nord EA, Chimungu J, Brown KM, Lynch JP (2014) Root cortical aerenchyma enhances nitrogen acquisition from low nitrogen soils in maize (Zea Mays L.) Plant Physiol 166:726–735
7. Saengwilai P, Tian X, Lynch JP (2014) Low crown root number enhances nitrogen acquisition from low nitrogen soils in maize (Zea Mays L.) Plant Physiol 166:581–589
8. Trachsel S, Kaeppler SM, Brown KM, Lynch J (2011) Shovelomics: high throughput phenotyping of maize (Zea Mays L.) root architecture in the field. Plant Soil 341:75–87
9. Das A, Schneider H, Burridge J, Ascanio AK, Wojciechowski T, Topp CN, Lynch JP, Weitz JS, Bucksch A (2015) Digital imaging of root traits (DIRT): a high-throughput computing and collaboration platform for field-based root phenomics. Plant Methods 11:51
10. Bucksch A, Burridge J, York LM, Das A, Nord EA, Weitz JS, Lynch JP (2014) Image-based high-throughput field phenotyping of crop roots. Plant Physiol 166:470–486
11. Burridge J, Jochua CN, Bucksch A, Lynch JP (2016) Legume shovelomics: high—throughput phenotyping of common bean (Phaseolus Vulgaris L.) and cowpea (Vigna Unguiculata

subsp, unguiculata) root architecture in the field. Field Crop Res 192:21–32

12. York LM, Galindo-Castaneda T, Schussler JR, Lynch JP (2015) Evolution of US maize (Zea Mays L.) root architectural and anatomical phenes over the past 100 years corresponds to increased tolerance of nitrogen stress. J Exp Bot 66:2347–2358
13. Hauck AL, Novais J, Grift TE, Bohn MO (2015) Characterization of mature maize (Zea Mays L.) root system architecture and complexity in a diverse set of ex-PVP inbreds and hybrids. Spring 4:424
14. Grift TE, Novais J, Bohn M (2011) High-throughput phenotyping technology for maize roots. Biosyst Eng 110:40–48
15. Abiven S, Hund A, Martinsen V, Cornelissen G (2015) Biochar amendment increases maize root surface areas and branching: a shovelomics study in Zambia. Plant Soil 395(1):45–55
16. Colombi T, Kirchgessner N, Le Marié CA, York LM, Lynch JP, Hund A (2015) Next generation shovelomics: set up a tent and REST. Plant Soil 388:1–20
17. Houle D, Govindaraju DR, Omholt S (2010) Phenomics: the next challenge. Nat Rev Genet 11:855–866
18. York L (2017) Root crown phenotyping (AKA shovelomics) cartoon. figshare

Chapter 3

Developmental Analysis of *Arabidopsis* Root Meristem

Riccardo Di Mambro and Sabrina Sabatini

Abstract

Plant postembryonic development takes place in region called meristems that represent a reserve of undifferentiated cells. In the root meristem of *Arabidopsis thaliana*, all tissues originate from a stem-cell niche. Stem-cell daughters undergo a finite number of cell divisions until they reach the transition zone where divisions cease and cells start to differentiate. For meristem maintenance, and therefore continuous root growth, the rate of cell differentiation must equal the rate of generation of new cells. How this balance is achieved is a central question in plant biology. In this chapter we described protocols to help the operator in approaching developmental studies on the *Arabidopsis* root meristem.

Key words Meristem, Cell division, Cell differentiation, Auxin, Cytokinin, Transition boundary

1 Introduction

In the *Arabidopsis* root, a pool of stem cells continuously provide precursors of more specialized cells to replace tissues and sustain growth. Stem cells are maintained in a specialized cellular microenvironment known as stem-cell niche [1]. This region provides the extracellular signals to maintain stem cell division and prevent their differentiation [2]. Longitudinal section of *Arabidopsis* root displays a set of concentric cylinders where four external layers (epidermis, cortex, endodermis, and pericycle) surround a central vascular cylinder. An external single cell layer, the root cap, and the lateral root cap protect the meristem while growing into the soil [3, 4]. The epidermis is the outermost layer of the root and consists of cell files with hair-bearing cells interspersed with files of hairless cells [5–7]. The epidermis is the only cell file where two different cell types can be distinguished; the cortex and endodermis layers are each composed of a single cell type [8, 9]. Internally to these three layers, in the middle of the root, there is the pericycle. Cell belonging to this cell file can initiate the formation of new root, the lateral roots [10]. The vascular tissue in the middle of the root has a bilateral symmetry. This consists of two types of cells: xilematic

Daniela Ristova and Elke Barbez (eds.), *Root Development: Methods and Protocols*, Methods in Molecular Biology, vol. 1761, https://doi.org/10.1007/978-1-4939-7747-5_3, © Springer Science+Business Media, LLC 2018

cells and floematic cells. The first one give rise to the xylem which transport water and inorganic nutrients from the root to the other organs, while floematic cells generate the phloem involved in the transport of organic nutrients, such as sugars, from leaves and parenchymas reserve [9].

Because of the simplicity of its structure, the *Arabidopsis thaliana* root meristem has become one of the best-studied model systems in plant biology [11]. The root is transparent allowing easy observation at the bright field and confocal microscopes. For this reasons, the *Arabidopsis* root is a powerful tool to study the molecular mechanisms controlling the balance between cell division and cell differentiation established in the root meristem.

The root meristem is located at the distal tip of the growing root and continuously gives rise to new cells [9].

Along the longitudinal axis, the root meristem can be divided, by morphology and cell developmental stage, into three distinct regions: [1] the stem-cell niche where stem cells continuously produce transit-amplifying cells, which enter [2] the division zone where they undergo a finite number of cell divisions until they reach [3] the elongation and differentiation zone where they lose the capacity to divide and rapidly expand and differentiate to reach maturity. Cell differentiation in each file starts at the transition boundary (TB); the position of the TB is different for each cell type generating the transition zone (TZ) encompassing the TB of the different cell files (Fig. 1) [12].

The coordinated activity of these three regions allows the establishment of a dynamic equilibrium between the dividing cells and those which differentiate, maintaining constant in time the size of the root meristem [13].

The balance between the rate of cell proliferation in the meristem and the extent of cell elongation/differentiation at the transition zone determines root meristem size and the overall rate of root growth [5, 12–15]. How this balance is achieved is a central question in plant developmental biology, and in the last years, the mechanisms underlying this process have begun to be clarified.

Upon seed germination, the root consists of few cells including the root stem-cell niche. The apical root meristem is built, thanks to a prevalence of cell division over cell differentiation, and eventually reaches its final size 5 days after germination [16]. At this stage a dynamic equilibrium between cell division and cell differentiation is reached, and a stable number of cells are established in the meristem [13, 17]. To determine the number of cell belonging to the meristem, it is possible to count the number of meristematic cortex cells in a file extending from the QC to the first elongated cell excluded (from white to black arrowheads in Fig. 1). The cortex is the best suitable tissue to count meristematic cells as it is composed of a single cell type and its cell number is maintained constant between different roots. The epidermis tissue, instead, is largely variable in meristematic cells number as cells giving rise to root hairs

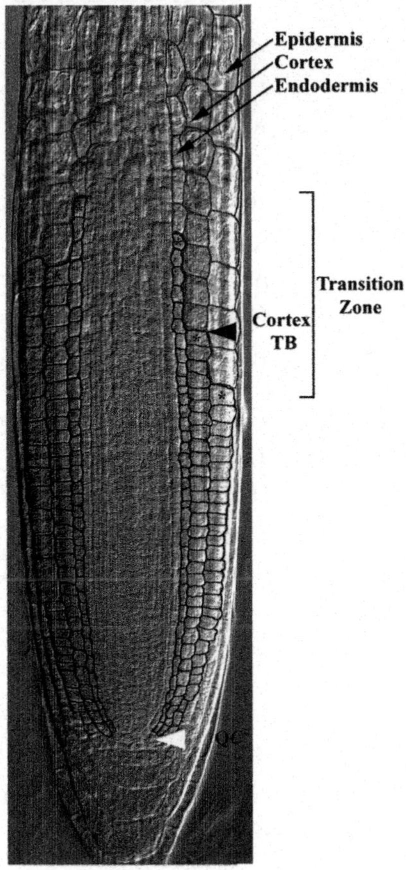

Fig. 1 Zonation of the *Arabidopsis* root. *Arabidopsis* root zonation at 5 days after germination. The meristematic and the differentiated cells of epidermis, cortex, and endodermis tissues are marked, respectively, in black and gray. The boundary between dividing and differentiating cells for each tissue is called transition boundary (TB), while the region including the different transition boundaries is called transition zone. White and black arrowheads indicate QC and cortex TB, respectively. Black stars indicate epidermis, cortex, and endodermis last meristematic cells. The meristematic cortex cells enclosed between the QC and the cortex TB correspond to the meristem size

divide more frequently than non-hair cells [6]. Since these different cell types are undistinguishable in the meristem, the number of cells can drastically change depending on which epidermis cell file has been counted. Cells belonging to inner tissues are difficult to count due to their small size. On the contrary, cortex meristematic cells are easily distinguishable from differentiating cells because of their size and morphology, clearly observable at the microscope. Meristematic cells, indeed, have a small size and a large central vacuole so that the cytoplasm looks denser than the differentiating cells one. Furthermore, differentiating cells undergo cell expansion along the longitudinal axis appearing with an elongated morphology.

2 Materials

2.1 Seed Sterilization

1. Hypochlorite solution: 1% final concentration of active Cl in water. Store at room temperature (*see* **Note 1**).
2. Sterilization solution: 50% bleach and 0.4% of polyethylene glycol sorbitan monolaurate (TWEEN® 20) (*see* **Notes 2** and **3**).
3. Sterile distilled water.
4. 0.1% agarose. Autoclave and store at room temperature (*see* **Note 4**).
5. Parafilm™ or masking tape or micropore tape.

2.2 Growth Conditions

1. MS medium: 0.5× Murashige and Skoog (MS) salt mixture, 1% sucrose, 0.5 g/L 2-(*N*-morpholino)ethanesulfonic acid (MES) at pH 5.8, 0.8% agar. Autoclave before use.
2. Square Petri dishes, 120 × 120 × 17 mm. Use 50 mL of medium each square Petri dishes.

2.3 Root Meristem Size Analysis

1. Chloral hydrate solution: 8:3:1 mixture of chloral hydrate/distilled water/glycerol [18]. Store at room temperature.
2. Glass slides (25.4 × 76.2 mm, 1.0–1.2 mm thick).
3. Cover glasses (24 × 50 mm).
4. Optical microscope with Nomarski optics, connected to a digital camera.

2.4 Root Meristem Confocal Analysis

1. Glass slides (25.4 × 76.2 mm, 1.0–1.2 mm thick) (*see* **Note 5**).
2. Cover glasses (24 × 50 mm) (*see* **Note 6**).
3. 10 μg/mL propidium iodide solution (dissolved in water) (*see* **Notes 7** and **8**).
4. SR staining solution: 0.1% (v/v) SR2200, 1% (v/v) DMSO, 0.05% (w/v) Triton X-100, 5% (w/v) glycerol, and 4% (w/v) paraformaldehyde dissolved in PBS buffer (pH 8.0). Fresh for each experiment.

2.5 Histochemical Staining

1. X-gluc solution: 100 mg/mL X-gluc (5-bromo-4-chloro-3-indolyl glucuronide) dissolved in *N*-*N*-dimethylformamide (100 mM Na_2HPO_4, 100 mM NaH_2PO_4, 0.5 mM $K_3Fe(CN)_6$, 0.5 mM $K_4Fe(CN)_6$, 0.1% Triton X-100, and 0.5 mg/mL X-gluc). Store at +4 °C or −20 °C. Fresh for each experiment (*see* **Note 9**).

3 Methods

3.1 Seed Dry Sterilization for Large Amount of Samples

1. Place the seeds in separate collecting tubes (1.5 or 2.0 mL). Use at least 90 seeds for each experiment, approximately 15 μL.
2. Place a rack with the tubes in a vacuum jar dryer placed under a chemical hood without applying the vacuum; leave the tubes opened.
3. Under a chemical hood, place a beaker with 100 mL of bleach in the vacuum jar dryer and then add 3 mL of 37% hydrochloric acid (HCl) and quickly close the vacuum jar dryer without applying the vacuum. Warning, the reaction determines the development of chlorine fumes (Cl_2 gas).
4. Leave the seeds in the chlorine fumes for at least 1 h (or for a maximum of 4 h).
5. Transfer quickly the rack with the tubes under a laminar hood, and vent the tubes 15 min before adding 300–500 μL of 0.1% agarose or sterile distilled water. The volume can be varied depending on the amount of seeds.
6. Place the tubes at 4 °C in the dark for 3–5 days (*see* **Note 10**).

3.2 Seed Liquid Sterilization from Mid to Low Amount of Samples

1. Place the seeds in separate collecting tubes (1.5 or 2.0 mL). Use at least 90 seeds for each experiment, approximately 15 μL.
2. Add 500 μL to 1 mL of hypochlorite solution to each tube, and gently mix for 10 min (*see* **Note 11**).
3. Shortly centrifuge the tubes to allow seeds deposition. From this step onward, use sterile solutions, and work under a laminar hood.
4. Remove the supernatant with pipette, and rinse the seeds three times for 10 min each with distilled water gently mixing.
5. Suspend the seeds in 300–500 μL of 0.1% agarose or sterile distilled water. The volume can be varied depending on the amount of seeds.
6. Place the tubes at 4 °C in the dark for 3–5 days.

3.3 Sterilization of Large Amount of Seeds

1. Place the seeds in separate 50 mL tubes. Use up to 5 mL of seeds each.
2. Add 50 mL of sterilization solution to each tube, and gently mix for 5 min.
3. Centrifuge the tubes at 410 rcf for 1 min. From this step onward, work under a laminar hood.
4. Remove the supernatant by inverting the tubes and rinse the seeds four times with sterile distilled water and gently mix the tubes.

5. Suspend the seeds in 50 mL of sterile distilled water.
6. Repeat **steps 4** and **5** three times.
7. Add 30–50 mL of sterile distilled water.
8. Place the tubes at 4 °C in the dark for 3–5 days.

3.4 Growth Conditions

1. Plate seeds by using a pipette with sterile tips on solid MS medium under a laminar hood. Lay down seeds on the medium at a 3–5 mm distance from each other to avoid any contact between seedlings (*see* **Notes 12** and **13**).
2. Incubate plates in a near vertical position at 22 °C and 60% humidity with long-day condition (16 h light and 8 h dark cycle).

3.5 Root Length Analysis

1. For root length analysis, draw on the plate a scale bar of 5 mm and then take a picture of the plate or use a scanner with a matte black background.
2. Use *ImageJ* software (or similar as *RootLM*) to track root on the acquired images. Use the segmented line tool to follow root curvatures.
3. Set the scale in accordance with the drawn scale bar on the plate. Calculate mean and standard deviation, and do statistical analysis (*see* **Note 14**).

3.6 Root Meristem Size Analysis

The cortex is the most suitable tissue to count meristematic cells, as its single cell type composition shows a conserved number of cells among different roots. The boundary between dividing and differentiating cells for each tissue is called transition boundary (TB), while the region including the different transition boundaries is called transition zone (TZ) [12, 13].

3.6.1 Microscope Observation

1. For meristem size analysis, prepare a glass slide with chloral hydrate solution.
2. Place the seedlings on the glass slide. When necessary, use a surgical blade or narrow-tip scissors to cut the root (approximately 1 cm from the root tip), and, using a tweezers, gently lay it in the chloral hydrate solution on the glass slide. Mount the cover glass and dry on paper towels to remove excess solution.
3. Use the manual focusing drive of the microscope to find the root longitudinal plane of interest, i.e., for meristem size analysis, focus until all tissue layers and the QC are clearly visible.
4. Use appropriate objective for the analysis. A 40× objective is recommended for meristem analysis.
5. To determine root meristem size, count the number of cortex cells in a file extending from the QC to the first elongated cell excluded (*see* **Note 15**).

6. Analyze a minimum of 30 plants for 2 biological replicates. Calculate mean and standard deviation, and do statistical analysis.

To strengthen the analysis performed with the optical microscope, the use of image analysis software that recognize the transition boundary of the cortex can help. *Cell-o-Tape* [19], a *Fiji* macro, is able to count and measure individual neighboring cells along the cortex file as well as to estimate the position of the last cortical cell of the meristem. Use 8-bit images obtained using a confocal laser scanning microscope. Threshold and stringency parameters of *Cell-o-Tape* can be set depending on the observation carried out by the operator and the quality of the image. For each experiment analysis of at least 30 plants has to be performed and in 2 biological replicates.

3.6.2 Root Meristem Size Analysis Over Time

Root meristem size gradually increases during development. Just upon germination root meristem growth is possible because the rate of cell division is higher compared to the rate of cell differentiation [16]. Five days after germination, the two rates are balanced, and root meristem size is set. At this stage a stable number of approximately 30 cells are established in the meristem, and it is maintained constant in time.

To perform the analysis of root meristem size over time, follow the steps described under Subheadings 3.4–3.6. Consider a number of plates equal to the number of points of the time course. For long-time course, consider "backup" plate in order to overcome potential contaminations. During the time course, avoid opening plates, which may easily get contaminated. However, if necessary (i.e., to remove condenses), open the plate under a laminar hood, and work with sterile materials.

3.6.3 Root Meristem Size Analysis After Treatment

The analysis of root meristem size after exogenous application of a hormone or any other substance is essential to determine, if this substance is involved in the control of root growth and root meristem size determination.

1. Perform the analysis on 5-day-old seedlings, grown as described under Subheading 3.4 (*see* **Note 16**).
2. Transfer seedlings to solid MS medium containing mock conditions or a suitable concentration of the desired substance (*see* Table 1 and **Note 17**). Some examples of substances concentrations used to investigate their role in meristem size control were reported in Table 1.
3. Calculate mean and standard deviation, and do statistical analysis.
4. Check root meristem size after several times of treatment, even just several hours (*see* **Note 17**).

Table 1
Substances concentrations usually used to investigate their role in controlling meristem size

Substance	Concentration
Trans-zeatin (tZ)/bihydrozeatin (DHZ)	1 nM–5 μM
Indole-3-acetic acid (IAA)	0.1 nM–5 μM
N-1-naphthylphthalamic acid (NPA)	2 μM
Ethylene	1–100 μM
Gibberellic acid (GA_3)	10 μM
Paclobutrazol (PAC)	10 μM

3.7 Sample Preparation and Confocal Settings for Root Meristem Analysis

1. Prepare a glass slide with dyes that stain the plasma membrane. Use propidium iodide (PI) (or FM4-64).
2. Place the seedlings on the glass slide. When necessary, use a surgical blade or narrow-tip scissors to cut the root (approximately 0.5 cm from the root tip), and, using a tweezers, gently lay it in the PI solution on the glass slide. Mount the cover glass and dry on paper towels to remove excess solution.
3. Use the manual focusing drive of the microscope to find the root longitudinal plane of interest, i.e., for meristem size analysis plane, focus until all tissue layers and the QC are clearly visible.
4. Use appropriate objective for the analysis. A 40× objective is recommended for meristem analysis.
5. Set the pinhole as smaller as possible (the smaller the pinhole, the better the device behaves). A diameter of the detection pinhole of 1 airy unit (AU) gives the best signal/noise ratio (you will find airy 1 button in the software that will set automatically the detection pinhole to the optimal value of 1 AU depending on the objective in use).
6. Set the photomultiplier (PMT) of PI channel to obtain a clear view of the different tissues and cells of the root.
7. Set the PMT of the other analyzed dyes in order to obtain a non-saturated fluorescence. It is strongly suggested observing at least five different roots to find the optimal setting before start the experiment (*see* **Notes 18–21**).

3.8 Sample Fixation and Confocal Analysis

In some cases it is necessary to fix sample before confocal analysis. The SR2200 dye labels cell walls (*see* **Note 22**). SR2200 can be used as fluorescence dye in combination with fluorescent proteins ranging in the emission spectrum from cyan to red (*see* **Note 23**). SR2200 is excited with a 405-nm laser line and emission recorded between 405 and 476 nm [20].

1. Prepare a glass slide with SR staining solution.
2. Place the seedlings on the glass slide. When necessary, use a surgical blade or narrow-tip scissors to cut the root (approximately 0.5 cm from the root tip), and, using a tweezers, gently lay it in the SR staining solution on the glass slide. Mount the cover glass and dry on paper towels to remove solution in excess.
3. Follow **steps 3**, **4**, **5**, and 7 described under Subheading 3.7 (*see* **Note 24**).

3.9 Fluorescence Quantification of Reporter in Root Confocal Images

The analysis of fluorescent reporter needs to be supported by a quantitative and statistical analysis. The method of analysis chosen depends on the type of reporter to be analyzed. Here two methods are described for the analysis of a reporter with no specific localization within the cell or a reporter localized in the nucleus (*see* **Note 25**).

3.9.1 Fluorescence Quantification of Reporters with No Specific Localization Signal

The plugin *MeasureRGB* of the software *ImageJ* can be used to quantify the fluorescence intensity of plants carrying a fluorescent reporter with no specific localization signal in the cell.

1. Acquire images of the roots with confocal laser scanning microscope.
2. Take as ROI the expression domain of the reporter expressed in the root.
3. Calculate with *MeasureRGB* the Σ of pixels of the channel (raw intensity density—*RawIntDen*).
4. Normalize the results for the chosen area.
5. Calculate mean and standard deviation, and do statistical analysis.

3.9.2 Fluorescence Quantification of a Reporter Targeted to the Nucleus

The software *ImageJ* can be also used to quantify the fluorescence intensity of plants carrying a fluorescent reporter with a nuclear localization signal. In this case it is highly recommended to measure the fluorescence of the entire nucleus of the cell to reduce bias.

1. Acquire Z-stack images of the roots with confocal laser scanning microscope. Make sure to include in the different planes of the Z-stack the entire nuclei area.
2. Create maximum intensity projection of the acquired Z-stack of root tip acquired with the confocal management software or *Fiji*.
3. Use the ROI manager tool of the software *Fiji* to select nuclei area.
4. Calculate mean and standard deviation of raw fluorescence intensity data, and do statistical analysis (*see* **Note 26**).

3.10 Histochemical Staining

The GUS histochemical staining is an established tool for qualitative and quantitative analysis of gene and protein activity.

1. Grow plants carrying the reporter *GUS* construct as described under Subheading 3.4.
2. Transfer seedling into well containing X-gluc solution. For each experiment use at least 60 plants.
3. Apply 10 min of vacuum to facilitate entry of the substrate into the cells (*see* **Note 27**).
4. Incubate for 1 h at 37 °C in the dark to activate β-glucuronidase (GUS) activity (*see* **Note 28**).
5. Prepare glass slides as described under Subheading 3.6.1.

4 Notes

1. Alternatively commercial bleach for household cleaning (containing 5% of active Cl) can be used, diluted at 1:5 in distilled water.
2. Alternatively bleach for household cleaning (containing 5% of active Cl) can be used.
3. Due to high density of Twin, prepare a 10% dilution solution in water.
4. To avoid contamination of 0.1% agarose, prepare it fresh for each experiment. This solution tends to form clumps, so include a magnet in the bottle, and after sterilization in autoclave, let it cool at room temperature under continuous stirring.
5. For inverted microscope you can use as glass slides cover glasses 24 × 50 mm, 1.0 mm thick.
6. For inverted microscope you can use cover glasses 18 × 18 mm, 1.0 mm thick.
7. Propidium iodide (PI) is light sensitive.
8. Prepare a 10 mg/mL starting propidium iodide solution, and keep it in aluminum at +4 °C.
9. X-gluc is light sensitive. This solution can form precipitates: make sure that no crystals are present in the solution before use.
10. This step, called stratification, is important to synchronize germination. Since old seeds might have problems to germinate, we recommend using seeds not older than 1 year. For dry seeds (stored at room temperature with drierite or silica gel), 3 days of stratification are usually sufficient. In contrast, fresh seeds (e.g., collected from a not yet completely dried plant)

need 5 days of stratification. We suggest keeping fresh seeds with drierite or silica gel for at least 8 days before sterilization.

11. This is an alternative method: place the seeds in separate collecting tubes (1.5 or 2.0 mL). Use at least 90 seeds for each experiment, approximately 15 μL. Add 1 mL of sterilization solution to each tube, and gently mix for 1 min. Shortly centrifuge the tubes to allow seeds deposition. From this step onward, work under a laminar hood. Remove the supernatant with a pipette, and rinse the seeds with 1 mL of EtOH (at least 96% EtOH); gently mix for 1 min. Shortly centrifuge the tubes to allow seed deposition. Remove the supernatant with a pipette, and rinse the seeds five times with sterile distilled water. Continue at **steps 5** and **6** of Subheading 3.2.
12. Use a pipette to plate the seeds. Place the seeds at about 1 cm from the top and the lateral board of the dish. Before closing the plate and sealing with Parafilm™, masking tape, or micropore tape, make sure that the agarose solution/water in which the seeds were embedded is dried.
13. For molecular analysis sometimes a large amount of sample is needed. In these cases the use of mesh is recommended in order to facilitate the cutting and the collecting of the roots. Cut 10 × 10 cm and autoclaved.
14. For root length analysis over time, add a step in which at each point of the time course mark with indelible marker the tip of the root on the plate in order to follow root growth during time.
15. In some cases it may be difficult to identify the boundary between the last meristematic cell and the first elongating cell. In these cases observe the other cortex file, the differentiated epidermis cells, and the lateral root cap height.
16. Analysis of 5-day-old seedlings, in which root meristem size is already established, allows understanding if the exogenous application of the tested substance controls meristem size. This analysis however can be performed at different developmental stage.
17. It is important to set up the experimental conditions, i.e., concentration and time of treatment, when treating a sample with a substance for the first time. Apply various concentrations of the substance, and follow the effect on root meristem size after several hours of treatment.
18. Do not change PMT settings during the experiment in order to avoid artifact in the result. Only, PMT of PI channel can be modified.

19. In order to reduce any bias of the instrument, alternate control and mutant/treated sample analysis or mutant/treated and at the end control sample analysis.
20. In some cases a transmitted light (Bright Field) PMT (T-PMT) combined with confocal PMT acquisition can improve image quality and highlight the observation, however, at the expense of fluorescence signal revelation.
21. If the experiment implies to carry out the analyses of a fluorescent reporter in different background (i.e., wild-type versus mutant or untreated versus treated root), set parameter settings based on control line.
22. Be careful; weak SR2200 signal can also be found in the nucleus, which should be taken into account when working with nuclear localized fluorescent proteins.
23. If the microscope allows recording spectral information of the fluorescent signal, SR2200 can even be combined with DAPI, which has a very similar emission spectrum.
24. To increase tissue penetration, soft vacuum can be applied for 5 min at RT. Afterward, the SR staining solution is replaced by water and again incubated under soft vacuum for 5 min. After replacing the water with 10% glycerol, the sample is mounted under a coverslip.
25. There are many fluorescence quantification methods that can be used, but your intuition and inventiveness will drive you to build up new methods.
26. Remember to subtract background noise selecting cytosolic area where no fluorescence is expected.
27. If the reporter is weakly expressed, extend vacuum treatment up to 1 h.
28. If the reporter is weakly expressed, extend the incubation at 37 °C in the dark up to an overnight.

Acknowledgment

The authors would like to thank Raffaele Dello Ioio, Rosangela Sozzani, Elena Pacifici, Laura Polverari, Emanuela Pierdonati, Elena Salvi, and Serena Perilli for helpful comment. This work was supported by The European Research Council.

References

1. Scheres B (2007) Stem-cell niches: nursery rhymes across kingdoms. Nat Rev Mol Cell Biol 8:345–354
2. van den Berg C, Willemsen V, Hendriks G et al (1997) Short-range control of cell differentiation in the Arabidopsis root meristem. Nature 390:287–289

3. Rougier M (1981) Secretory activity of the root cap. In: Tanner W, Loewus FA (eds) Plant carbohydrates II. PLANT, vol 13B. Springer, Berlin, Heidelberg, pp 542–574
4. Sievers A, Hensel W (1991) Root cap. In: Waisel Y, Eshel A, Kafkfi U (eds) Plant roots: the hidden half. Marcel Dekker, New York, pp 53–74
5. Dolan L, Janmaat K, Willemsen V et al (1993) Cellular organisation of the *Arabidopsis thaliana* root. Development 119:71–84
6. Berger F, Hung CY, Dolan L et al (1998) Control of cell division in the root epidermis of *Arabidopsis thaliana*. Dev Biol 194:235–245
7. Galway ME, Masucci JD, Lloyd AM et al (1994) The TTG gene is required to specify epidermal cell fate and cell patterning in the Arabidopsis root. Dev Biol 166:740–754
8. Helariutta Y, Fukaki H, Wysocka-Diller J et al (2000) The *SHORT-ROOT* gene controls radial patterning of the *Arabidopsis* root through radial signaling. Cell 101:505–517
9. Leyser O, Day D (2005) Mechanism in plant development. Blackwell Publishing, Malden, USA
10. De Smet I, Vanneste S, Inzè D et al (2006) Lateral root initiation or the birth of a new meristem. Plant Mol Biol 60:871–887
11. Benfey PN, Scheres B (2000) Root development. Curr Biol 10:R813–R815
12. Di Mambro R, De Ruvo M, Pacifici E, Salvi E, Sozzani R, Benfey PN, Busch W, Novak O, Ljung K, Di Paola L, Marée AFM, Costantino P, Grieneisen VA, Sabatini S (2017) Auxin minimum triggers the developmental switch from cell division to cell differentiation in the root. Proc Natl Acad Sci U S A 114(36):E7641–E7649
13. Dello Ioio R, Linhares FS, Scacchi E et al (2007) Cytokinins determine Arabidopsis root-meristem size by controlling cell differentiation. Curr Biol 17:678–682
14. Beemster GTS, Baskin TI (1998) Analysis of cell division and elongation underlying the developmental acceleration of root growth in Arabidopsis thaliana. Plant Physiol 116:1515–1526
15. Scheres B, Wolkenfelt H, Willemsen V et al (1994) Embryonic origin of the Arabidopsis primary root and root meristem initials. Development 120:2475–2487
16. Moubayidin L, Perilli S, Dello Ioio R et al (2010) The rate of cell differentiation controls the Arabidopsis root meristem growth phase. Curr Biol 20:1138–1143
17. Dello Ioio R, Nakamura K, Moubayidin L et al (2008) A genetic framework for the control of cell division and differentiation in the root meristem. Science 322:1380–1384
18. Mayer U, Torres Ruis R, Berleth T et al (1991) Mutations affecting body organization in the Arabidopsis embryo. Nature 353:402–407
19. French AP, Wilson MH, Kenobi K et al (2012) Identifying biological landmarks using a novel cell measuring image analysis tool: cell-o-tape. Plant Methods 8:7
20. Musielak TJ, Schenkel L, Kolb M, Henschen A, Bayer M (2015) A simple and versatile cell wall staining protocol to study plant reproduction. Plant Reproduction 28(3-4):161–169

Check for updates

Chapter 4

Genetic and Phenotypic Analysis of Lateral Root Development in *Arabidopsis thaliana*

Selene Napsucialy-Mendivil and Joseph G. Dubrovsky

Abstract

Root system formation to a great extent depends on lateral root (LR) formation. In *Arabidopsis thaliana*, LRs are initiated within a parent root in pericycle that is an external tissue of the stele. LR initiation takes place in a strictly acropetal pattern, whereas posterior lateral root primordium (LRP) formation is asynchronous. In this chapter, we focus on methods of genetic and phenotypic analysis of LR initiation, LRP morphogenesis, and LR emergence in *Arabidopsis*. We provide details on how to make cleared root preparations and how to identify the LRP stages. We also pay attention to the categorization of the LRP developmental stages and their variations and to the normalization of the number of LRs and LRPs formed, per length of the primary root, and per number of cells produced within a root. Hormonal misbalances and mutations affect LRP morphogenesis significantly, and the evaluation of LRP abnormalities is addressed as well. Finally, we deal with various molecular markers that can be used for genetic and phenotypic analyses of LR development.

Key words Root branching, Root system architecture, Root primordium, Lateral root, Pericycle, Root initiation, Developmental stages, Founder cells, Quantitative analysis, Morphogenesis

1 Introduction

The most dramatic changes in plant shape take place during post-embryonic development. The growth of aerial plant organs depends, to a great extent, on how the roots expand within the soil. Germination starts with the root emergence. Any root formed on another root, independent of its origin, is considered a LR. An extensive root system is formed as a result of the iterative process of de novo LR formation. LRs are initiated within a parent root. Therefore, LR production depends on the growth rate and, thus, on the cell production within the root apical meristem of the parent root.

In *Arabidopsis thaliana*, as in most angiosperms, LRs start their formation in pericycle [1–6], an external tissue of the stele, which maintains cell proliferation activity outside the root apical meristem

Daniela Ristova and Elke Barbez (eds.), *Root Development: Methods and Protocols*, Methods in Molecular Biology, vol. 1761, https://doi.org/10.1007/978-1-4939-7747-5_4, © Springer Science+Business Media, LLC 2018

[5, 7]. In this species, soon after pericycle cells leave the root apical meristem, some of them become primed to form founder cells from which a lateral root primordium (LRP) originates. Priming, which takes place within the transition domain of the root apical meristem and the distal part of the elongation zone, is accompanied by time-dependent expression of the genes involved in LR initiation [8–11]. After priming, pericycle founder cells are specified [12], and when these cells divide anticlinally (i.e., divisions that result in new cell walls oriented perpendicularly to the nearest organ surface), LR initiation starts [2, 7, 13]. The time between founder cell specification and LR initiation is very short [14], and both processes take place in the differentiation zone where cells acquire their final developmental fate. Pericycle maintains its cell proliferation capacity even after LR initiation, and it eventually forms lateral meristems (i.e., cambium and phellogen), which permit root growth in girth.

During root growth, the root apex is displaced away continuously from the more proximal root portion as a result of cell division and rapid cell elongation. Therefore, the position where cell differentiation starts is also continuously displaced toward the root apex (acropetally), and thus, LR initiation continues in the same direction as well. In *A. thaliana*, LR initiation takes place in a strictly acropetal pattern, but posterior LRP development is asynchronous [15]. This is why younger LRP stages are found among later developmental stages and why LRPs are present among emerged LRs.

In this chapter, we provide details of methods that permit genetic and phenotypic analysis of LR development in *Arabidopsis*. First, we describe methods of root clearing required for quantitative analysis of LR initiation and formation (Fig. 1) in wild-type and mutant lines of *Arabidopsis*. Next, we provide details on how to identify the LRP stages in these preparations. The main parameters and logics of the analysis have been described earlier [16, 17], and here we focus on the respective methods and practical approaches that permit to obtain a number of parameters to evaluate how a treatment or a gene mutation affects LR formation. Genetic and hormonal regulation of LR development is complex [18–20]. Hormonal regulation of root branching and its study is not straightforward. As LRs are formed within the parent root, any external substance treatment that affects the parent root development will have an effect on LR formation, and identifying specific outcome (s) of a treatment can be a difficult task. To this end, we designed experiments that imply transferring *Arabidopsis* seedlings from normal growth medium to a medium supplemented with a substance of interest [21–23]. In this chapter, we are going to describe methodological specifications of these experiments. Hormonal misbalances or mutations affect LRP morphogenesis significantly [6, 24–26], and we are going to describe, shortly, how

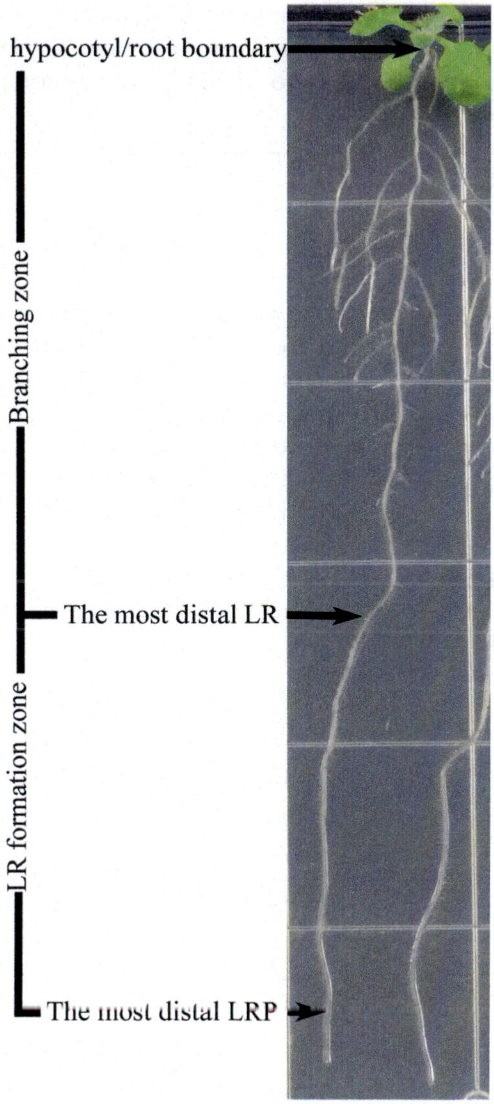

Fig. 1 The primary root portions of *Arabidopsis* seedling, corresponding to the lateral root formation and branching zones. Arrows indicate the root and hypocotyl boundary (root base), the most distal (rootward) emerged lateral roots, and the most distal lateral root primordium. Position of the most distal lateral root primordium is indicated arbitrarily as its precise location can be determined only on cleared root preparation. The root portion from the root base to the most distal lateral root is defined as branching zone; in this zone lateral roots are growing. The root portion from the most distal lateral root to the most distal lateral root primordium is defined as lateral root formation zone; in this zone lateral root initiation and posterior lateral root morphogenesis take place until lateral root is emerged. Distance between horizontal lines of a Petri dish is 13.4 mm

developmental abnormalities can be evaluated as well. Finally, we address how molecular markers of a number of processes can be used for genetic and phenotypic analysis of LR development.

2 Materials

Plants are grown in soil or on Petri dishes oriented vertically and containing 0.2× Murashige and Skoog (MS) medium [27], pH 5.7, supplemented with vitamins (0.1 mg/L pyridoxine, 0.1 mg/L nicotinic acid, 1% sucrose, and 0.8% agar (w/v, Bacto™ Agar; BD Difco, Sparks, MD, USA). Seeds are sterilized for 10 min in 20% commercial bleach and 0.08% Triton X-100, and then washed four times with sterile distilled water, and imbibed at 4 °C for 2 days. Plants are grown at 21 °C, under a 16/8 h light/dark photoperiod with a light intensity of 105 μmol photons/m^2/s.

2.1 Root Clearing

2.1.1 Methanol-NaOH-Based Clearing of Roots

1. Small Petri dishes (35 mm in diameter) or ELISA 6-well plates, each well of 34 mm in diameter (*see* **Note 1**).
2. Forceps.
3. Parafilm.
4. Glass microscope slides.
5. Micropipette for 200 μL or 1 mL.
6. Microscope with an ocular micrometer and an object-micrometer (*see* **Note 2**).
7. An incubator at 62 °C.
8. Acidified methanol solution (0.24 N HCl, 20% methanol). For 125 mL of the solution: 25 mL of methanol, 2.5 mL of conc. HCl, 97.5 mL of distilled water.
9. Basic solution (7% NaOH, 60% ethanol). For 50 mL: 30 mL of absolute ethanol, 20 mL of distilled water, 3.5 g of NaOH (*see* **Note 3**).
10. Ethanol series: 40, 20, and 10% ethanol in distilled water.
11. 50% glycerol.

2.1.2 NaI-Based Clearing of Roots

1. Forceps.
2. Paraformaldehyde fixation solution (4% paraformaldehyde, 0.025 M phosphate buffer, pH 7.2). For 50 mL: 20 mL of 10% paraformaldehyde, 1.25 mL of phosphate buffer and 28.75 mL of distilled water (*see* **Note 4**).
3. 30% glycerol and 2% DMSO solution. For 50 mL of the solution: 15 mL of glycerol, 1 mL of DMSO, and 34 mL of distilled water.

4. NaI-based clearing-and-mounting solution (saturated NaI solution prepared in 65% glycerol (v/v) and 2% DMSO (v/v). For 20 mL of the solution: 13 mL of glycerol, 0.4 mL of DMSO, 6.6 mL of distilled water, 17 g of NaI, 0.04 g of $Na_2S_2O_3$.

2.1.3 Mounting

1. Microscope slides.
2. Coverslips No. 1 of different size.
3. Parafilm.
4. A dissecting needle.
5. Fine forceps.
6. Fine scissors.
7. Nail polish.
8. A razor blade.
9. Slide trays or slide boxes.

2.2 Identification of LRP Stages

1. A microscope equipped with differential interference contrast (DIC) microscopy or Nomarski optics.
2. An ultrafine-point pen marker.

2.3 Quantitative Analysis of Lateral Root Development

1. Square Petri dish, 100 × 100 × 15 mm or 120 × 120 × 17 mm.
2. 3 M Micropore tape.
3. Supports for Petri dishes to maintain them in vertical position.
4. A desktop scanner.
5. A binocular microscope (or magnifying glasses).

3 Methods

3.1 Root Clearing

LR initiation takes place in an internal root tissue, the pericycle. Chemical clearing is an alternative to histological sectioning for analyzing LRP development in *Arabidopsis*. General steps and mounting of samples are outlined in Fig. 2. We describe here two clearing protocols of roots (*see* **Note 5**): one with the use of methanol and NaOH (Malamy and Benfey (1997)-modified protocol [15]) and another one with the use of NaI [17]. Both protocols can be used for quantitative analysis of LR initiation, development, emergence, and pattern formation. However, for quantitative analysis, we preferentially use the Malamy and Benfey-modified protocol (*see* **Note 6**) as it permits easier identification of early LRP developmental stages (compare Figs. 3 and 4). NaI-based clearing protocol (*see* **Note 7**) also permits identification of LRP developmental stages, but confident identification of early

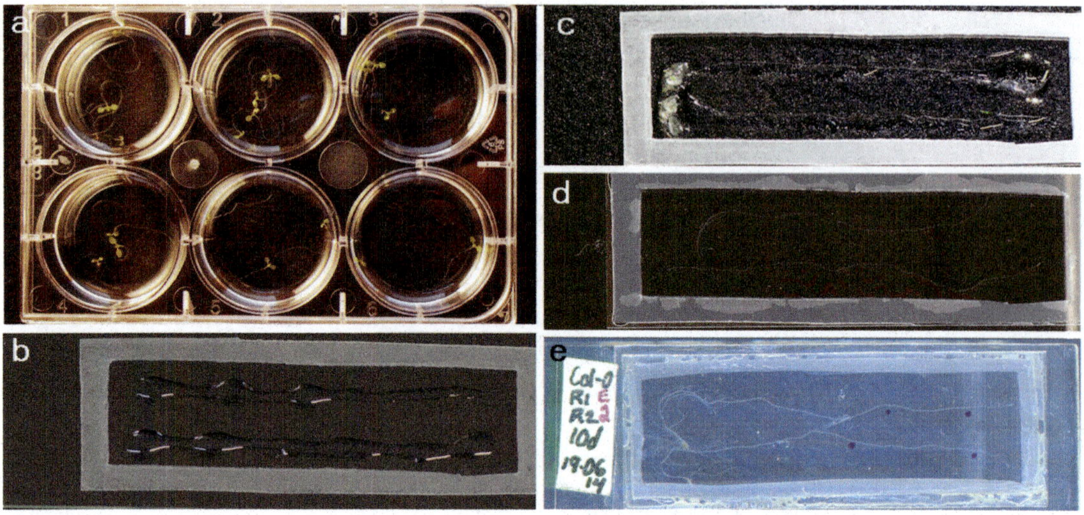

Fig. 2 The most essential steps of microscope slide preparation of cleared root specimen. (**a**) ELISA 6-well plates with clearing roots and solutions. (**b**) A parafilm frame attached to a microscope slide. Note two lines of 50% glycerol solution inside the parafilm frame. (**c**) The same frame with seedlings transferred from the last clearing solution. (**d**) A freshly made preparation of the cleared roots; the parafilm frame is covered with a coverslip. Note that the shoots are eliminated. (**e**) A nail polish-sealed microscope slide image obtained after scanning. The attached label indicates genotype (Col-0), plant age (d), root (R) number, experiment (E) number, and date

developmental stages requires certain amount of experience because the thin cell walls in early-stage LRPs are not easy to identify as the procedures used in this protocol do not induce partial plasmolysis contrary to the Malamy and Benfey-modified protocol (Fig. 3). However, NaI-based protocol has an advantage in the analysis of pattern formation in LRPs at later developmental stages and in the root apical meristem of emerged LRs (Fig. 4).

3.1.1 Tissue Fixation

The material processed with the NaI-based clearing protocol requires root fixation. The Malamy and Benfey-modified protocol does not require fixation. However, if there is no time available to perform clearing using Malamy and Benfey-modified protocol immediately after termination of the experiment, the material can also be fixed (*see* **Note 8**).

1. Place the root samples in small containers with 4% paraformaldehyde prepared in 0.025 M phosphate buffer.
2. Store the fixed roots at 4 °C for 24 h (*see* **Note 8**).

3.1.2 Clearing of Roots

Methanol-NaOH-Based Clearing of Roots

1. Place one to four wild-type seedlings (*see* **Note 9**) to each well of a 6-well ELISA plate (Fig. 2a) containing 4 mL of acidified methanol solution (0.24 N HCl in 20% methanol) per well (*see* **Note 10**). To diminish methanol evaporation, wrap the ELISA plate or the containers with parafilm.

2. Incubate the material at 62 °C for 40 min (*see* **Note 11**).
3. Using micropipette replace the acidified methanol solution with basic solution (7% NaOH in 60% ethanol). Avoid touching the samples with micropipette tip.
4. Incubate samples in the basic solution at room temperature for 20 min.

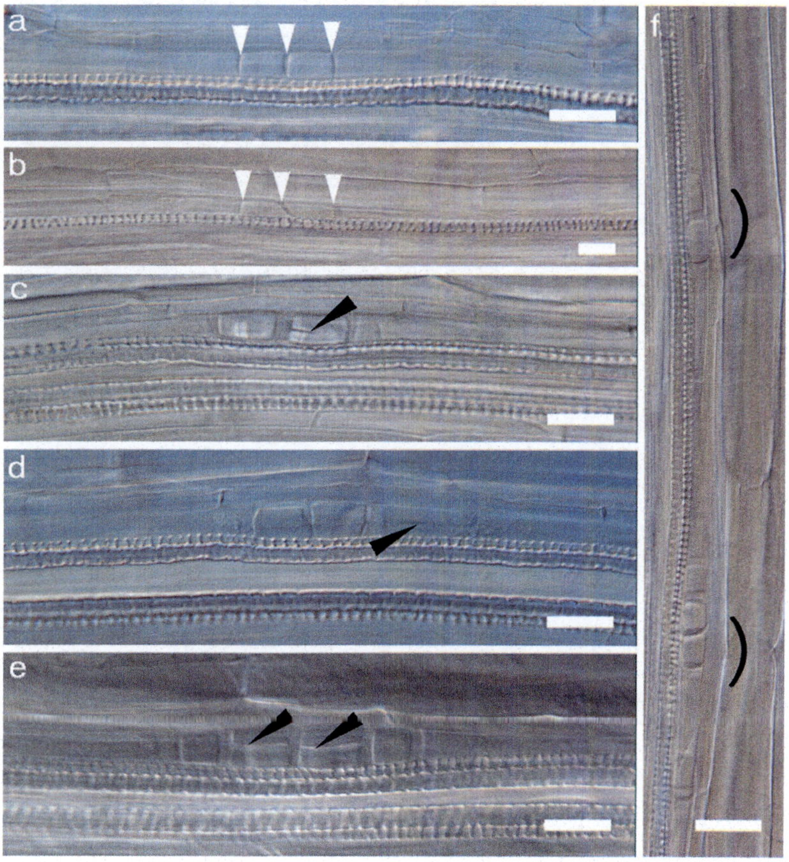

Fig. 3 Lateral root primordia cleared with methanol and NaOH (modified protocol of Malamy and Benfey (1997) [13]). (**a**) Stage I LRP. (**b**) Stage I LRP. Note that the first pericycle division can be anticlinal (a) or oblique (b); white arrowheads indicate cell walls resulted from anticlinal or oblique cells division. (**c**) Stage II LRP. (**d**) Atypical Stage II LRP. Note that periclinal divisions only in a single cell are sufficient to qualify this LRP stage as Stage II; cell walls resulted from the periclinal divisions are indicated with black arrowheads; in (**c**) periclinal cell division is not completed. (**e**) A more common case of Stage II primordium. Black arrowheads indicate cell walls resulted from periclinal division. (**f**) Closely spaced Stage I LRPs, marked with brackets. (**g**) Emerged lateral root. Col-0 (**a**, **d**, **g**) and Ws (**b**, **c**, **e**, **f**) seedlings 8 days after germination. Scale bar: 20 μm. The same magnification in panels (**a**) and (**c**–**f**)

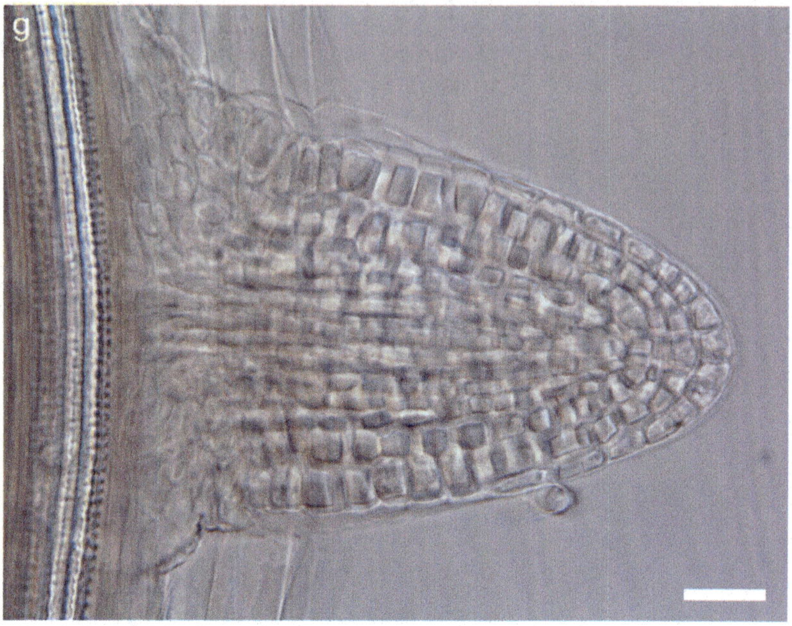

Fig. 3 (continued)

5. Rehydrate the samples sequentially with ethanol series (40, 20, and 10%). Incubate the seedlings in each solution for at least 2 h (*see* **Note 12**).
6. Finally, add to wells containing 10% of ethanol equal volume of 50% glycerol (final concentration is 25% glycerol and 5% ethanol) and incubate for at least 2 h (*see* **Note 12**).

NaI-Based Clearing of Roots

1. Fix the roots with 4% paraformaldehyde in 0.025 M phosphate buffer, pH 7.2, for 4 h to overnight at 4 °C.
2. Replace the fixative solution with 30% glycerol supplemented with 2% DMSO (final concentration) and incubate for at least 30 min at room temperature (*see* **Note 13**).
3. Mount the roots in NaI-based solution and leave for at least 1 h before observation.

3.1.3 Mounting

1. Cut out a rectangle of parafilm (Fig. 2b) using coverslip as a mold. The rectangle size is 1 mm less on each side than the coverslip used. For wild type, 8 days after germination (dag) seedlings, use a 22 × 60 mm coverslip.
2. Prepare slides with parafilm chambers. Place a parafilm frame on a clean slide, cover with a sheet of paper, and press with a finger to permit parafilm frame to stick to the slide.
3. If roots are cleared with methanol and NaOH, place the 50% glycerol solution inside the parafilm frame over the glass

microscope slide in a form of two lines of solution (Fig. 2b). If the other protocol is used, place NaI-based clearing-and-mounting solution in a similar manner.

4. Take the cleared seedling shoots with forceps and carefully transfer them into the lines of solution (Fig. 2c; *see* **Note 14**).
5. Excise the seedling shoot with a razor blade or fine scissors and remove it from preparation.
6. Add more solution (*see* **Note 15**).
7. Gently cover the parafilm frame with a coverslip (*see* **Note 16**).
8. Seal the coverslip with nail polish (*see* **Notes 17** and **18**).
9. Store the slides with cleared roots in slide trays in horizontal position. If slide boxes are used, maintain them in vertical position so that the slides are kept in the horizontal position.

3.2 Identification of LRP Stages

Arabidopsis LRP developmental stages are commonly classified in accordance with Malamy and Benfey (1997) [13]. In their study, the seedlings were grown in full MS medium supplemented with 4.5% sucrose. Currently, most laboratories use 0.2×–0.5× MS medium supplemented with 1% sucrose. This is one of the reasons why the pattern of LRP development can be more variable than what has been described in this classical work. Some examples of variability are given in [28] and in Figs. 3, 4, and 5. In spite of this variability, the most essential features of the developmental stages stay the same as what has been initially defined [13]. Overall, the detection of LRPs, especially early-stage LRPs, is recommended without the use of any transgenic marker lines. As an example, the auxin response reporter, *DR5::GUS*, is an excellent tool to monitor auxin response [29]. In some laboratories this line is used to count LRP stages. However, it was shown that up to 25% of LRP do not show DR5 activity [30]. Similarly, cell cycle progression, G2-M, marker *CycB1;1*$_{DB}$*::GUS* [31] is expressed in many but not all LRPs. The lack of expression of these markers is explained by asynchrony in LRP development as some LRPs become slowly developed or arrested in their development. Therefore, it is essential to obtain good-quality cleared root preparations and to learn how to detect LRPs.

Here, we pay attention to identifying LRPs, including those formed within the LR formation and branching zones, without any markers for any LRP developmental stages. In the branching zone, LRPs are present between already-emerged LRs. In the shootward portion of this zone, secondary growth is in progress [32]. A hallmark of the beginning of the secondary growth is the formation of a relatively short pericycle cells (Fig. 5a), which are longer than Stage I LRP cells (compare Fig. 5a and Fig. 3a–c). These relatively short pericycle cells divide periclinally and form piles of the cells (Fig. 5b), which are derivatives of pericycle, and show increased

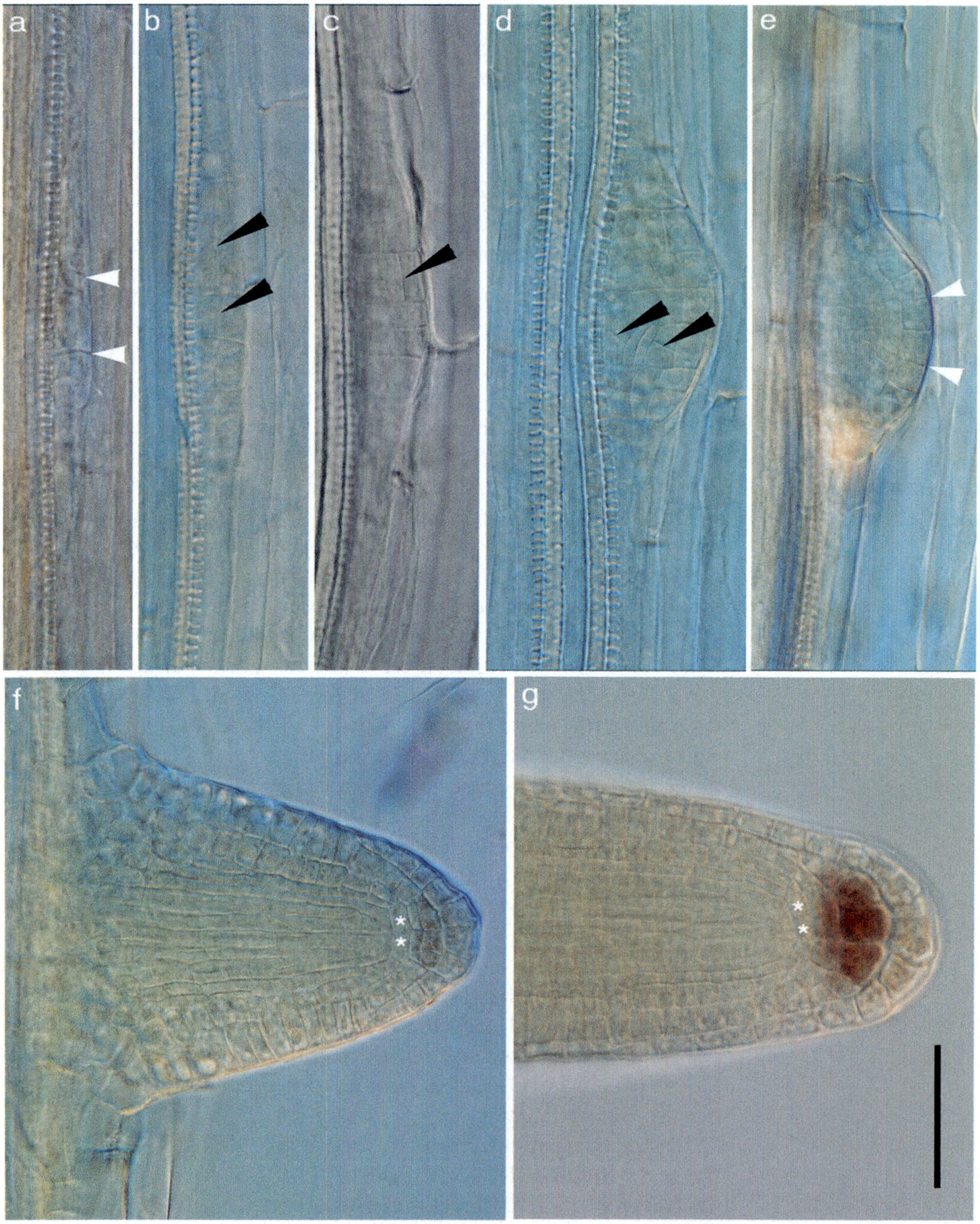

Fig. 4 Lateral root primordia and emerged lateral roots cleared with NaI-based protocol. (**a**) Stage I LRP; white arrowheads indicate transversal cell walls of a single relatively short cells resulted from LR initiation. (**b**) LRP stage II. Note that not all LRP cells passed through periclinal divisions indicated by black arrowheads. (**c**) LRP stage III. Note that a periclinal division in a single cell (indicated by black arrowhead) resulted in the beginning of the formation of the third layer within the LRP is sufficient to qualify it as Stage III. (**d**) Stage IV LRP. Note that in this LRP four cell layers in the core portion can be recognized and in the lower core portion the fifth layer is

birefringence (Fig. 5b–e). Some slowly developing or arrested primordia are frequently found in this root portion, and here we define how to classify their developmental stages as well.

It is essential to use differential interference contrast (DIC) microscopy, or Nomarski optics. Without DIC, it is practically impossible to identify early developmental stages. It is very common for beginners to underscore early LRP developmental stages. Even when DIC optics is used, students that start to identify LRPs are able to find approximately 50% more LR initiation events on the same preparations after a month or two of practicing. A common distribution of LRP stages within the LR formation zone is shown in Fig. 6. Note that more than half of LRPs within the LR formation zone of Col-0 and Ler wild-type seedlings at 8–10 dag are at the Stages I and II and that the average number of LRPs within LR formation zone at this age is close to 16. If this distribution is changed under a treatment or in a mutant, this is a clear sign that either the LR initiation or LR emergence or both are changed. Indirectly, it indicates that the rate of LR formation (time from initiation to emergence) is also affected.

To identify the developmental stage of each LRP, we use a conservative approach, i.e., in the cases where the most essential feature of the subsequent developmental stage is not present yet, the assigned stage is the previous one. Similarly, if the most essential feature of a stage is already present, an LRP is categorized as of this stage. The examples are given below.

Stage I of LRP development. The most essential feature: one-cell-layer LRP is formed as a result of anticlinal divisions of pericycle founder cells. At this stage, an LRP can contain eight to ten cells [13]. When the founder cells start to form LRPs, a single short cell within pericycle cell layer is a clear indication of LRP initiation (Fig. 4a). The presence of one or two short cells is sufficient to qualify them as Stage I LRP [7] (Figs. 3a, b and 4a). In addition to anticlinal divisions, oblique divisions in founder cells can also lead to Stage I LRP formation (Fig. 3b and [28]).

Stage II of LRP development. The most essential feature: two cell layers within an LRP. Commonly, after completion of the anticlinal divisions in the founder cells, the Stage I LRP cells start periclinal divisions (Figs. 3e and 4b), resulting in the formation of outer layer (OL) and inner layer (IL) of cells. However, periclinal

Fig. 4 (continued) formed. In spite of this, anticlinal division in the outer cell layer did not start yet; this feature does not permit to categorize this primordium as Stage V. Note the presence of oblique divisions (black arrowheads). (**e**) Stage VI LRP. Note that this LRP pattern in the external layer is 4-2-4. White arrowheads indicate those two outer layer cells that passed through periclinal division. (**f**) Recently emerged LR. (**g**) Root apical meristem of recently emerged LR. Asterisks indicate the quiescent center. Note that in (**f**) and (**g**) the root apical meristem organization and cell types can be clearly recognized. All panels show Col-0 roots at 8 (**a**) and 12 (**b–g**) days after germination. Scale bar: 20 μm. The same magnification for all panels

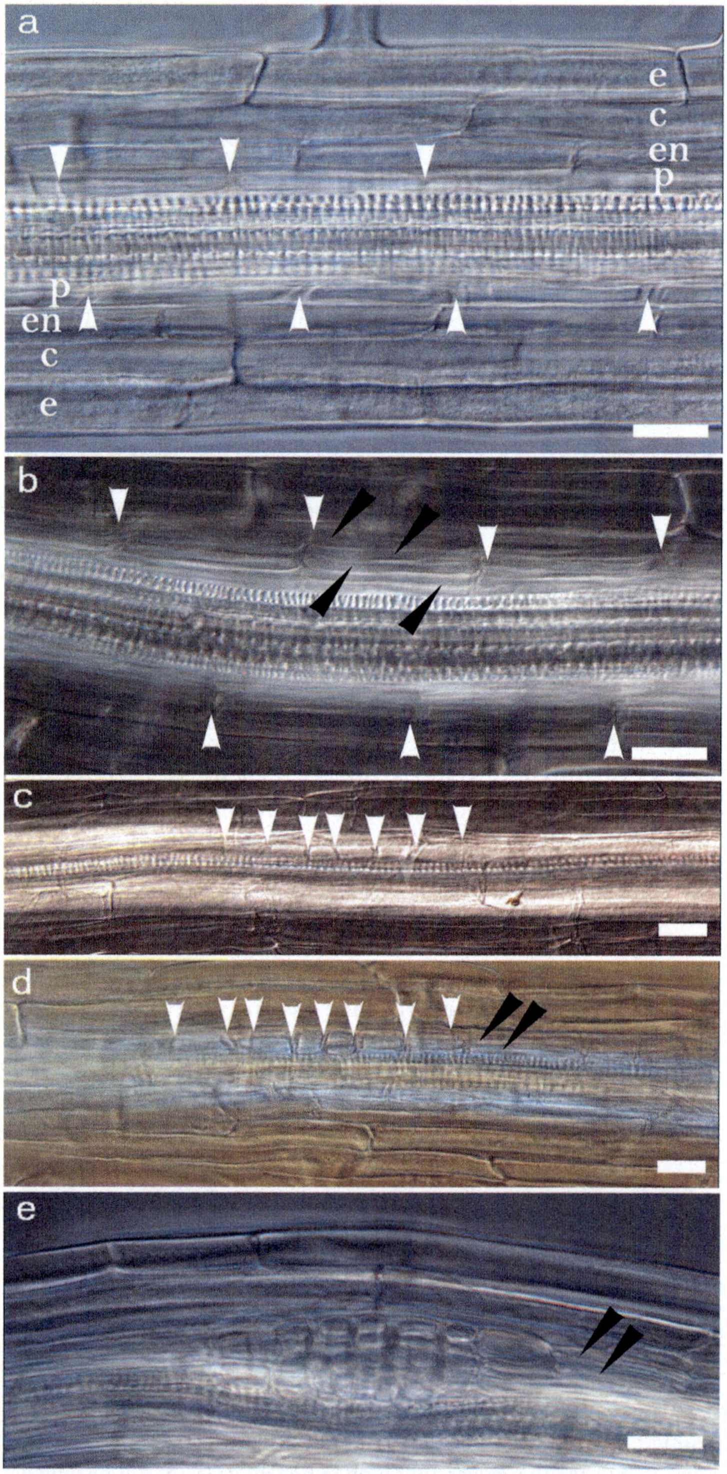

Fig. 5 Pericycle cells and its derivatives in the primary root portion within the root branching zone cleared with methanol and NaOH. (**a**) Relatively short pericycle cells which do not participate in LRP formation; e, epidermis; c, cortex; en, endodermis; p, pericycle. (**b**) Stacks of pericycle derivative cells resulted from

division can start much earlier when only one [28] or two short cells of Stage I LRP cells (Fig. 3c) are formed. Periclinal division in only one cell of Stage I LRP is sufficient to qualify it as a Stage II LRP (Fig. 3c, d). DIC optics permits optical sectioning. To unambiguously establish that the periclinal division took place, it is of paramount importance to ensure that the optical plane in which periclinal cell wall observed is real and from the cell. For example, Fig. 4a shows a Stage I LRP, in which a white line parallel to the protoxylem can be observed in a cell marked with arrowheads. This line is a shade produced by a subtending cell, and not a cell wall resulted from periclinal cell division. So, this is not a Stage II but Stage I LRP. Similar approach is used during the recognition of other developmental stages.

Stage III of LRP development. The most essential feature: presence of three cell layers within an LRP. Periclinal division of a single cell of Stage II LRP is sufficient to qualify it as Stage III LRP (Fig. 4c).

Stage IV of LRP development. The most essential feature: presence of four cell layers formed, IL1, IL2, OL1, and OL2. In some cases more than four layers can be formed, and still the stage should be considered Stage IV (Fig. 4d). In spite of the fact that more than four cell layers are recognized, yet the most essential feature of Stage V is not present in this LRP. So, in accordance with a conservative approach described, this LRP is categorized as Stage IV. In this LRP, anticlinal division in the outer cell layer, which is a characteristic of Stage V, has not started yet. By this reason, the LRP in Fig. 4d cannot be categorized as Stage V LRP.

Stage V of LRP development. The most essential feature: cells in OL1 and OL2 divide anticlinally. Up to 10–12 small cuboidal cells can be formed in the OL1 layer.

Stage VI of LRP development. The most essential feature: the central cells of OL1 divide periclinally. Such divisions in one or two cells are sufficient to categorize it as Stage VI (Fig. 4e). When this stage is completed, 4-4-4 pattern is formed [13].

Fig. 5 (continued) periclinal divisions of pericycle cells depicted on panel (**a**). Note an increased birefringence of the derivative cells; black arrowheads indicate individual cells. (**c**) Delayed or arrested Stage I LRP present among pericycle derivatives. (**d**) Delayed or arrested Stage II LRP present among pericycle derivatives resulted from periclinal divisions; pericycle derivatives are indicated with black arrowheads. Note the roots on (**c**) and (**d**) are laying on a slide in the protophloem plane. (**e**) Atypical delayed or arrested Stage IV LRP present among pericycle derivatives. Note that in this LRP more than four cell layers are formed; as no anticlinal divisions in outer layers OL1 and OL2 are present, this stage cannot be categorized as Stage V. Black arrowheads show derivatives of pericycle cells contiguous to the LRP (**d**, **e**). White arrowheads indicate the end walls of pericycle cells (**a**), their derivatives (**b**), or LRP cells (**c**, **d**). Col-0 (**c**) and Ws (**a**, **b**, **d**, **e**) seedlings 8 days after germination. Scale bar: 20 μm. The same magnification in panels (**a**, **b**, and **e**) and panels (**d** and **c**)

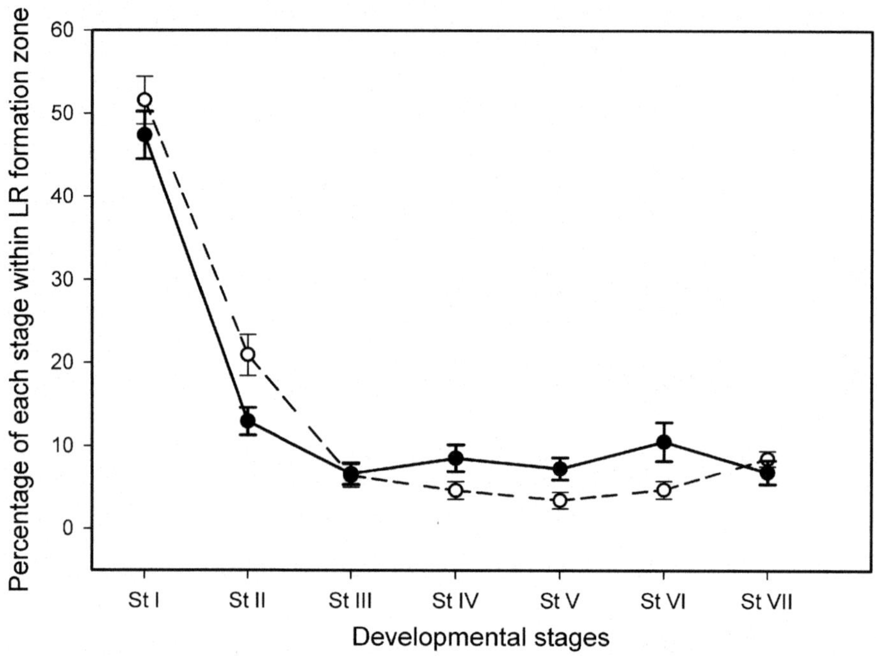

Fig. 6 Distribution of lateral root primordia stages within the LR formation zone in Col-0 (open circles) and Ler (closed circles) 8 and 10 dag roots, respectively; mean ± SE. Average of total number of LRPs per LR formation zone in a seedling is 16.3 ± 0.8 (Col-0) and 16.4 ± 1.0 (Ler), $n = 16$–18. Seedling grew in Petri dishes on 0.2× MS growth medium supplemented with 1% sucrose and solidified with 0.8% agar

Stage VII of LRP development. The most essential feature: an LRP just before emergence with a pattern more advanced than 4-4-4. When this stage is completed, 8-8-8 pattern is formed [13].

Emerged LR. The most essential feature: appearance of the LR above the surface of the epidermis when viewed under the microscope or with a magnifying glass (Figs. 3g and 4f, g) (*see* **Note 19**).

Lateral root primordia within the branching zone. If Stage I or II LRP were arrested in their development, they still can be found in the proximal portion of the branching zone where secondary growth has been started (Fig. 5c, d). Frequently arrested LRPs in this root portion are located contiguously with pericycle cell derivatives (Fig. 5d, e). *A. thaliana* root has a diarch radial organization. As LRP initiation and development takes place in pericycle cells adjacent to protoxylem, to define an LRP developmental stage, it is desirable that the primary root is positioned so that the protoxylem plane is parallel to the slide surface (Fig. 5a). When roots are positioned obliquely, it is still feasible to define an LRP stage (*see* **Notes 20** and **21**).

1. Place a slide with a preparation on a stage of a microscope equipped with DIC optics. Using a 10× objective, find the appropriate focal plane.

2. Focus on the specimen and establish Köhler illumination (*see* **Note 22**).
3. Change magnification to 20× and 40× or greater. To achieve maximum contrast, adjust positions of Wollaston prism, polarizers, and condenser diaphragm (*see* **Note 23**).
4. Locate the primary root tip and slowly move the field of view basipetally (toward the root base or shootward) along the root. Using a micro-adjustment knob, focus on pericycle cell layer and locate the first LRP closest to the root tip.
5. Move the objective revolver of the microscope to a position without an objective or to the lowest magnification objective.
6. Mark over the slide the location of the light path with an ultra-fine-point pen marker. This mark will correspond to the most distal LRP location. Return to the high magnification objective, and scan the root up and down through the tissues and field by field in a shootward direction.
7. Record in a logbook each LRP developmental stages and LR found in order of appearance (*see* **Note 24**). If LRP stages are closely spaced (as in Fig. 3f), consider them as two different LR initiation events, if the distance between them is greater than the diameter of the LRP.
8. When the most distal LR is found, repeat **step 5** and mark over the slide its position with ultra-fine-point pen marker (Fig. 2e).
9. Continue to record all LRs and LRPs in the branching zone. Finally, the record of all the LR initiation events will be completed from the latest LRP initiation to the root base.

3.3 Quantitative Analysis of Lateral Root Development

General approaches to quantitative analysis of LR development in *Arabidopsis* have been described [16, 17, 22]. Here, we focus on the most important parameters that permit to compare different genotypes and/or treatments quantitatively. Each of the described parameters is obtained for each individual root. Therefore, the roots have to be numbered, and values of the length of each root portion and number of LRPs or LRs in a respective portion should be associated within an individual root. It is recommended to perform at least two independent experiments in which various parameters are estimated. If the results of these experiments show similar trends, the data can be combined. If no concordance exists between two independent experiments, more experiments have to be performed. First, we describe here how to estimate the number of LRs and branching density in seedlings without clearing. In the following sections, we will explain how to obtain various LR development parameters on cleared root preparations. It is important to keep in mind that estimation made without clearing gives only preliminary evaluations; some young recently emerged LR are not visible by the naked eye.

3.3.1 Estimation of Number of LRs and Branching Density in Seedlings Without Clearing

Seedlings of 8–10 dag are used for quantitative analysis. Usually 10–12 seeds are sown in 100 × 100 × 15 mm square Petri dishes with 38 mL of MS medium (*see* **Note 25**), and the dishes are sealed with 3M Micropore tape, which is important to not inhibit gas exchange.

1. Score the number of the first-order LRs on the primary root either by the naked eye or with a binocular microscope. A magnifying glass can be used as well.
2. Scan the Petri dishes with seedlings.
3. Launch the ImageJ program (http://rsb.info.nih.gov/ij/) and open the seedlings image (*see* **Note 26**).
4. In the main menu, select "analyze" tab and choose the "set scale" window.
5. If the images were scanned at 800 dpi, in the "set scale" window, in the box "distance in pixels," write 800 and "known distance" as 25.4 mm, corresponding to 1 in. (*see* **Note 27**); indicate "pixel aspect ratio" as 1, "unit of length" as mm, check "Global," and finally click "OK" to save the values.
6. Select "segmented line" option and start to measure the length of the branching zone, which comprises the root portion from the hypocotyl and root boundary (root base) to the most distal emerged LR (Fig. 1).
7. To calculate branching density for each individual seedling, divide the number of LRs by the length of the root branching zone, expressed in mm (Fig. 7a; *see* **Notes 28** and **29**).

3.3.2 Density of All LR Initiation Events Detected on Cleared Root Preparations

Density of all LR initiation events refers to the density of both, LRs and LRPs, along the root portion that comprises LR formation and branching zones (Fig. 1; *see* **Note 30**).

1. After completing recording all the LRPs and LRs, with the aid of desktop image scanner, scan the microscope slides with root preparations on which the position of the most distal LRP and LR is marked (*see* **Note 31**).
2. On scanned images of root preparations, measure by ImageJ, as described in Subheading 3.3.1, the following distances in mm: from the most distal LRP to the most distal LR (LR formation zone) and from the most distal LR to the root base (branching zone).
3. From the records obtained, as described in Subheading 3.2, **step 9**, count the combined total number of LRPs and LRs of the entire root.

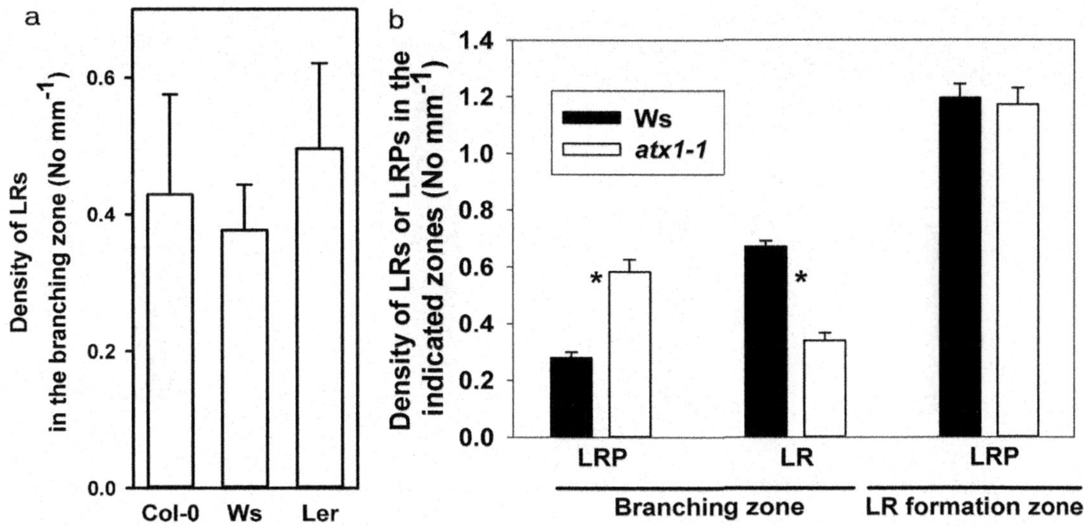

Fig. 7 Density of lateral roots (LRs) and lateral root primordia (LRPs) in wild type and in *Arabidopsis homolog of trithorax1* (*atx1-1*) mutant. (**a**) Lateral root density estimated without clearing in three wild-type accessions; 8–9 dag seedlings. Mean ± SD, $n = 12–14$. Counts and measurements were performed based on scanned seedling grown in Petri dishes. (**b**) LR and LRP density in the branching and LR formation zones in Ws wild type and *atx1-1* mutant 8 dag. Mean ± SD; $n = 22$; $*P < 0.001$, Student's *t*-test. Note that LR density in branching zone in Ws estimated without clearing (**a**) is lower than the same parameter evaluated in cleared root. In *atx1-1* mutant, LR initiation is not affected (see bars of the densities of LRP in the LR formation zone). Note that in the mutant LR density in the branching zone is decreased but LRP density is increased; this indicates that some LRPs are not developing into LRs and the emergence is affected in the mutant. Panel (**b**) is reproduced with some modifications from Napsucialy-Mendivil S, Alvarez-Venegas R, Shishkova S, Dubrovsky JG (2014) *ARABIDOPSIS HOMOLOG of TRITHORAX1* (*ATX1*) is required for cell production, patterning, and morphogenesis in root development. Journal of Experimental Botany, 65: 6373–6384 with permission of Journal of Experimental Botany

4. To obtain the density of all LR initiation events, divide the total number of LRPs and LRs by the length of the primary root from the most distal (rootward) LRP to the root base and express it as number per millimeter.
5. Create a graph representing the data on the density of all LR initiation events similar to that in Fig. 8.

3.3.3 LRP Density in the Lateral Root Formation Zone Determined on Cleared Root Preparations

1. From the records obtained, as described in Subheading 3.2, **step 9**, count the number of LRPs localized between the most distal (rootward) LRP and the most distal emerged LR (i.e., LRPs within the LR formation zone; Fig. 1).
2. Obtain the LRP density in the LR formation zone dividing the number of LRPs found in this zone by the length (in mm) of the LR root formation zone obtained in Subheading 3.3.2, **step 2**.

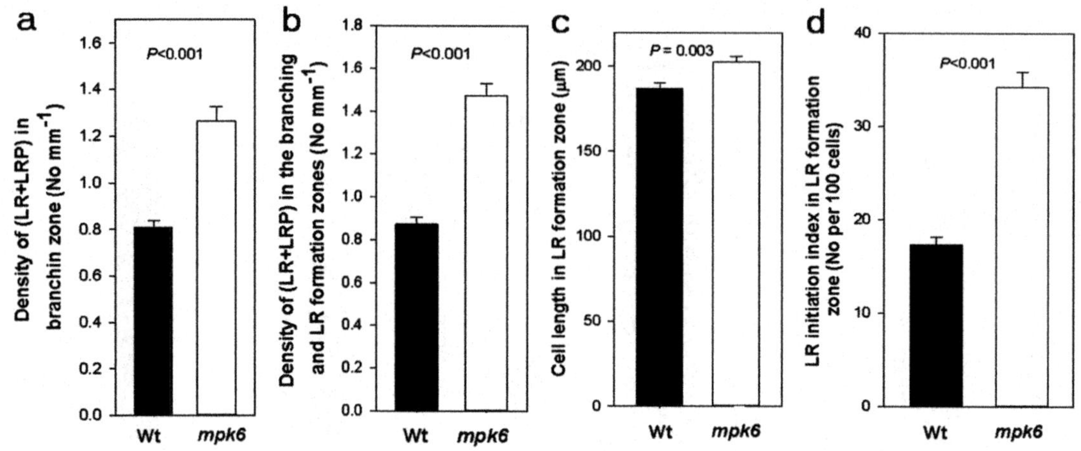

Fig. 8 An example on increased lateral root initiation caused by a gene mutation. Density of lateral roots (LRs) and lateral root primordia (LRPs) and LR initiation index (I_{LRI}) in wild type (Col-0) and in *mitogen-activated protein kinase 6* (*mpk6*) mutant. Density of all LR initiation events in the branching zone (**a**) and in the branching and LR formation zone (**b**). Note that in both panels values are very similar. (**c**) Cell length in the LR formation zone. For each individual root, ten fully elongated cortical cells were measured. (**d**) Lateral root initiation index (I_{LRI}) in the LR formation zone. Note that LR initiation in *mpk6* mutant is increased based both on density and LR initiation index. Error bars represent mean ± SE, $n =$ of 23 roots from two independent experiments. Asterisks mark Student's *t*-test significant differences at the *P* values indicated. Reproduced with some modifications from López-Bucio JS, Dubrovsky JG, Raya-González J, Ugartechea-Chirino Y, López-Bucio J, de Luna-Valdez LA, Ramos-Vega M, León P, Guevara-García AA (2014) *Arabidopsis thaliana* mitogen-activated protein kinase 6 is involved in seed formation and modulation of primary and lateral root development. Journal of Experimental Botany, 65: 169–183 with permission of Journal of Experimental Botany

3.3.4 LR Density in the Root Branching Zone Determined on Cleared Root Preparations

1. From the records obtained, as described in Subheading 3.2, **step 9**, count the number of LRs starting from the most distal emerged LR to the root base.
2. To obtain the branching density, divide the number of LR by the length of the branching zone of the primary root obtained in Subheading 3.3.2, **step 2**.

3.3.5 Percentage of LRs or LRPs Identified on Cleared Root Preparations

These parameters are useful to determine whether LR initiation, LR emergence, or both are affected in a genotype or under a treatment.

1. From the records obtained, as described in Subheading 3.2, **step 9**, estimate the percentage of LRs and LRPs out of all LR initiation events.
2. From the same records, estimate the percentage of LRs and LRPs only in the branching zone.

3.3.6 Distribution of LRPs by Developmental Stage Within the LR Formation Zone

1. From the records obtained, as described in Subheading 3.2, **step 9**, count the number of LRPs of each developmental stage within the LR formation zone (Fig. 1).

2. Estimate the percentage of each LRP developmental stage with respect to the total number of LRP in the LR formation zone.
3. Create a graph representing the data on the percentages of each of the LRP developmental stage similar to that in Fig. 6 (*see* **Note 32**).

3.3.7 LR Initiation Index

The LR initiation index evaluates the number of LR initiation events that take place in a root segment equivalent to the average length of 100 cortical cells [17, 22]. The evaluation of this parameter is very important because it permits normalization of the density data by the length of fully elongated cells. Any genotype, different from wild type, and/or any treatment may cause differences in cell length. Hence, simple density data may not be sufficient to appreciate the effect of a genotype or a treatment. For example, sodium nitroprusside does not change the density of all LR initiation events, but as cells become shorter during the treatment, compared to untreated roots, LR initiation index diminishes almost two times [22] showing clear effect on LR initiation (Fig. 9). It is known that protoxylem- and protophloem-adjacent pericycle cells in the differentiation zone differ in length

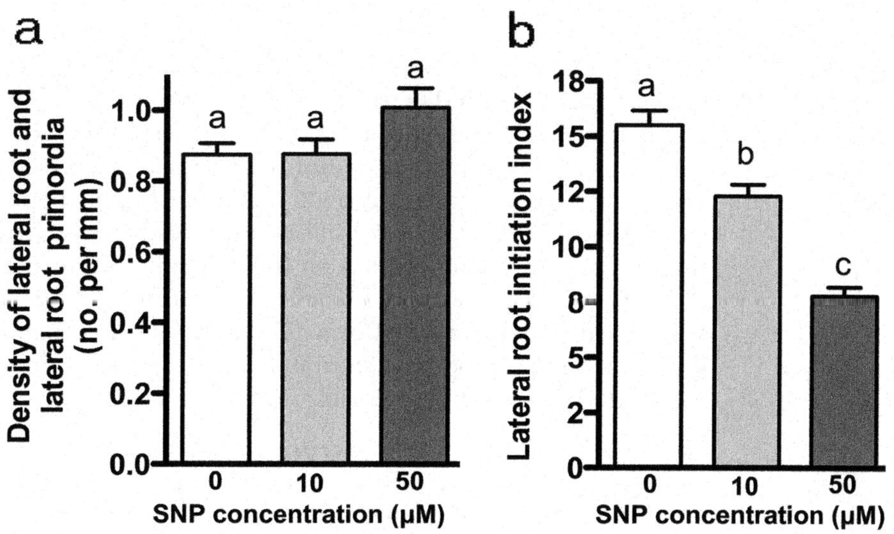

Fig. 9 An example of a decrease in lateral root initiation caused by a treatment. Roots grew for 5 days on growth medium and then were transferred either to the same medium or to the same medium supplemented with sodium nitroprusside (SNP) and grew for 5 additional days. The data shown are for the distal root portion formed during treatment. (**a**) Density of all lateral root initiation events, including lateral roots and primordia. (**b**) Lateral root initiation index estimated as the number of all initiation events along the length of root portion that comprises 100 cortical cells of average length. Combined data of at least three experiments ($n = 25$–38), mean $\pm$ SE. Different letters above graph bars indicate statistical differences among treatments at $P < 0.05$ in Tukey's multiple comparison test. Reproduced with some modifications from Lira-Ruan V, Napsucialy-Mendivil S, Dubrovsky JG (2013) Heuristic aspect of the lateral root initiation index: A case study of the role of nitric oxide in root branching. Applications in Plant Sciences, 1(10): 1300029, doi:10.3732/apps.1300029

[2, 7]. Additionally, before cambium formation, pericycle cells become shorter (Fig. 5a). By this reason, normalization by pericycle cell length is not recommended.

1. With an ocular micrometer, measure the length of ten cortical cells in the same file within the differentiation zone (where the cells are fully elongated) and estimate the average cell length per root (*see* **Note 33**).
2. Using an object-micrometer with a known distance, convert units of an ocular micrometer into micrometers. Express average length of measured cortical cells in mm.
3. Calculate the density of all LR initiation events as described in Subheading 3.3.2.
4. Calculate the LR initiation index (I_{LRI}) for each individual root using the equation $100dl$, where d is the density of lateral root initiation events (number/mm) and l is the average length of the cortical cell (mm) for the same root.
5. Compare averages of I_{LRI} values for each group, and perform statistical analysis.
6. Create a graph with the data of the LR initiation index similar to that shown in Figs. 8d and 9b.

3.4 Quantitative Analysis of Lateral Root Development Before and After a Treatment

Genetic and phenotypic analysis of lateral root development frequently includes studies of hormonal response of a mutant in comparison with wild type. To establish the response(s) of a mutant to a hormonal treatment (i.e., normal response, increased sensitivity, or acquired resistance), growth medium is supplemented with a substance of interest, and responses are investigated. Growth regulators have differential effects on the root portions formed before and during a treatment [21–23]. Therefore, it is advisable to study in the same experiment effects of a hormone treatment on LR initiation and development separately for the root portions formed before and during the treatment.

1. Sterilize the seeds and plate them in Petri dishes in a medium not supplemented with a substance of interest (*see* Subheading 3.3.1). Let the seedlings grow for 5 days.
2. Prepare Petri dishes with a growth medium supplemented with a substance of interest.
3. With a sterile scalpel, excise the upper 8–10 mm segment of agar supplemented with a substance of interest.
4. Take with sterilized forceps a seedling from a Petri dish where seedlings germinated, and gently transfer it to a dish with medium supplemented with a substance of interest. Place the seedling in such a way that its aerial part is located above the agar layer where the excised segment was taken away and its

root is arranged vertically. Avoid direct contact of leaves with agar (*see* **Note 34**). Repeat the same with other seedlings.

5. Immediately after transfer, mark with an ultra-fine-point pen marker a position of the root tip and number each seedling.
6. Return the Petri dish to a growth chamber and let them grow for 5 days more.
7. Using a razor blade, cut gently the root at the marked position and separate the root portion above the mark (proximal root portion formed before the transfer) and below the mark (distal root portion formed after the transfer).
8. Collect the root portions formed before and after termination of the treatment in containers with either a fixative or the first step solution of a Malamy and Benfey-modified clearing protocol.
9. Clear roots and make microscope preparations as described in Subheadings 3.1.2 and 3.1.3.
10. Analyze the preparations as described in Subheading 3.3.
11. Obtain data on the length of the proximal and distal root portions, LR density, LRP density, and LR initiation index in these portions.

3.5 Analysis of Abnormalities in LR Primordium Morphogenesis

One aspect of phenotypic characterization of LR formation is directly related to the LRP and LR morphogenesis. A number of mutants have been identified which have various abnormalities in their development [24]. These mutants include those affected in LR initiation, LRP positioning, and LRP morphogenesis (e.g., [14, 33, 34]). Analysis of disturbed cell division pattern in a mutant LRP is based on counts of number of cells in the outermost layer and measurements of LRP width [35]. Depending on the type of abnormality, the analysis can be different and no unified protocol can be proposed. Here, we describe a straightforward analysis of asymmetry of LRP formed.

3.5.1 Asymmetry in LRP Shape at Late Developmental Stages

To analyze the symmetry of the LRP at Stages V–VII, it is essential to perform good-quality root clearing. Only roots positioned in the protoxylem plane can be used for this analysis [36] (Fig. 10a, b).

1. Place a root preparation on the microscope stage and follow the steps described in Subheading 3.2.
2. Find Stage V–VII LRPs of the parent root laying in the protoxylem plane and take a photo of each one with a camera.
3. Using ImageJ software, open a file with an LRP image and draw a line at the base of the LRP where it consists of more than one cell layer as shown in Fig. 10a (line b).

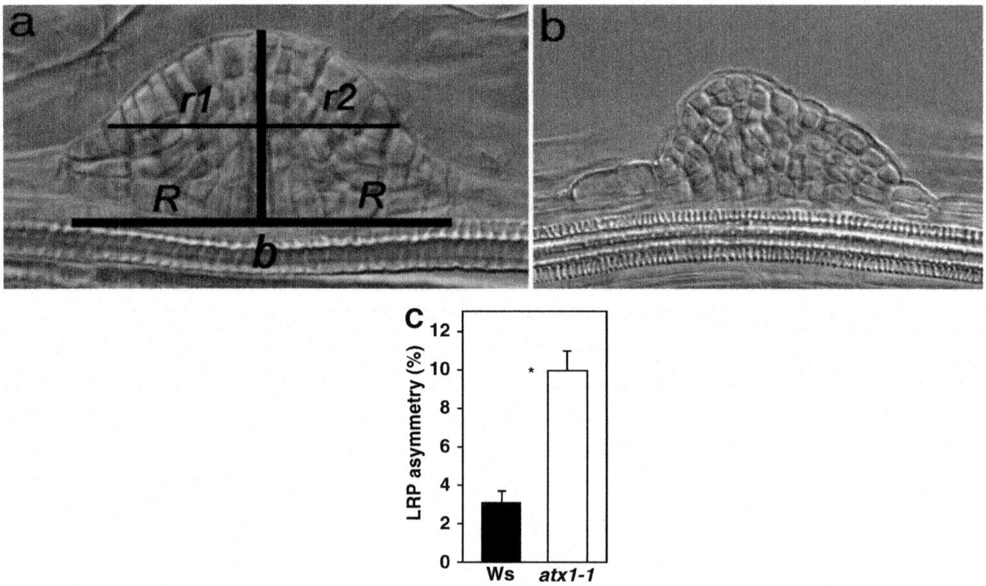

Fig. 10 Analysis of asymmetry of lateral root primordia. (**a**) A scheme showing how primordium asymmetry is evaluated. A primordium base (*b*) length is measured; from the center of the primordium base, a perpendicular line corresponding to the longitudinal axis of a developing primordium is drawn. From the center of this longitudinal axis, two radii, *r1* and *r2*, are drawn and measured. When the radii are of different lengths, the longer radius is considered as r_l and the shorter as r_s. Ws and *atx1-1* plants. (**b**) Asymmetric primordium in *atx1-1* mutant. (**c**) Quantification of asymmetry in wild type (Ws) and in *atx1-1* primordia; $^*P < 0.001$, Student's *t*-test; mean $\pm$ SE, $n = 25$ LRPs in Ws and 39 LRPs in *atx1-1*. Reproduced with some modifications from Napsucialy-Mendivil S, Alvarez-Venegas R, Shishkova S, Dubrovsky JG (2014) *ARABIDOPSIS HOMOLOG of TRITHORAX1* (*ATX1*) is required for cell production, patterning, and morphogenesis in root development. Journal of Experimental Botany, 65: 6373–6384 with permission of Journal of Experimental Botany

4. Using the same software, measure, as described in Subheading 3.3.1, the length of the LRP base.
5. Dividing this distance by 2, find the center of the LRP base. Draw a perpendicular line (an axis of the nascent root) from the center of the LRP base line as shown in Fig. 10a.
6. Using ImageJ software, measure the longitudinal axis of the LRP and find its center. Draw two radii parallel to the LRP base from the center of this axis and measure the radii *r*1 and *r*2 length. When the length of the two radii is the same, the LRP is considered symmetrical. If the lengths of the radii are different, a longer radius (r_l) and a shorter (r_s) radius are determined.
7. Estimate the percentage of asymmetry (A) for each primordium using the equation $A = r_l - r_s (r_l + r_s)^{-1}$ 100, where r_l is the longer radius and r_s is the shorter radius.
8. Create a graph with the data of the percentage of asymmetry (*see* Fig. 10c).

3.6 Marker Lines Used for Characterization of LR Development

Depending on particular processes affected in certain mutants, different marker lines can be used. The most frequently used lines are as follows:

Auxin response: *DR5::GUS* [29] and *DR5rev::GFP* [37].

Auxin transport: *AUX1::GUS* [38], *AUX1::AUX1:YFP* [39], *PIN1::GUS*, *PIN2::GUS*, *PIN3::GUS* [40], *PIN1::PIN1:GFP*, and *PIN2::PIN2:GFP* [41].

Cell cycle markers: $CycB1;1_{DB}$*::GUS* [31].

Cell type markers: protoxylem-adjacent pericycle J0121 line [42], endodermis and quiescent center marker *SCR::GFP* [43], and quiescent center marker *WOX5::GUS* and *WOX5::GFP* [44].

Plasma membrane marker: *UB10::WAVE131:YFP* [45].

To characterize loss-of-function mutants, a marker is commonly introgressed by genetic crosses. In the recessive mutants with strong phenotype, seedlings with specific phenotype segregate in F2 generation in ratio 3 (wild-type phenotype) to 1 (mutant phenotype). Selected seedlings with a mutant phenotype are verified on the presence of expression of the reporter. In mutants with less obvious phenotype, F2 seedlings should be genotyped by the polymerase chain reaction. When mutant genotype is confirmed in F2 generation, verify that the marker is expressed. From the confirmed seedlings, obtain F3 generation and maintain the pure line.

4 Notes

1. The size of the ELISA plates depends of the length of the seedling's primary root. Six-well ELISA plate is sufficient for 8 dag wild-type seedlings.
2. To convert the units of the ocular micrometer to micrometers, use an object-micrometer.
3. It is recommended to use this solution fresh. The solution can be kept at −20 °C.
4. To prepare 10% stock, weigh 10 g of paraformaldehyde and dissolve it with 100 mL of hot distilled water; add drops of NaOH until the solution becomes transparent. All the paraformaldehyde solutions should be handled under the fume hood. Aliquots of the stock solution can be stored −20 °C for months. Either phosphate or PIPES buffer, pH 7.2, can be used to prepare the fixation solution.
5. Many different clearing protocols are available [46–48]. In *Arabidopsis* and other species, many protocols use chloral hydrate [49–52]. In some countries, e.g., in the USA [53], this compound is a regulated one and requires special permits

for handling. In certain protocols chloral hydrate is used together with phenol [49, 51] which is harmful for human health.

6. Original Malamy and Benfey (1997) protocol [13] works well for wild-type seedlings. In our experience some mutants are characterized by more dense cytoplasm than in wild type and their clearing represents a challenge, e.g., *moots koom 1* (*mko1*) mutant [54]. The modified protocol of Malamy and Benfey (1997) described here includes some modifications reported earlier [15]. With these modifications material becomes more transparent. However, sometimes even these modifications are not sufficient for good clearing of mutants with dense cytoplasm. In these cases, it is preferable to use NaI-based protocol.
7. NaI-based clearing-and-mounting solution [17] has a high refractive index, close to 1.5 [55], and is an easy one-step clearing procedure.
8. Only freshly prepared fixative solution can be used. The duration of the fixation can be extended. We did not note the deterioration of samples up to 1 month of fixation at 4 °C.
9. Avoid taking for analysis those seedlings whose roots grew inside the agar or touched the bottom of vertically oriented Petri dish.
10. Ensure that the seedlings are not floating. If this is the case, use micropipette drop of the same solution over the seedling until they sink.
11. For some mutants with thicker roots, the time at 62 °C should be increased to 80 min.
12. The time of incubation in each step can be increased, and usually we incubate 24 h per step.
13. For better clearing results, the incubation during this step can be increased to 24 h. For mutants with thicker roots and/or denser cytoplasm, the samples can be incubated over a week.
14. Be careful with the roots at this point because these are very fragile. Do not fold the roots; if the roots are longer than the frame prepared, curve the primary root on a slide (Fig. 2d).
15. If a 22 × 60 mm coverslip is used, the total solution volume is approximately 150 μL. If the mounting solution is applied in excess, it will be leaking out of the coverslip. In this case, use absorbing paper to clean microscope slide from any solution outside the coverslip.
16. To avoid bubbles under coverslip, maintain a coverslip at angled position above the parafilm frame supporting coverslip's shorter side with a single forceps branch. To avoid its slipping, support other shorter coverslip side with a needle.

Gently and slowly lower a forceps branch until the coverslip completely covers the parafilm frame.

17. It is recommended to apply nail polish twice to the same preparation. During the first application, place nail polish generously at the sides of the coverslip and let it to dry. During the second application, lower amount of nail polish is used to seal possible openings.
18. Any brand of nail polish can be used. We prefer transparent nail polish. We also successfully use a nitrocellulose lacquer.
19. Some authors propose Stage VIII of LRP formation [56] for recently emerged LRs. By definition an LRP is converted to a LR when it emerges. By this reason we do not consider Stage VIII of LRP formation.
20. When most of the root length is positioned in the protophloem plane parallel to the slide surface (as in Fig. 5c,d), many early-stage LRP can be missed, and thus, the LRP number can be underscored. Therefore, if most of the primary root is in the protophloem plane, roots should be discarded from the analysis. We discard 1/10 or less of all the roots in a sample.
21. When LRP stage identification is uncertain, we indicate that as "nd" (not defined).
22. Establishment of the Köhler illumination is an obligatory step as it permits to set up a uniform illumination of a sample, which is critical for good contrast. To establish Köhler illumination, refer to microscopy technique basic books, e.g., [57].
23. A greater magnification gives better results. Be mindful of not to squash the preparation with the microscope objective. This depends on a working distance of a particular objective. Remember that the whole mounts of cleared roots are thick preparations. If inappropriate objective is used, this can damage the lens of the objective. We commonly use 40× and 60× dry or 63× water-immersion objectives (depending on microscope model).
24. It is useful to score the LRP developmental stages and count the LRPs in order of their appearance over the primary root starting from the most distal LRP. An example of a complete record of LRP stages (Roman numbers) and LRs (marked as E: emerged) for a single primary root of an 8-dag Col-0 seedling is the following: I, I, II, III, I, I, I, V, I, II, VI, II, VII, I (14 LRPs)*, E, II, E, E, I, E, I, E, IV, I, E, II, I, E, E, II, E, E, E (11 LRs + 8 LRPs)**. One asterisk (*) refers to the LRPs in the LR formation zone only; ** refers to the number of LR initiation events in the branching zone only.
25. When quantitative parameters have to be evaluated, the volume of the growth medium has to be standard. Otherwise loss of

water by growth medium due to drying will be different in different Petri dishes. These differences may affect the resultant data.

26. Images of seedlings plates have to be scanned at 800 dpi (dots per inch) resolution or higher.
27. If the resolution is higher, indicate the number of pixels corresponding to 1 in.
28. It is important to take into account that in very young seedlings in which LR starts to emerge, the branching zone is very short, and many of LRPs initiated have not emerged yet. The more reliable results are obtained when one third (or more) of the length of the primary root bear LRs. Commonly, for LR number scoring, we use 7–8 dag seedlings at earliest. As branching zone is defined as the zone where LRs are present, it is very important to recognize the most distal recently emerged LR. Any LR that protrudes epidermis can be considered emerged LR. Position of the most distal LR is important both for LR scoring and for measuring the length of the branching zone.
29. Estimation of the number of LRs and the branching density without clearing is the easiest and fastest way to evaluate LR formation. In spite of this simplicity, we very rarely use this method. Instead, we clear roots and count emerged lateral roots on cleared root preparations. This approach simultaneously permits unambiguous determination of the position of the most distal emerged LR and scoring all LRPs within the primary root. The position of the most distal LR is of great importance also for correct measurement of the length of the LR formation zone (Fig. 1).
30. In the literature sometimes LRPs and LRs are referred to as "lateral organs" [15]. Strictly speaking as LRPs are not yet formed organs but primordia, we preferentially use the term "LR initiation events."
31. Images of slides have to be scanned at 1200 dpi resolution or higher.
32. When the percentage of each LRP developmental stage is analyzed, it is easier to perceive the differences between genotypes or treatments if they are present.
33. As fully elongated cells are long (around 180–200 μm), when the cell length is measured, it is important to ensure that the transversal cell walls are in the same focal plane as the longitudinal walls of the same cell. This permits to avoid a possible error of measuring distance between transversal walls of two different cells (e.g., of one cell in a focal plane at one side and of another side a wall of a subtending cell in the same focal plane).

By this reason we always measure cell lengths directly on a root preparation with an ocular micrometer and never on photos.

34. A contact of seedling shoots with agar facilitates sugar [58] and a substance-of-interest uptake by leaves which affect LR formation. Excised agar segment approach permits to avoid this problem [59]. It also is important to ensure that the roots of transferred seedlings are in direct contact with agar; otherwise, they can dry out.

Acknowledgments

We thank K. Nikoumanesh and H. H. Torres-Martínez for the critical reading of the manuscript, N. Doktor for the help with illustrations, and S. E. Ainsworth, J. M. Hurtado-Ramírez, and A. Ocádiz-Ramírez for their help with logistics. The work in the corresponding author laboratory is supported by the Mexican Consejo Nacional de Ciencia y Tecnología (CONACyT: Grant 237430) and Universidad Nacional Autónoma de México-DGAPA-PAPIIT (IN205315).

References

1. Charlton WA (1996) Lateral root initiation. In: Waisel Y, Eshel A, Kafkafi U (eds) Plant roots: hidden half, 2nd edn. Marcel Dekker, New York, pp 149–173
2. Laskowski MJ, Williams ME, Nusbaum HC, Sussex IM (1995) Formation of lateral root meristems is a two-stage process. Development 121:3303–3310
3. Malamy JE, Benfey PN (1997) Down and out in *Arabidopsis*: the formation of lateral roots. Trends Plant Sci 2:390–396
4. Dubrovsky JG, Rost TL, Colón-Carmona A, Doerner PW (2001) Early primordium morphogenesis during lateral root initiation in *Arabidopsis thaliana*. Planta 214:30–36
5. Beeckman T, Burssens S, Inzé, D (2001) The peri-cell-cycle in *Arabidopsis*. J Exp Bot 52:403–411
6. Jansen L, Demeulenaere M, Beeckman T, Eshel A, Beeckman T (2013) Lateral root development. In: Eshel A, Beeckman T (eds) Plant roots-the hidden half, 4th edn. CRC Press, Boca Raton, pp 6–1
7. Dubrovsky JG, Doerner PW, Colón-Carmona A, Rost TL (2000) Pericycle cell proliferation and lateral root initiation in *Arabidopsis*. Plant Physiol 124:1648–1657
8. Moreno-Risueno MA, Van Norman JM, Moreno A, Zhang J, Ahnert SE, Benfey PN (2010) Oscillating gene expression determines competence for periodic *Arabidopsis* root branching. Science 329:1306–1311
9. De Smet I, Tetsumura T, De Rybel B, Frei dit Frey N, Laplaze L, Casimiro I, Swarup R, Naudts M, Vanneste S, Audenaert D, Inzé D, Bennett MJ, Beeckman T (2007) Auxin-dependent regulation of lateral root positioning in the basal meristem of *Arabidopsis*. Development 134:681–690
10. Möller BK, Xuan W, Beeckman T (2017) Dynamic control of lateral root positioning. Curr Opin Plant Biol 35:1–7
11. Van Norman JM, Xuan W, Beeckman T, Benfey PN (2013) To branch or not to branch: the role of pre-patterning in lateral root formation. Development 140(21):4301–4310
12. Dubrovsky JG, Sauer M, Napsucialy-Mendivil-S, Ivanchenko MG, Friml J, Shishkova S, Celenza J, Benková E (2008) Auxin acts as a local morphogenetic trigger to specify lateral root founder cells. Proc Natl Acad Sci U S A 105:8790–8794
13. Malamy JE, Benfey PN (1997) Organization and cell differentiation in lateral roots of Arabidopsis thaliana. Development 124:33–44
14. Dubrovsky JG, Napsucialy-Mendivil S, Duclercq J, Cheng Y, Shishkova S, Ivanchenko MG, Friml J, Murphy AS, Benková E (2011)

Auxin minimum defines a developmental window for lateral root initiation. New Phytol 191:970–983
15. Dubrovsky JG, Gambetta GA, Hernández-Barrera A, Shishkova S, González I (2006) Lateral root initiation in *Arabidopsis*: developmental window, spatial patterning, density and predictability. Ann Bot 97:903–915
16. Dubrovsky JG, Forde BG (2012) Quantitative analysis of lateral root development: pitfalls and how to avoid them. Plant Cell 24(1):4–14
17. Dubrovsky JG, Soukup A, Napsucialy-Mendivil S, Jeknic Z, Ivanchenko MG (2009) The lateral root initiation index: an integrative measure of primordium formation. Ann Bot 103:807–817
18. Dubrovsky JG, Laskowski M (2017) Lateral root initiation. In: Murray BG, Murphy DJ (eds) Encyclopedia of applied plant sciences, 2nd edn. Academic Press, Oxford, pp 256–264
19. Fukaki H, Tasaka M (2009) Hormone interactions during lateral root formation. Plant Mol Biol 69:437–449
20. Jansen L, Hollunder J, Roberts I, Forestan C, Fonteyne P, Van Quickenborne C, Zhen R-G, McKersie B, Parizot B, Beeckman T (2013) Comparative transcriptomics as a tool for the identification of root branching genes in maize. Plant Biotechnol J 11:1092–1102
21. Ivanchenko MG, Muday GK, Dubrovsky JG (2008) Ethylene-auxin interactions regulate lateral root initiation and emergence in *Arabidopsis thaliana*. Plant J 55:335–347
22. Lira-Ruan V, Mendivil SN, Dubrovsky JG (2013) Heuristic aspect of the lateral root initiation index: a case study of the role of nitric oxide in root branching. Appl Plant Sci 1 (10):1300029. https://doi.org/10.3732/apps.1300029
23. Ivanchenko MG, Napsucialy-Mendivil S, Dubrovsky JG (2010) Auxin-induced inhibition of lateral root initiation contributes to root system shaping in *Arabidopsis thaliana*. Plant J 64:740–752
24. Szymanowska-Pułka J (2013) Form matters: morphological aspects of lateral root development. Ann Bot 112:1643–1654
25. De Smet I (2012) Lateral root initiation: one step at a time. New Phytol 193(4):867–873
26. Fukaki H, Okushima Y, Tasaka M (2007) Auxin-mediated lateral root formation in higher plants. In: Kwang WJ (ed) International review of cytology, a survey of cell biology, vol 256. Academic, San Diego, pp 111–137
27. Murashige T, Skoog F (1962) A revised medium for rapid growth and bio assays with tobacco tissue cultures. Physiol Plant 15 (3):473–497
28. Białek J, Potocka I, Szymanowska-Pułka JM (2014) Various scenarios of the cell pattern formation in *Arabidopsis* lateral root. Acta Soc Bot Pol 83(1):85–89
29. Ulmasov T, Murfett J, Hagen G, Guilfoyle TJ (1997) Aux/IAA proteins repress expression of reporter genes containing natural and highly active synthetic auxin response elements. Plant Cell 9:1963–1971
30. Manzano C, Ramirez-Parra E, Casimiro I, Otero S, Desvoyes B, De Rybel B, Beeckman T, Casero P, Gutierrez C, C del Pozo J (2012) Auxin and epigenetic regulation of *SKP2B*, an F-Box that represses lateral root formation. Plant Physiol 160(2):749–762
31. Colón-Carmona A, You R, Haimovitch-Gal T, Doerner P (1999) Spatio-temporal analysis of mitotic activity with a labile cyclin-GUS fusion protein. Plant J 20:503–508
32. Dubrovsky JG, Rost TL (2012) Pericycle. Wiley. New York doi:https://doi.org/10.1002/9780470015902.a0002085.pub2
33. De Smet I, Lau S, Voá U, Vanneste S, Benjamins R, Rademacher EH, Schlereth A, De Rybel B, Vassileva V, Grunewald W, Naudts M, Levesque MP, Ehrismann JS, Inzé, D, Luschnig C, Benfey PN, Weijers D, Van Montagu MCE, Bennett MJ, Jürgens G, Beeckman T (2010) Bimodular auxin response controls organogenesis in *Arabidopsis*. Proc Natl Acad Sci U S A 107:2705–2710
34. Van Damme D, De Rybel B, Gudesblat G, Demidov D, Grunewald W, De Smet I, Houben A, Beeckman T, Russinovaa E (2011) Arabidopsis a aurora kinases function in formative cell division plane orientation. Plant Cell 23:4013–4024
35. Hirota A, Kato T, Fukaki H, Aida M, Tasaka M (2007) The auxin-regulated AP2/EREBP gene *PUCHI* is required for morphogenesis in the early lateral root primordium of *Arabidopsis*. Plant Cell 19:2156–2168
36. Napsucialy-Mendivil S, Alvarez-Venegas R, Shishkova S, Dubrovsky JG (2014) ARABIDOPSIS HOMOLOG of TRITHORAX1 (ATX1) is required for cell production, patterning, and morphogenesis in root development. J Exp Bot 65:6373–6384
37. Friml F, Vieten A, Sauer M, Weijers D, Schwarz H, Hamann T, Offringa R, Jürgens G (2003) Efflux-dependent auxin gradients establish the apical-basal axis of *Arabidopsis*. Nature 426:147–153
38. Marchant A, Kargul J, May ST, Muller P, Delbarre A, Perrot-Rechenmann C, Bennett

MJ (1999) AUX1 regulates root gravitropism in *Arabidopsis* by facilitating auxin uptake within root apical tissues. EMBO J 18 (8):2066–2073
39. Swarup R, Kargul J, Marchant A, Zadik D, Rahman A, Mills R, Yemm A, May S, Williams L, Millner P (2004) Structure-function analysis of the presumptive *Arabidopsis* auxin permease AUX1. Plant Cell 16 (11):3069–3083
40. Benková E, Michniewicz M, Sauer M, Teichmann T, Seifertová D, Jürgens G, Friml J (2003) Local, efflux-dependent auxin gradients as a common module for plant organ formation. Cell 115(5):591–602
41. Xu J, Hofhuis H, Heidstra R, Sauer M, Friml J, Scheres B (2006) A molecular framework for plant regeneration. Science 311:385–388
42. Laplaze L, Parizot B, Baker A, Ricaud L, Martiniere A, Auguy F, Franche C, Nussaume L, Bogusz D, Haseloff J (2005) GAL4-GFP enhancer trap lines for genetic manipulation of lateral root development in Arabidopsis thaliana. J Exp Bot 56:2433–2442
43. Sabatini S, Heidstra R, Wildwater M, Scheres B (2003) SCARECROW is involved in positioning the stem cell niche in the *Arabidopsis* root meristem. Genes Dev 17:354–358
44. Sarkar AK, Luijten M, Miyashima S, Lenhard M, Hashimoto T, Nakajima K, Scheres B, Heidstra R, Laux T (2007) Conserved factors regulate signalling in *Arabidopsis thaliana* shoot and root stem cell organizers. Nature 446:811–814
45. Geldner N, Dénervaud-Tendon V, Hyman DL, Mayer U, Stierhof YD, Chory J (2009) Rapid, combinatorial analysis of membrane compartments in intact plants with a multicolor marker set. Plant J 59:169–178
46. O'Brien TP, McCully ME (1981) Whole mounts of cleared preparations. The study of plant structure principles and methods. Termarcaphi Pty Ltd, Melbourne, pp 614–630
47. Lersten NR (1967) An annotated bibliography of botanical clearing methods. Iowa State J Sci 41:481–486
48. Haseloff J (2003) Old botanical techniques for new microscopes. BioTechniques 34 (6):1174–1183
49. Herr JM Jr (1971) A new clearing-squash technique for the study of ovule development in angiosperms. Am J Bot 58:785–790
50. Geldner N, Richter S, Vieten A, Marquardt S, Torres-Ruiz RA, Mayer U, Jürgens G (2004) Partial loss-of-function alleles reveal a role for GNOM in auxin transport-related, post-embryonic development of Arabidopsis. Development 131:389–400
51. Beeckman T, Engler G (1994) An easy technique for the clearing of histochemically stained plant tissue. Plant Mol Biol Rep 12 (1):37–42
52. Grunewald W, De Smet I, De Rybel B, Robert HS, Van De Cotte B, Willemsen V, Gheysen G, Weijers D, Friml J, Beeckman T (2013) Tightly controlled WRKY23 expression mediates Arabidopsis embryo development. EMBO Rep 14 (12):1136–1142
53. Villani TS, Koroch AR, Simon JE (2013) An improved clearing and mounting solution to replace chloral hydrate in microscopic applications. Appl Plant Sci 1(5):1300016
54. Hernández-Barrera A, Ugartechea-Chirino Y, Shishkova S, Napsucialy-Mendivil S, Soukup A, Reyes-Hernández BJ, Lira-Ruan V, Dong G, Dubrovsky JG (2011) Apical meristem exhaustion during determinate primary root growth in the *moots koom 1* mutant of *Arabidopsis thaliana*. Planta 234:1163–1177
55. Soukup A, Tylová E (2014) Essential methods of plant sample preparation for light microscopy. In: Žárský V, Cvrčková F (eds) Plant cell morphogenesis, Methods in molecular biology, vol vol 1080. Humana, New York, pp 1–23
56. Casimiro I, Beeckman T, Graham N, Bhalerao R, Zhang H, Casero P, Sandberg G, Bennett MJ (2003) Dissecting *Arabidopsis* lateral root development. Trends Plant Sci 8:165–171
57. Ruzin SE (1999) Plant microtechnique and microscopy. Oxford University Press, Oxford, 322 pp
58. MacGregor DR, Deak KI, Ingram PA, Malamy JE (2008) Root system architecture in *Arabidopsis* grown in culture is regulated by sucrose uptake in the aerial tissues. Plant Cell 20:2643–2660
59. Walch-Liu P, Forde BG (2008) Nitrate signalling mediated by the NRT1.1 nitrate transporter antagonises L-glutamate-induced changes in root architecture. Plant J 54:820–828

Check for updates

Chapter 5

Adapting the Lateral Root-Inducible System to *Medicago truncatula*

Violaine Herrbach, Fabienne Maillet, and Sandra Bensmihen

Abstract

Almost all legume plants have the capacity to form two types of root organs: lateral roots and nodules (that will host rhizobia that fix nitrogen). Transcriptomic analyses are useful to understand both the similarities and differences between nodule and LR formation and to compare the LR developmental programs used by *Arabidopsis* and model legumes such as *Medicago truncatula*. However, in *M. truncatula* as in *Arabidopsis*, root cells "committed" to LR formation programs are scattered along the primary root and localized in the inner most layers of the root. To gain access to these cells, a lateral root-inducible system (LRIS) was first developed in *Arabidopsis*. This LRIS was recently shown to be effective in maize as well. Here we present a LRIS protocol adapted to the model legume *Medicago truncatula*. Using the same auxin transporter inhibitor and permeant auxin molecules used for *Arabidopsis* and maize but with slight modifications in their concentrations, we obtained very efficient enrichment and synchronization in LR development stages in *M. truncatula*.

Key words Root development, Synchronization, Auxin, Legumes, Transcriptomics, In vitro culture

1 Introduction

Legume plants, or Fabaceae, are the third most important family of flowering plants in terms of number of species. This evolutionary success is believed to be largely due to the capacity of most of its 20,000 species to establish a symbiotic interaction, called nodulation, with soil bacteria called rhizobia [1]. The rhizobia provide the plant with an assimilable source of nitrogen, inside a root organ called a nodule. Thanks to this symbiotic capacity, legume plants are rich in proteins and provide an important source of animal feed and human food. *Medicago truncatula* is a model legume plant used for many years. We recently performed a histological approach to describe *M. truncatula* lateral root formation (LRF) [2]. In angiosperms, the primary root is formed of different concentric cell layers: the innermost is the stele (comprising the xylem and phloem vessels), then comes the pericycle, surrounded by the endodermis, the

Daniela Ristova and Elke Barbez (eds.), *Root Development: Methods and Protocols*, Methods in Molecular Biology, vol. 1761, https://doi.org/10.1007/978-1-4939-7747-5_5,

cortex, and the root epidermis [3]. In *M. truncatula*, there are several layers of cortex and the layer closest to the endodermis is called the inner cortex. We have shown that, as for *Arabidopsis thaliana* [4], *M. truncatula* LRF also starts with divisions in the pericycle, but in contrast to *Arabidopsis*, divisions of the endodermis and the inner cortex also contribute to the formation of the lateral root primordium. Interestingly, symbiotic signals such as Nod factors (produced by rhizobia and necessary for the nodulation process) and Myc-LCOs (produced by arbuscular mycorrhizal fungi from the *Glomeromycota* phylum) induce LRF in *M. truncatula* [5, 6]. However, the molecular mechanisms controlling LRF in *M. truncatula* and, more particularly, controlling the stimulation of LRF by symbiotic signals are not yet understood. Auxin is a major regulator of LRF in angiosperm species [7]. Based on the prime role of auxin in controlling LRF, Himanen and collaborators have developed a "lateral root-inducible system" (LRIS) in *Arabidopsis* [8]. The LRIS does not only allow to synchronize the development stages of the LR primordia along the primary root but also leads to an enrichment in LR forming cells. This is achieved through the sequential treatments by, first, the auxin transport inhibitor 1-*N*-naphthylphthalamic acid (NPA) to disrupt auxin gradients and block LRF, followed by a (hours scale) treatment with the auxin permeant analog 1-naphthaleneacetic acid (NAA) to restart the LRF program along all the primary root. In that way, synchronization and enrichment in early LRF stages can be achieved. Since the original *Arabidopsis* work, this method was already adapted to maize and used as well for maize transcriptomic studies [9, 10]. In order to identify LRF regulators potentially regulated by symbiotic signals with no a priori, we recently successfully adapted the LRIS to perform transcriptomic analysis in *Medicago truncatula* [11]. The same concentration of NPA as the one used for *Arabidopsis* (10^{-5} M) was effective to block LRF in *M. truncatula*. After testing several concentrations of the auxin permeant analog NAA, we found 10^{-6} M as most appropriate concentration to restart the LRF program (Fig. 1). This is ten times less than the concentration used in *Arabidopsis* [10]. The full method we used in our transcriptomic approach to achieve such enrichment and synchronization in LRF is detailed below.

2 Materials

2.1 Seed Sterilization

1. 95% concentrated sulfuric acid.
2. 3.2% sodium hypochlorite solution.
3. Sterile deionized water.
4. Round petri dishes (diameter 92 mm × 16 mm height).

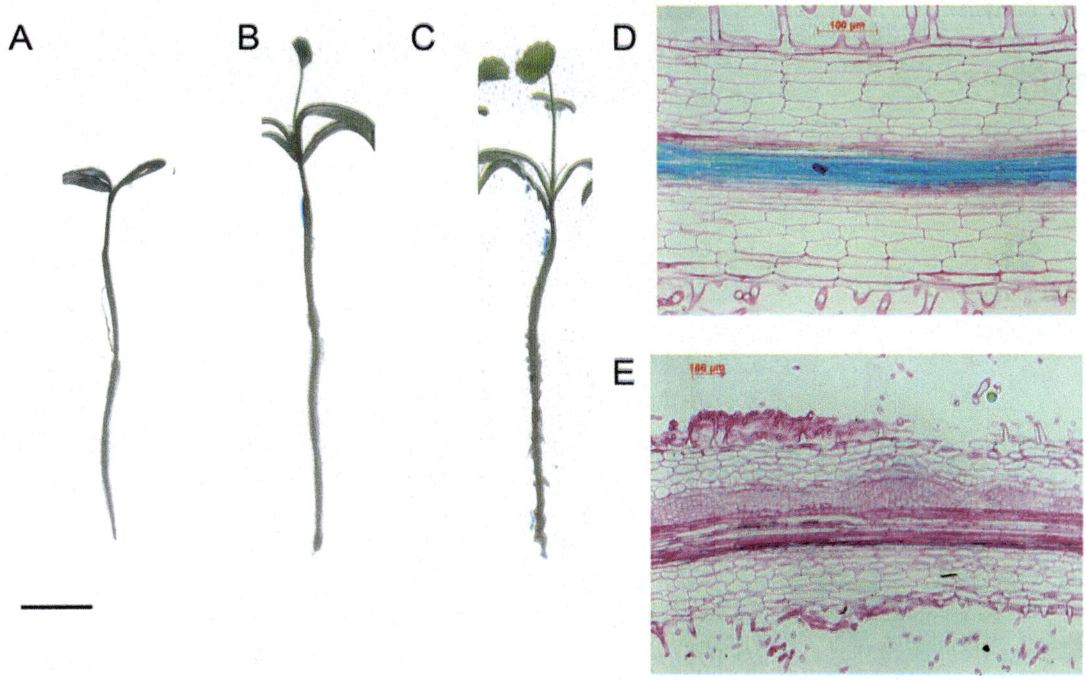

Fig. 1 Aspects of *M. truncatula* seedlings upon NPA and NPA+ auxin treatments. (**a**) A 2-day-old *M. truncatula* seedlings grown on M medium prior to NPA transfer; (**b**) *M. truncatula* seedling grown for 2 days on M medium and then for 2 days on 10^{-5} M NPA; (**c**) *M. truncatula* seedling grown for 2 days on M medium, 2 days on NPA, and 8 days on 10^{-6} M NAA. Scale bar is 1 cm. (**d**) Longitudinal cross section of a DR5:GUS *M. truncatula* transgenic seedling grown for 2 days on M medium and then for 2 days on 10^{-5} M NPA; (**e**) longitudinal cross section of a DR5:GUS *M. truncatula* transgenic seedling grown for 2 days on M medium and then for 2 days on 10^{-5} M NPA and 48 h on 10^{-6} M NAA. Scale bars are 100 μm

5. Parafilm.
6. 0.5× gelose medium (*see* **Note 1**).
7. 70% ethanol.
8. Metal lab spatula.

2.2 M Medium Solidified with Phytagel

M medium composition (*see* **Note 2**): macronutrients (80 mg/L KNO_3, 731 mg/L $MgSO_4{\cdot}7H_2O$, 65 mg/L KCl, 4.8 mg/L KH_2PO_4), 176.4 mg/L $CaCl_2{\cdot}2H_2O$, 8 mg/L NaFe-EDTA, micronutrients (6 mg/L $MnCl_2{\cdot}4H_2O$, 1.5 mg/L H_3BO_3, 2.65 mg/L $ZnSO_4{\cdot}7H_2O$, 0.0024 mg/L $Na_2MoO_4{\cdot}2H_2O$, 0.13 mg/L $CuSO_4{\cdot}5H_2O$), 0.75 mg/L KI. Adjust pH to 5.7 with diluted KOH (*see* **Note 3**). Gelify medium with 4 g/L of Phytagel (*see* **Notes 4** and **5**).

2.3 NPA, NAA Stock Solutions

1. 10^{-2} M stock solution of naphthylphthalamic acid (NPA) solubilized in dimethyl sulfoxide (DMSO) (*see* **Note 6**).
2. 10^{-2} M stock solution of naphthalene acetic acid (NAA) (*see* **Note 7**).

2.4 Growth Conditions and Transfers

1. Sterile forceps.
2. Square petri dishes (15.8 × 120 × 120 mm).
3. Growth cabinet: 21 °C, 16 h light/8 h dark photoperiod.

3 Methods

3.1 Seed Scarification and Sterilization

1. Put the necessary amount of *Medicago* seeds in a clean Erlenmeyer flask (*see* **Note 8**).
2. Add concentrated 95% sulfuric acid in the flask to cover the seeds for 5 min under a fume hood. This is a chemical scarification step.
3. Remove the concentrated sulfuric acid under a fume hood (*see* **Note 9**) and then proceed to a sterile environment (laminar hood or Bunsen burner proximity). Rinse one or two times rapidly with sterile water, and transfer to a sterile Erlenmeyer flask.
4. Rinse five times with sterile deionized water.
5. Add 3.2% sodium hypochlorite solution for 2 min (*see* **Note 10**).
6. Rinse five times with sterile deionized water.
7. Remove the excess of water.
8. Using a lab spatula, take a small amount (approximately 20–30) of scarified and sterilized seeds, and put them on round petri dishes containing 0.5× gelose medium.
9. Using the lab spatula, spread evenly the seeds on the 0.5× petri dish (*see* **Note 11**).
10. Close the petri dishes with Parafilm, cover in foil, and put for 3–4 days at 4 °C (*see* **Note 12**). This step is to synchronize germination.
11. Germinate the seeds by an overnight incubation at 25 °C.

3.2 Transfer and Growth on M Medium

1. Prepare the necessary amount of M medium (as indicated above), and count on 70 mL of M medium per plate (*see* **Note 13**).
2. After overnight germination, transfer the germinated seedling on M medium plates, typically putting seven to eight seedlings per plate (*see* **Note 14**).
3. Seal the plates on three sides (letting the upper part of the plate not sealed) with Parafilm.
4. Cover the plates with a kraft and aluminum foil cover (*see* **Note 15**) to protect roots from light.
5. Put the plates nearly vertically, in a "crystal box", for instance, in a growth chamber at 21 °C.

6. Let the seedlings grow for 2 days before transferring them to NPA plates.

3.3 Lateral Root Induction Treatment

1. Prepare plates for NPA and NAA treatments: prepare one bottle of medium per NPA and NAA (and possibly control) treatment. Let the M medium cool down but not solidify before adding the necessary amount of NPA or NAA stock solution. Prepare NPA plates to a final concentration of 10^{-5} M and NAA plates to the final concentration of 10^{-6} M from the 10^{-2} respective stock solutions (*see* **Note 16**).
2. After 2 days of growth on M medium, transfer the seedlings onto NPA plates (in sterile condition under a laminar hood). Be careful to pick seedlings with a well-developed primary root but without any lateral root. Pick seedlings with a homogenous growth phenotype (*see* **Note 17**).
3. Seal the plates on three sides with Parafilm as explained above. Put back the kraft and foil cover and put back to 21 °C growth cabinet in a near-vertical position.
4. After 2 days on growth on NPA, seedlings should show an inflated meristem and reduced growth. Transfer the seedlings showing no emerged lateral root to the 10^{-6} M NAA medium (as described for the transfer onto NPA).
5. Seal the plates on three sides with Parafilm as explained above. Put back the kraft and foil cover, and put back to 21 °C growth cabinet in a near-vertical position for the desired time. Typically, a 10 h NAA treatment will correspond to seedlings synchronized for the first periclinal cell divisions (Stage II as described in [2]). Stage III should be reached after approximately 20 h of NAA treatment (*see* ref. 11).

4 Notes

1. 0.5× gelose medium is obtained by diluting with sterile deionized water 2× stock solution.0.5× gelose medium is poured in round petri dishes.
2. Use MilliQ (micro-filtrated) water; make 100× stock (for macronutrients, $CaCl_2$ and NaFe-EDTA) solution and 1000× stock (for micronutrients and KI) solutions. Stock solutions can be kept at 4 °C and not autoclaved, while every fresh medium at 1× concentration should be autoclaved (typically 16 min at 121 °C).
3. This is not a buffered solution and a very low amount of diluted KOH is required. pH should not be too far from 5.6 in the original 1× medium solution.

4. Stock of Phytagel should be checked first for gelifying properties as small variations from one stock to another could influence these solidifying properties.
5. Keep medium at 65 °C until use since solidified Phytagel containing medium cannot be melt again.
6. This stock can be kept at −20 °C for a few months.
7. NAA powder should be solubilized in a few μL of 95% ethanol, then add a few μL of KOH 0.1 N (typically 500 μL for 10 mL of final water volume) before adding MilliQ water. Filtrate with a 22 μm mesh filter for sterilization. Store 100 μL aliquots at −20 °C covered in foil. This should be stable for several months.
8. 100 *Medicago truncatula* seeds have an average weight of 0.4 g.
9. Beware to put the concentrated sulfuric acid in a dedicated waste bin.
10. Do not leave in sodium hypochlorite solution for more than 2 min as it will make the seeds more permeable to water and difficult to rinse.
11. First, sterilize the spatula with ethanol and a Bunsen burner, and then cool it not to burn the seeds.
12. Seeds should not be too close from each other on the plate as they will dry rapidly once germinated during the transfer step.
13. Pour plates "inclined" by putting a 2 mL pipette underneath the plate. Be careful if you use this kind of "inclined" medium to always incline the plates in the orientation of the slope when you put them vertically. Beware that the medium does not slide too much by pouring from the upper corners of the plates.
14. When picking germinated seedlings, pay attention to picking only seedlings with a comparable well-elongated radicule. Put seedlings at approximately 2 cm for the top of the petri dish.
15. This handmade cover is made from regular kraft paper used for parcel packaging, covered by an extra layer of aluminum foil. These covers should be 10 cm width × 10 cm height.
16. NPA- and NAA-containing plates can be poured at the same time and kept in the dark at 4 °C until use. They can be stored for a few days before use but preferably not more than 2 days.
17. Pick the seedlings under the cotyledons with a curved forceps not to damage the root part. Seedlings should be small enough to be manually transferred using forceps from one medium to another during the course of the experiment.

References

1. Doyle JJ (2011) Phylogenetic perspectives on the origins of nodulation. Mol Plant-Microbe Interact 24(11):1289–1295
2. Herrbach V, Remblière C, Gough C, Bensmihen S (2014) Lateral root formation and patterning in *Medicago truncatula*. J Plant Physiol 171(3–4):301–310
3. Bensmihen S (2015) Hormonal control of lateral root and nodule development in legumes. Plants (Basel) 4(3):523–547
4. Malamy JE, Benfey PN (1997) Organization and cell differentiation in lateral roots of Arabidopsis thaliana. Development 124(1):33–44
5. Olah B, Briere C, Becard G, Denarie J, Gough C (2005) Nod factors and a diffusible factor from arbuscular mycorrhizal fungi stimulate lateral root formation in Medicago truncatula via the DMI1/DMI2 signalling pathway. Plant J 44(2):195–207
6. Maillet F, Poinsot V, Andre O, Puech-Pages V, Haouy A, Gueunier M, Cromer L, Giraudet D, Formey D, Niebel A, Martinez EA, Driguez H, Becard G, Denarie J (2011) Fungal lipochitooligosaccharide symbiotic signals in arbuscular mycorrhiza. Nature 469(7328):58–63
7. Overvoorde P, Fukaki H, Beeckman T (2010) Auxin control of root development. Cold Spring Harb Perspect Biol 2(6):a001537
8. Himanen K, Boucheron E, Vanneste S, de Almeida Engler J, Inze D, Beeckman T (2002) Auxin-mediated cell cycle activation during early lateral root initiation. Plant Cell 14(10):2339–2351
9. Jansen L, Hollunder J, Roberts I, Forestan C, Fonteyne P, Van Quickenborne C, Zhen RG, McKersie B, Parizot B, Beeckman T (2013) Comparative transcriptomics as a tool for the identification of root branching genes in maize. Plant Biotechnol J 11(9):1092–1102
10. Jansen L, Parizot B, Beeckman T (2013) Inducible system for lateral roots in Arabidopsis thaliana and maize. Methods Mol Biol 959:149–158
11. Herrbach V, Chirinos X, Rengel D, Agbevenou K, Vincent R, Pateyron S, Huguet S, Balzergue S, Pasha A, Provart N, Gough C, Bensmihen S (2017) Nod factors potentiate auxin signaling for transcriptional regulation and lateral root formation in Medicago truncatula. J Exp Bot 68(3):569–583

Check for updates

Chapter 6

Characterization of Root Epidermal Cell Patterning and Differentiation in *Arabidopsis*

Jorge E. Salazar-Henao, Girish Mokkapati, Eunice Han Xian Khor, Yi-Chia Chou, Wann-Neng Jane, and Wolfgang Schmidt

Abstract

The root epidermis of *Arabidopsis thaliana* has been established as a model system for elucidating the mechanisms which govern the spatial patterningAbstract and morphogenesis of plant cells. Investigations into root hairs focus on various aspects of the biology of epidermal cells, using methods specifically developed to dissect the biological question under study. Despite the large number of studies related to epidermal cell differentiation, a survey of methods to analyze the phenotypic readout resulting from environmental conditions or the genetic background of the plant has not been provided so far. This protocol describes how to analyze the spatial arrangement and morphologic characteristics of cells in the root epidermis based on whole mount roots or cross sections, using confocal, scanning electron and light microscopy. This comparison of methods aids in selecting the most suitable strategy to examine the differentiation of root epidermal cells at different developmental stages.

Key words Root hairs, Scanning electron microscopy, Cell length, Cell patterning

1 Introduction

Root hairs are tubular outgrowth of root epidermal cells derived from specialized cells referred to as trichoblasts that are formed in the maturation zone of the root [1]. The morphogenesis of root epidermal cells is modulated by interactions between genetically determined, intrinsic developmental programs and environmental cues, which defines the pattern, number, and length of root hairs. Several genes with critical functions in root hair morphogenesis have been identified by molecular genetic studies that affect fate assignment, initiation, or the tip growth of root hairs [2, 3]. Similarly, root hair morphogenesis is strongly affected by the availability of mineral nutrients and other environmental conditions [4–8]. In *Arabidopsis thaliana* (Arabidopsis), root hairs are organized in cell files that are arranged in alternation with non-hair cell files [9, 10]. Epidermal cells that are in contact with two cortical cells,

Daniela Ristova and Elke Barbez (eds.), *Root Development: Methods and Protocols*, Methods in Molecular Biology, vol. 1761, https://doi.org/10.1007/978-1-4939-7747-5_6, © Springer Science+Business Media, LLC 2018

i.e., positioned over the clefts of two radial cortical cell walls (H position), develop into a root hair cell; cells that are in contact with only one cortical cell, i.e., over a tangential cortical cell walls (N position), enter the non-hair cell developmental program [10, 11]. Owing to their shorter cycle time, trichoblasts are generally shorter, are less vacuolated, and have denser cytoplasm than non-hair cells (atrichoblasts) [4, 12].

Root epidermal cells have been widely used to study fate assignment and differentiation of plant cells. Besides intrinsic, genetically determined developmental programs, environmental cues can affect cell fate decisions and morphogenetic programs, widening the model character toward an understanding of phenotypic plasticity [3].

Root hairs greatly expand the absorptive surface area of the root and improve their capacity to take up water and mineral nutrients. Due to their extreme polar growth, root hairs extend the reach of roots radially, thereby minimizing the generation of depletion zones and increasing nutrient absorption by interception. In particular, root hairs have critical role for the uptake of mineral nutrients with limited phyto-availability such as phosphorous, taken up as inorganic phosphate, manganese, or iron [3]. In Arabidopsis, root hair density increases up to fivefold upon Pi deficiency, as a result of an increase in the number of epidermal cells that differentiate into hair cells [6, 13] and a reduction in epidermal cell length with decreasing Pi availability in the medium [14, 15]. Also, root hairs of Pi-deficient plants are significantly longer than those formed by plants grown in the presence of sufficient Pi [5, 6].

In the following protocol, we provide a comprehensive survey that gathers the methodologies used to analyze the patterning and morphology of root epidermal cells from plants subjected to different treatments or mutants defective in the expression of genes with critical roes in cell fate determination of root hair morphogenesis, using light microscopy, confocal microscopy, scanning electron microscopy (SEM), and analysis of the arrangement of root hairs and non-hair cells in the epidermis using cross sections. Additionally, for some cases we describe how one can perform root hair analysis using different techniques in order to select the most appropriate tool for the question under study.

2 Materials

2.1 Plant Growth

1. Seed sterilizing solution: 35% (v/v) bleach. Store at room temperature.
2. Solid half-strength MS medium: 2.151 g/L MS, 1% (w/v) sucrose, 0.05% (w/v) MES. Adjust the pH at 5.5 (with

KOH) and 4.5 g/L Gelrite pure. Autoclave the medium at 121 °C for 20 min.

3. Petri dishes (120 × 120 × 12 mm).

2.2 Fluorescence Microscopy

- Glass microscope slides (75 × 25 × 1 mm) Superfrost Plus (+).
- Glass cover slips (22 × 22 × 1.5 mm).
- Propidium iodide (PI) 1000× concentrated stock: 10 mg/mL PI in water (avoid light).
- Sterile water.
- Confocal microscope: inverted confocal laser scanning microscope in combination with a super resolution imaging system.
- Software for analysis of confocal microscope images.

2.3 Cross Sections

- A plastic embedding system based on MMA (methyl methacrylate).

2.4 Scanning Electron Microscope

- Scanning electron microscope.
- Liquid nitrogen.
- Razor blades.

3 Methods

3.1 Plant Growth Conditions

1. Sterilize Arabidopsis seeds with seed-sterilizing solution for 5 min, followed by five washes with sterile water for 5 min each. Stratify sterilized seeds at 4 °C in the dark for 3 days before plating.
2. Grow plants on the surface of horizontally positioned plates at 22 °C with constant illumination (55 μmol/m^2/s) for 5 days.

3.2 Analysis of Root Epidermal Cell Patterning in Arabidopsis

3.2.1 Confocal Microscopy

The pattern of root epidermal cell types, i.e., the position of trichoblasts and atrichoblasts, is analyzed as described in [16–18] with some modifications. Roots from 5-day-old plants are stained with PI, followed by examination with a fluorescence microscope to determine the type and location of epidermal cells. The files are determined by their relative positions to the underlying cortical cells. Root epidermal cells are defined to be in the H position if they are located above an anticlinal wall of a cortical cell and to be in the N position if they are located above a periclinal cortical cell wall.

1. Place seedlings in 10 μg/mL propidium iodide in water for 1 min (*see* **Note 1**).
2. Rinse seedlings gently in water for 1 min.
3. Mount roots in freshwater.

4. Observe roots with an inverted confocal laser scanning microscope in combination with a super resolution imaging system.
5. The peak excitation λ and emission λ for PI is 536 nm and 620 nm, respectively.
6. Root hair frequency is determined using images captured 4–6 mm above the root tip (*see* **Note 2**). For each seedling root, five consecutive epidermal cells from the same file are observed, and a total of 20 cells from two hair cell files and the adjacent two non-hair cell files are counted. The proportion of hair and non-hair cells in the root epidermis is determined by examining at least ten 5-day-old seedlings (for a total of 200 cells). An epidermal cell is scored as a root hair if a protrusion is visible regardless of its length (Fig. 1).

3.2.2 Cross Sections

Alternatively, root epidermal patterning can be inferred from the analysis of cross sections.

1. Fix, dehydrate, and embed root samples in resin.
2. Cut transverse sections (30 μm) using a microtome. Dry and stain sections with toluidine blue (0.05%) on glass slides, and examine the sections on a bright-field microscope.
3. Count the number of cortical and epidermal cells and the percentage of root hairs in the H and N positions in 50 cross sections from 10 roots of each genotype or growth type. To avoid counting the same root hair twice, analyze only every fourth section from serial cuttings.

3.3 Root Hair Density

3.3.1 Relative Root Number

The relative hair number indicates the number of root hairs formed on a segment of the root in which epidermal cells have readily elongated. The following protocol is based on [19].

1. Prepare and mount specimens for confocal microscopy as in subheading 3.2.1, **steps 1–5**.
2. Root hair density and cell lengths are determined using images captured 4–6 mm behind the root tip.
3. The length of the cells is determined using the ImageJ (Fiji) software (https://imagej.net/Fiji) [20].
4. Calculate the relative root number as the numbers of root hairs per mm × average length of epidermal cell μm/1000 (Fig. 1).

3.3.2 Percentage of Root Hairs that Develop Beyond the Transition to Tip Growth

The proportion of elongated root hairs that successfully passed the initiation stage (i.e., root hairs that are longer than>40 μm) is calculated as an indication if root hairs of a certain growth type or genotype can develop beyond the transition to tip growth, which takes place when hairs are 20–40 μm long [21]. Root hairs longer than 40 mm are presumed to be tip-growing, and their length at maturity is determined to assess their ability to elongate [22]. The

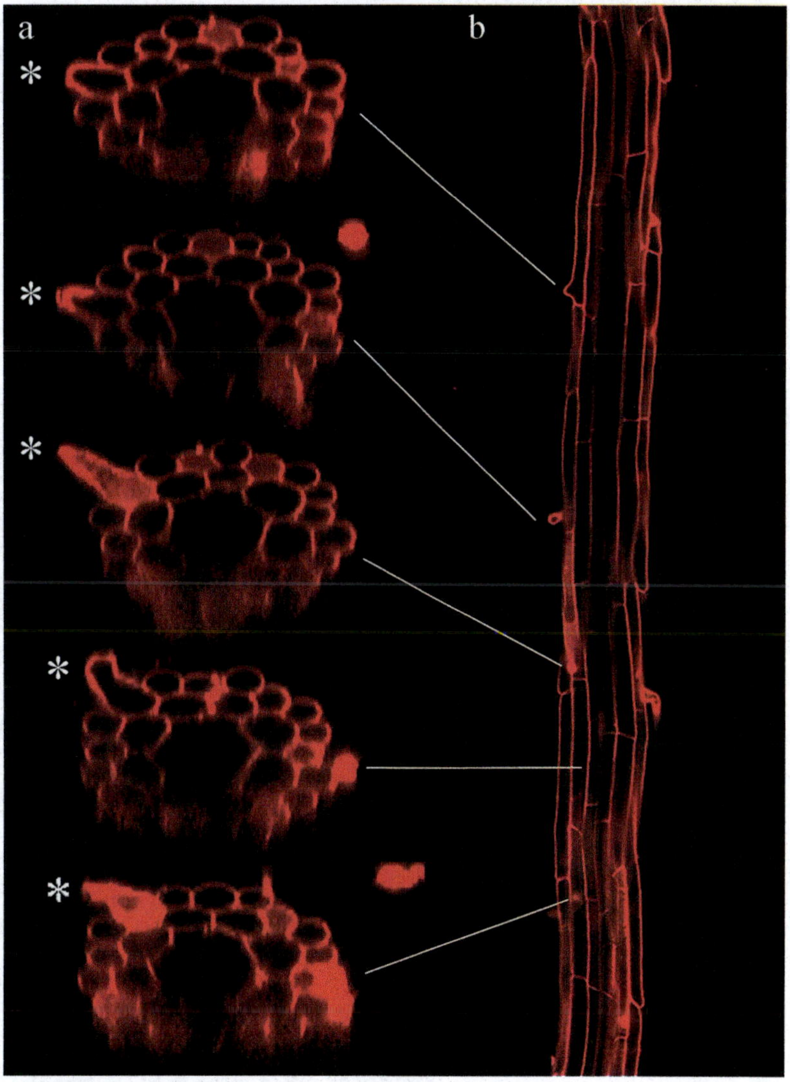

Fig. 1 Confocal micrograph of a 5-day-old Arabidopsis (Col-0) root. (**a**) Transverse sections showing the trichoblast (H) position, indicated by an asterisk. (**b**) Longitudinal section showing an emerging root hair at each corresponding section in **a**, indicated by a white line

percentage of root hairs that grow longer than 40 μm is determined using images captured with a scanning electron microscope (SEM) at 80× magnification. Alternatively, light microscopy can be employed to determine this parameter.

1. Dissect the terminal 1 cm of the primary root and load the root on stub.
2. Freeze samples with liquid nitrogen slush and transfer the roots to a sample preparation chamber cooled to −160 °C. After

5 min, raise the temperature to −95 °C, and allow sublimation to proceed for 23 min (*see* **Notes 3** and **4**).

3. Coat the samples with platinum (Pt) at −130 °C, transfer specimen to a cryo-stage in the SEM chamber, and observe the sample under −160 °C on a cryo-SEM with 20 kV at 80× magnification.
4. Count the number of root hairs from one slide of a 1 mm segment at the beginning of the differentiation region of the roots in 5-day-old plants (Fig. 2a, c).

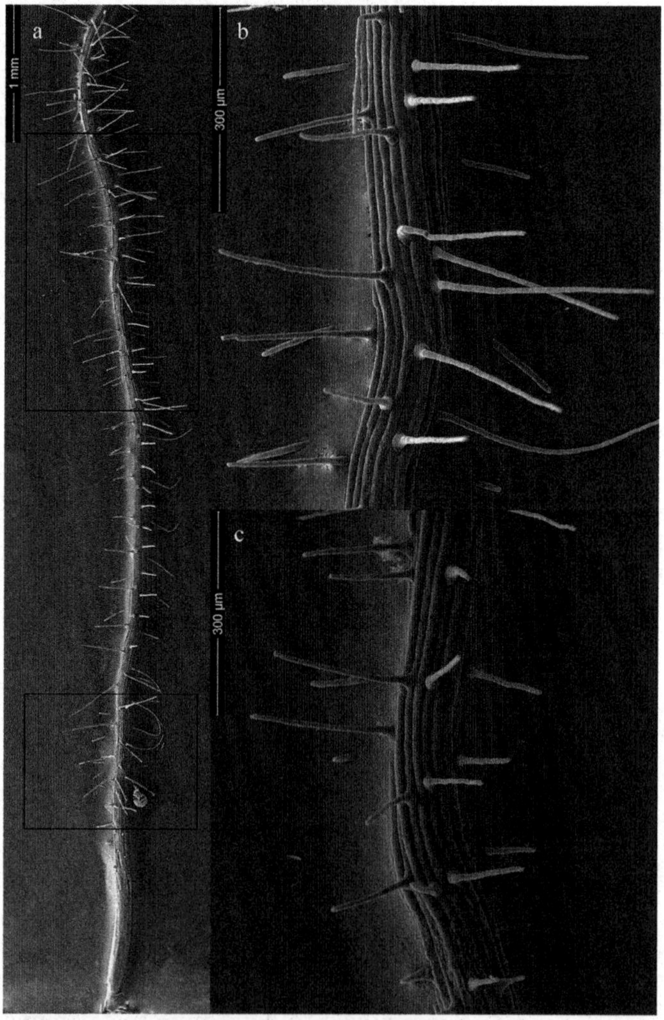

Fig. 2 Scanning electron microscope (SEM) images of an Arabidopsis (Col-0) root. (**a**) Root hair patterning of 5-day-old seedling at 80× magnification. (**b**) Upper outlined area in **a** at 400× magnification showing mature root hairs (2–4 mm above the root tip). (**c**) Magnification (400×) of the lower outlined area in **a** showing root hairs at the beginning of the differentiation zone

3.3.3 Root Hair Density and Length

Root hair length is calculated from micrographs acquired with SEM at 80× magnification and analyzed with the image analysis software ImageJ (Fiji) [20].

1. Prepare and mount specimens for SEM as described in subheading 3.3.2, **steps 1–3**.
2. Count the number of root hairs from one side within 1 mm segment from a zone 4 mm behind the root tip (Fig. 2a, b).
3. Select mature root hairs elongated horizontally to the agar surface from a zone 4–6 mm behind the root tip for length determination. Immature root hairs or bulges with lengths <5 μm should be omitted.
4. Quantify the length of the root hairs using ImageJ (Fiji) [20] (Fig. 2a, b).

Alternatively, root hair length can be determined using a light microscope equipped with a scale in the eyepiece. One hundred root hairs from five roots are measured for each genotype and growth type [23].

3.4 Analysis of Longitudinal Cell Length of Trichoblasts and Atricoblasts

The longitudinal length of epidermal cells in the differentiation zone determines the number of root hairs per unit root length. Cell length is genetically determined but responsive to environmental cues. Longitudinal cell length is determined from compiled confocal micrographs.

1. Prepare and mount specimens for confocal microscopy as described in subheading 3.3.2, **steps 1–6**. Compile micrographs spanning the first 6–8 mm of the root from the tip. Calculate the position of each cell from the cumulative length of all cells in the respective cell file from the quiescent center (QC).
2. Determine the length of trichoblasts and atrichoblasts using the ImageJ (Fiji) software [20]. The data sets can be smoothed and interpolated into 25-mm-spaced data points using a Kernel-smoothing routine [24]; this is done using a Microsoft Excel macro, which enables the average calculation between replicate roots. Plot the longitudinal cell length of trichoblasts and atrichoblasts against their distance from the root tip (QC).

3.5 Root Hair Morphology

To analyze root hair morphology, root hairs from 5-day-old seedling are examined with a SEM at 400× magnification in a zone between 4 and 6 mm behind the root tip.

1. Prepare and mount specimens for SEM as described in subheading 3.3.2, **steps 1–6**.

Fig. 3 Scanning electron micrographs of Arabidopsis root hairs. (**a–c**) Representative images of normally developed root hairs and hairs with altered morphology (e.g., by mutation). Normally developed (**a**), jagged (**b**), and branched (**c**) root hair cell

2. Determine percentages of each root hair class (i.e., wild type, branched, wavy, jagged, or other defects) for the growth or genotypes under investigation (Fig. 3).

4 Notes

1. A 1000× concentrated stock solution can be stored at 4 °C (10 mg/mL PI in water, avoid light).
2. Between 4 and 6 mm above the root tip, root hairs are mature, and their length is characteristic of the respective growth or genotype.
3. The critical step in freeze-drying techniques is to rapidly freeze the sample to avoid the formation of ice crystals that may damage the root hairs.
4. Preferably take the liquid nitrogen from the storage system just before starting the experiment to avoid the presence of water resulting from condensation.

References

1. Leavitt RG (1904) Trichomes of the root in vascular cryptogams and angiosperms. Proc Boston Soc Nat Hist 31:273–313
2. Grierson C, Schiefelbein J (2009) Genetics of root hair formation. In: Emons AMC, Ketelaar T (eds) Root hairs. Springer, Berlin, pp 1–25
3. Salazar-Henao JE, Vélez-Bermúdez IC, Schmidt W (2016) The regulation and plasticity of root hair patterning and morphogenesis. Development 143(11):1848–1858
4. Grierson C, Nielsen E, Ketelaarc T, Schiefelbein J (2014) Root hairs. The Arabidopsis book 12:e0172
5. Bates TR, Lynch JP (1996) Stimulation of root hair elongation in *Arabidopsis thaliana* by low phosphorus availability. Plant Cell Environ 19 (5):529–538
6. Müller M, Schmidt W (2004) Environmentally induced plasticity of root hair development in Arabidopsis. Plant Physiol 134(1):409–419
7. Schmidt W, Schikora A (2001) Different pathways are involved in phosphate and iron stress-induced alterations of root epidermal cell development. Plant Physiol 125(4):2078
8. Ma JF, Ryan PR, Delhaize E (2001) Aluminium tolerance in plants and the complexing

role of organic acids. Trends Plant Sci 6 (6):273–278

9. Bünning E (1951) Über die Differenzierungsvorgänge in der Cruciferen Wurzel. Planta 391 (1):126–153
10. Dolan L, Costa S (2001) Evolution and genetics of root hair stripes in the root epidermis. J Exp Bot 52(spec issue):413–417
11. Schiefelbein JW (2000) Constructing a plant cell. The genetic control of root hair development. Plant Physiol 124(4):1525–1531
12. Schiefelbein J, Huang L, Zheng X (2014) Regulation of epidermal cell fate in Arabidopsis roots: the importance of multiple feedback loops. Front Plant Sci 5(47):1–4
13. Ma Z, Bielenberg DG, Brown KM, Lynch JP (2001) Regulation of root hair density by phosphorus availability in *Arabidopsis thaliana*. Plant Cell Environ 24(4):459–467
14. Savage N, Yang TJ, Chen CY, Lin KL, Monk NA, Schmidt W (2013) Positional signaling and expression of ENHANCER OF TRY AND CPC1 are tuned to increase root hair density in response to phosphate deficiency in *Arabidopsis thaliana*. PLoS One 8(10):e75452
15. Sánchez-Calderón L, López-Bucio J, Chacón-López A, Gutiérrez-Ortega A, Hernández-Abreu E, Herrera-Estrella L (2006) Characterization of low phosphorus insensitive mutants reveals a crosstalk between low phosphorus-induced determinate root development and the activation of genes involved in the adaptation of Arabidopsis to phosphorus deficiency. Plant Physiol 140(3):879–889
16. Lee MM, Schiefelbein J (2002) Cell pattern in the Arabidopsis root epidermis determined by lateral inhibition with feedback. Plant Cell 14 (3):611–618
17. Simon M, Lee MM, Lin Y, Gish L, Schiefelbein J (2007) Distinct and overlapping roles of single-repeat MYB genes in root epidermal patterning. Dev Biol 311(2):566–578
18. Wieckowski Y, Schiefelbein J (2012) Nuclear ribosome biogenesis mediated by the DIM1A rRNA dimethylase is required for organized root growth and epidermal patterning in Arabidopsis. Plant Cell 24(7):2839–2856
19. Wada T, Tachibana T, Shimura Y, Okada K (1997) Epidermal cell differentiation in Arabidopsis determined by a Myb homolog, CPC. Science 277(5329):1113–1116
20. Schindelin J, Arganda-Carreras I, Frise E, Kaynig V, Longair M, Pietzsch T, Preibisch S, Rueden C, Saalfeld S, Schmid B, Tinevez J-Y, White DJ, Hartenstein V, Eliceiri K, Tomancak P, Cardona A (2012) Fiji: an open-source platform for biological-image analysis. Nat Methods 9(7):676–682
21. Dolan L, Duckett CM, Grierson CS, Linstead P, Schneider K, Lawson E, Dean C, Poethig S, Roberts K (1994) Clonal relationships and cell patterning in the root epidermis of Arabidopsis. Development 120:2465–2474
22. Parker JS, Cavell AC, Dolan L, Roberts K, Grierson CS (2000) Genetic interactions during root hair morphogenesis in Arabidopsis. Plant Cell 12(10):1961–1974
23. Li W-F, Perry PJ, Prafulla NN, Schmidt W (2010) Ubiquitin-specific protease 14 (UBP14) is involved in root responses to phosphate deficiency in Arabidopsis. Mol Plant 3(1):212–223
24. Beemster GTS, Baskin TI (1998) Analysis of cell division and elongation underlying the developmental acceleration of root growth in *Arabidopsis thaliana*. Plant Physiol 116 (4):1515–1526

Chapter 7

In Vitro Assay for Induction of Adventitious Rooting on Intact *Arabidopsis* Hypocotyls

Hoang Khai Trinh, Inge Verstraeten, and Danny Geelen

Abstract

Adventitious roots (AR) are de novo formed roots that emerge from any part of the plant or from callus in tissue culture, except root tissue. The plant tissue origin and the method by which they are induced determine the physiological properties of emerged ARs. Hence, a standard method encompassing all types of AR does not exist. Here we describe a method for the induction and analysis of AR that emerge from the etiolated hypocotyl of dicot plants. The hypocotyl is formed during embryogenesis and shows a determined developmental pattern which usually does not involve AR formation. However, the hypocotyl shows propensity to form de novo roots under specific circumstances such as removal of the root system, high humidity or flooding, or during de-etiolation. The hypocotyl AR emerge from a pericycle-like cell layer surrounding the vascular tissue of the central cylinder, which is reminiscent to the developmental program of lateral roots. Here we propose an easy protocol for in vitro hypocotyl AR induction from etiolated *Arabidopsis* seedlings.

Key words Etiolated seedlings, Hypocotyls, Adventitious roots

1 Introduction

Adventitious roots (AR) develop primarily, but not exclusively, from tissue types showing the capacity to engage in post-embryonic cell divisions [1]. In the hypocotyl it is the pericycle-like cell layer surrounding the vascular cylinder that shows high propensity to engage in the formation of new adventitious root organs [2]. In more mature stems, it is the cambium and surrounding tissue from the vascular bundle that initiate formation of ARs [3]. Under regular light conditions, dicotyledonous plants rarely form adventitious roots spontaneously. Typically an extended period of darkness promotes the induction of AR formation, which is often implemented for the industrial cloning of cuttings. Similarly, AR induction in hypocotyls requires a dark period exposure but additionally depends on the capacity of the plant to de-etiolate upon reexposure to light [4]. Exposure to light is hence a critical component of the

Daniela Ristova and Elke Barbez (eds.), *Root Development: Methods and Protocols*, Methods in Molecular Biology, vol. 1761, https://doi.org/10.1007/978-1-4939-7747-5_7,

AR induction process. The formation of AR is subdivided in different developmental stages similar to the developmental steps taking place during lateral root formation. Microscopic observation distinguishes VIII developmental phases. For routine analysis, typically three stages are defined. In stage I the pericycle has started to divide, and a site of AR initiation is established. Stage II reveals a primordium, which is a dome-shaped group of small meristematic cells that resides with its base attached to the central cylinder and with the top pushing aside the cortex cells. In stage III, the size of the primordium exceeds the cortex layer and has emerged [5]. In addition to light, other environmental factors such as temperature stress, flooding, and interaction with microbes influence AR formation and can lead to an arrest of AR development [2, 4–6]. The environmental cues are converted into signaling events during which the plant hormone auxin plays a central role. External application of auxin strongly promotes AR formation, and this depends on functional auxin signaling [6]. Multiple steps in AR development are controlled by auxin signaling as different chemicals with auxin-like structure alter the number of AR primordia, their position, and capacity of AR to elongate.

The protocol presented here is developed to determine the effect of exogenously applied chemicals on the induction and development of AR in *Arabidopsis* hypocotyls. Compounds can be applied to the growth medium, either from the start during seed germination or at later stages by transferring plantlets to new agar plates. The protocol is also suitable for other dicot plants, albeit that some adaptations may be needed, in particular regarding the light quality and regime.

2 Materials

1. Col-0 wild-type and WOX5-GFP seeds.
2. Kl medium [MS salts (basal salt mix), D-sucrose, 2-(N-morpholino) ethanesulfonic acid monohydrate (MES), agar (MC029, LAB M)].
3. Autoclave.
4. pH meter.
5. Milligram scale.
6. 500 mL glass bottle.
7. Squared petri dishes (12 × 12 cm).
8. Growth room.
9. Cold room.
10. Laminar air flow cabinet.
11. Flow hood.

12. Desiccator.
13. Heater.
14. Surgical knives.
15. Needles.
16. Beaker/flask/jar.
17. Binoculars system.
18. Fluorescence microscope.
19. ImageJ/Fiji software.
20. 10 mM NAA stock solution (dissolved in DMSO).
21. Bleach.
22. Distilled water (dH_2O).
23. Hydrochloric acid (HCl).
24. Glass slides.
25. Cover slips (22 mm × 22 mm).
26. Aluminum foil.
27. Autoclave tape.
28. Desinfectol spray bottle.
29. Hot bead sterilizers.

3 Methods

3.1 K1 Medium Preparation

1. In a 1 L beaker, weight all K1 ingredients (*see* **Note 1** below). Add 1 L distilled water and a magnetic stirrer in the beaker. Put the pH electrode in the medium, and set the pH at 5.7 using 1 M KOH (or 0.1 M KOH to adjust pH to reach exactly at 5.7) by adding approximately 750 μL of 1 M KOH. In case the solution becomes too basic, use 0.1 M HCl to adjust it (*see* **Notes 2** and **3**).
2. Add 4.0 g agar and 0.5 L of the medium into 0.5 L reagent bottles with a screw cap (borosilicate glass, VWR), and label bottles using autoclave tape (label medium name in case there are different types of chemicals to be added after autoclaving).
3. Mix well and put the bottle in the autoclave at 112 °C for 30 min (leave the screw cap a little loose when autoclaving).
4. When the autoclave is finished, put the medium bottles 30 min–1 h in the 60 °C oven. Start a laminar flow cabinet 30 min before you want to start working in it and fully sterilize the surface with 96% ethanol/desinfectol.
5. Get 12 × 12 cm squared petri dishes, and label the cover using a permanent maker.

6. Get the medium flasks and pour 50 mL medium into the 12 × 12 cm petri dishes. Keep the plates open to dry in the laminar flow for 30 min. Close with a lid, and store the plates upside down in the bags or in aluminum foil at 4 °C in cold room until use. To make medium flasks containing 10 μM NAA (*see* **Notes 4** and **5)**

3.2 Seed Sterilization

1. Bring 200–300 *Arabidopsis* seeds into a 1.5 mL Eppendorf tube.
2. Label the tube using a permanent marker, then cover the label with a piece of transparent tape and leave the open tubes in a rack or open box in a desiccator in the hood.
3. Add 100 mL bleach in a 250 mL beaker, and then place the beaker/jar/flask in the desiccator next to the seed rack. Pipette 3 mL hydrochloric acid (fuming 37%) into the beaker with bleach, and close the desiccator as fast as possible because the chlorine gas starts forming upon mixing the two ingredients.
4. Keep the desiccator closed to gas the seeds for 4 h.
5. After 4 h, sterilize your hands using desinfectol sprayer, open the desiccator, remove the seed rack, and close the Eppendorf tubes as fast as possible (still inside the desiccator).
6. Bring the tubes to the laminar flow, and leave them to ventilate in the laminar flow for 15 min before spreading the seeds on agar plates.

3.3 Sowing Seeds and Vernalization

1. Wear gloves and clean your hands with desinfectol.
2. Open agar square plate (12 × 12 cm) and open the Eppendorf tube with seeds. Spread the seeds over the agar surface medium by keeping the tube horizontally and gentle shaking. Pay attention not to contaminate the seeds/plates and work aseptically.
3. Close the plate and label the bottom of the plate with date.
4. Seal the plate using Micropore tape.
5. Vernalize the seeds on the plates for 4 days in a dark cold room at 4 °C to ensure a high and synchronized germination rate.

3.4 Germination and Etiolation

1. To initiate germination, the plates are put for 8 h in the light in the growth room (room temperature at 21.4 °C, humidity at 40–60%, light intensity at 70 μm/m^2/S, photo period: 14 h light/10 h dark).
2. To etiolate the seedlings, the plates then are wrapped in aluminum foil (three layers to ensure complete coverage and darkness) or put in a dark box for 3 days (seedlings will elongate to about 1.5–2 cm after this time of etiolation). Make sure that no light reaches the seedlings as this will affect the root induction process.

3.5 Root Induction on Vertical-Positioned Plates

1. Start the laminar flow 30 min before use, and sterilize surgical knives at 300 °C hot bead sterilizers. Leave the knives to cool before use.
2. Unwrap the plates and remove the tape.
3. Take a cooled sterilized surgical knife and gently lift the seedling by placing the blade end underneath the junction (the connection between hypocotyl and primary root). Transfer the seedling gently to fresh Kl medium or Kl medium containing your treatment of choice (avoid damaging seedlings when transferring to new medium because wounding can both induce and hamper AR formation).
4. Place the seedlings in a row, 8 mm apart, and push gently down to ensure contact with medium. Mark the position of the apical hook and the junction between hypocotyl and root with a marker (often, the seed coat remains at this junction). Depending on the size of your seedlings and the evaluation time, two rows can be used for one petri dish (*see* **Note 6**).
5. Seal this plate using Micropore tape and put the plate vertically in the growth room (cotyledons upward).
6. Leave the seedlings in the growth room for 10 days to evaluate the number of newly initiated AR. The first signs of AR initiation can be observed after 4 days.

3.6 Hypocotyl AR and ARP Scores

1. Observe the hypocotyls using a binocular at 10× magnification, and count the number of emerged AR (there are the root structures that surpassed the epidermis of the hypocotyl) and the adventitious root primordia (ARP; these are the root protuberances that are still within the borders of the hypocotyl tissue) (*see* Fig. 2).
2. The average and standard deviation of AR and ARP numbers are calculated using Excel. Further statistical analysis can be performed with any software of choice, e.g., S-plus or R.

3.7 Hypocotyl AR Length and Position

1. The position of the AR and ARP in the hypocotyl is not random. Typically in Col0-WT plants, AR appear at the lower end of the hypocotyl. Depending on the mutants analyzed or on the incubation conditions, AR may also emerge at the upper part of the hypocotyl. This part typically corresponds to the former position of the apical hook. Upon transfer to light, the etiolated hypocotyl will further expand for about 20–30%, mainly at the upper end which will accumulate chlorophyll. To record these features and to obtain positional information, make pictures of the full plates on a flatbed scanner/camera, and include a ruler to calibrate the pictures.
2. The position and the length of AR are measured by using ImageJ/Fiji software [7] (*see* **Notes 7** and **8**).

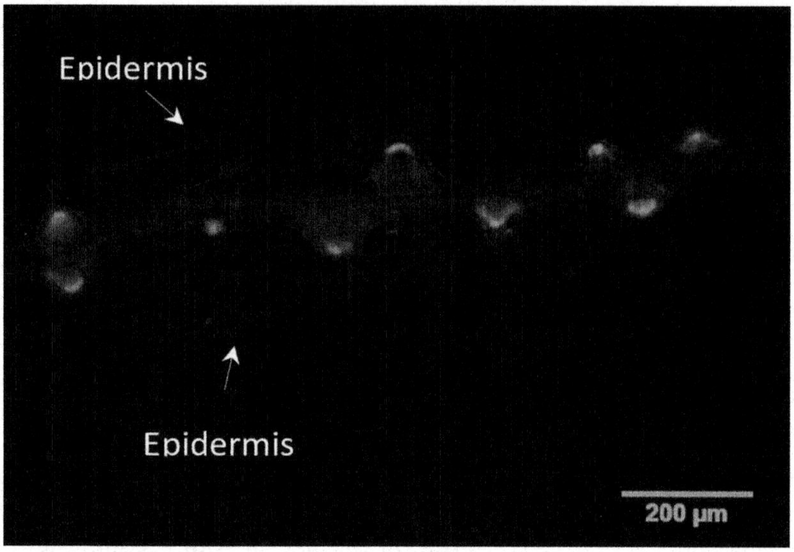

Fig. 1 WOX5-GFP signal in hypocotyl ARP upon 72 h 10 μM NAA treatment

3.8 Using WOX5-GFP Marker to Determine Early AR and ARP Induction Patterns

1. To massively induce AR formation, we apply 10 μM NAA (1-naphthaleneacetic acid) to the Kl medium at the transfer (*see* Subheading 3.1).
2. Transfer 3-day etiolated WOX5-GFP *Arabidopsis* seedlings to medium containing 10 μM NAA to evaluate root induction, and record the AR and ARP numbers (*see* Subheading 3.5).
3. Upon 72 h of incubation in the light, the first signs of primordia formation can be observed. WOX5-GFP fluorescence first appears near the vascular cylinder, presumably in the pericycle, and then migrates radially along with the growing tip of the formed primordia.
4. To image WOX5-GFP, prepare slides with 100 μL water to submerge the seedling. Gently put the cover slip on the seedling, while you avoid bubbles to form between the glass slide and cover slip. Leave the slide under microscope, count the number of fluorescent dots, and determine their position along the hypocotyl axis (*see* **Note 9** and Fig. 1).

4 Notes

1. Kl medium ingredients for 1 L: 1.5 g MS salts (basal salt mix, Duchefa), 5.0 g D-sucrose, 0.5 g MES (Duchefa), 8.0 g agar (Product MC029 of LAB M), set pH at 5.7 using KOH 1 M.
2. To make 100 mL 1 M KOH, weight 5.612 g KOH and dissolve in deionized water. Dilute this solution 1:10 in water to have 0.1 M KOH.

3. To make 100 mL HCl (stock solution of hydrochloric acid is calculated to be 12.178 M based on a density of 1.2 g/mL, a formula weight of 36.46 g/mol, and a concentration of 37% w/w), slowly add 8.212 mL of the 37% stock solution to 25 mL deionized water. Adjust the final volume to 100 mL with deionized water. Dilute this solution 1:10 to become 0.1 M HCl.
4. To make 2 mL 10 mM NAA stock solution (MW: 186.21 g/mol), weigh 3.724 mg NAA and dissolve in dimethyl sulfoxide, DMSO (SIGMA), and store the stock solution at −20 °C for long-term use.
5. For the NAA treatment plates, prepare K1 as in **Note 1** and let it cool down to 65 °C. Add 500 μL of the 10 mM NAA stock to 500 mL K1 medium, shake well, and pour 50 mL into square petri dish (12 × 12 cm). Leave the plates open in laminar flow for 30 min to dry out. Plates can be stored in the cold room for max 1 week.
6. Regular spacing of seedlings is done by using a guide/template paper with ten dots/row, two rows/plate, which is put underneath the plate.
7. ImageJ/Fiji [7]. Download and install ImageJ/Fiji following the guidance from the link https://imagej.net/Fiji/Downloads.
8. Measuring AR length using Fiji.
   (a) *Setting measurement scale:* Select ***Straight line*** and draw a line between two points of known distance, usually a ruler or scale bar on the photo. Go to ***Analyze\Set scale***. The *Set scale* will be displayed, type unknown distance and units in the appropriate boxes, and check *Global* to apply this scale to the other images taken the same scale. Every time the software is restarted, calibration is required.
   (b) *Root length measurement*: Select ***Freehand line*** to draw a line following the root shape from start to end of the root. Go to ***Analyze\Measure*** or simply ***Press M*** to get the result. The measurement value will be in the exportable *Results* table. Copy all values into an Excel file for further analysis.
9. AR and ARP identification (Fig. 2)
   (a) AR refers to a mature adventitious root emerged out of the epidermis layer.
   (b) ARP refers to the adventitious root formed but still located inside the epidermis layer.

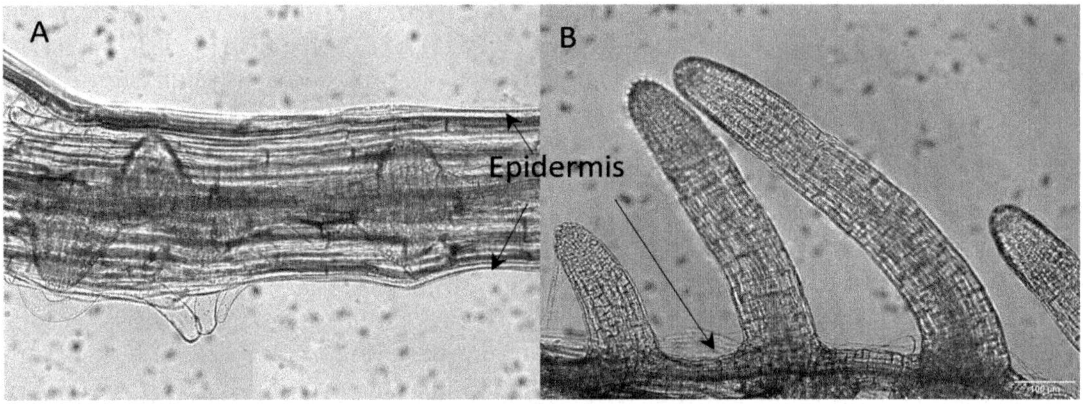

Fig. 2 ARP (**a**) and AR (**b**) formation on *Arabidopsis thaliana* hypocotyl

References

1. Bellini C, Pacurar DI, Perrone I (2014) Adventitious roots and lateral roots: similarities and differences. Annu Rev Plant Biol 65:639–666
2. Verstraeten I, Schotte S, Geelen D (2014) Hypocotyl adventitious root organogenesis differs from lateral root development. Front Plant Sci 5(9):1–13
3. da Rocha Correa L, Troleis J, Mastroberti AA, Mariath JEA, Fett-Neto AG (2012) Distinct modes of adventitious rooting in Arabidopsis thaliana. Plant Biol 14(1):100–109
4. Sorin C et al (2005) Auxin and light control of adventitious rooting in Arabidopsis require ARGONAUTE1. Plant Cell 17:1343–1359
5. de Klerk G-J, van der Krieken W, de Jong JC (2012) The formation of adventitious roots: new concepts , new possibilities. In Vitro Cell Dev Biol Plant 35(3):189–199
6. Verstraeten I, Beeckman T, Geelen D (2013) Adventitious root induction in arabidopsis thaliana as a model for in vitro root organogenesis. Methods Mol Biol 959:159–175
7. Schindelin J et al (2013) Fiji - an open source platform for biological image analysis. Nat Methods 9(7):676–682

Chapter 8

Root Gravitropism: Quantification, Challenges, and Solutions

Lukas Muller, Malcolm J. Bennett, Andy French, Darren M. Wells, and Ranjan Swarup

Abstract

Better understanding of root traits such as root angle and root gravitropism will be crucial for development of crops with improved resource use efficiency. This chapter describes a high-throughput, automated image analysis method to trace *Arabidopsis* (*Arabidopsis thaliana*) seedling roots grown on agar plates. The method combines a "particle-filtering algorithm with a graph-based method" to trace the center line of a root and can be adopted for the analysis of several root parameters such as length, curvature, and stimulus from original root traces.

Key words Root gravitropism, RootTrace, Image analysis, *Arabidopsis*

1 Introduction

Environmental signals can have profound impact on plant growth and development. Plants have evolved sophisticated mechanisms to respond to these signals. For example, in response to gravity, shoots and roots of higher plants modify their direction of growth by employing tropic responses. Root gravitropism enables plants to explore the soil environment and acquire anchorage and resources and therefore has a direct bearing on crop productivity [1]. Better understanding of root traits such as root angle and root gravitropism will be crucial for development of crops with improved resource use efficiency [1].

Root gravitropism can be divided into three spatially and temporally distinct phases: gravity perception, gravitropic signal transmission, and gravitropic response [2–4]. Laser ablation experiments suggest that the columella cells in the root cap are the site of gravi-perception [5]. A large body of genetic and physiological evidence suggests a crucial role for starch-filled plastids (termed statoliths) in columella cells for gravity perception [2–5].

Daniela Ristova and Elke Barbez (eds.), *Root Development: Methods and Protocols*, Methods in Molecular Biology, vol. 1761, https://doi.org/10.1007/978-1-4939-7747-5_8,

Upon gravity perception, a signal is generated in the columella cells which is then transmitted through the lateral root cap cells to the gravi-responsive tissues of the elongation zone. Genetic and cell biology approaches have now established that *this signal* is plant hormone auxin [2, 4, 6, 7].

Plastid sedimentation results in asymmetric release of auxin in the columella cells [8–10]. The resulting auxin gradient is transported from columella cells via the lateral root cap to the gravitropic responsive tissues in the elongation zone [7]; here auxin inhibits cell expansion on the lower side, thus causing differential growth resulting in root bending.

Pioneering work from Knight first demonstrated that roots can bend downward, irrespective of the angle that the germinating seed was positioned [11]. Since then several different approaches have been used to measure and quantify root gravitropism. They usually involve placing the roots horizontally and then measuring the rate of root bending as the roots reposition themselves toward the gravity. Blancaflor et al. [5] measured root bending by acquiring images every 20 min for a period of 12 h, whereas Swarup and Peret [12] acquired images every 30 min over an 8-h period.

With advances in imaging techniques and refinement in image acquisition software, several software packages have been developed for root architectural trait analysis including root gravitropism [13–15]. Many of these packages required the use of high-resolution images or considerable user intervention.

Later French et al. [16] developed a high-throughput, automated image analysis package termed RootTrace to trace *Arabidopsis* seedling roots grown on agarose plates. The method combines a "particle-filtering algorithm with a graph-based method" to trace the center line of a root. Additional software can quantify several root parameters including length and curvature from original root traces. This package is simple, fast, and robust and does not require high-resolution images. It does however require suitably captured images of light roots on a dark background and the roots to not be overlapping. It also makes assumptions about the growth of the root, namely, that roots should be generally growing from top to bottom on the image plane. It does not measure architecture—for this we refer the reader to our RootNav tool [17]. For detailed requirements *see* French et al. [16] and accompanying software documentation.

In our laboratory, RootTrace is the method of choice for root gravitropic analysis. This chapter describes the use of RootTrace in analysis of root gravitropism.

2 Materials

1. Growth medium. Any suitable agar-based growth medium can be used for these assays. We typically use full-strength Murashige and Skoog (MS) media [18] (4.3 g/L) supplemented with 1% sucrose. However, different strength of MS medium with or without sucrose can also be used.
2. Petri plates (preferably square-125 × 125 mm).
3. Laminar flow cabinet.
4. Bleach for seed surface sterilization. We typically use 50% sodium hypochlorite solution for surface sterilization of *Arabidopsis* seeds. However, lower sodium hypochlorite solutions (5–25%) have also been reported.
5. 70% (v/v) ethanol.
6. Sterile Whatman filter paper.
7. Suitable growth room. Typically for *Arabidopsis* we use 20 ± 2 °C with 16 h photoperiod; 100 μE (light intensity).
8. Camera with tripod or automated image acquisition system.
9. Computer suitable for running RootTrace (*see* **Note 1**).

3 Methods

3.1 Growth Media

To prepare full-strength MS medium supplemented with 1% sucrose, dissolve 4.3 g of MS salts and 10 g of sucrose in approximately 800 mL of distilled water. Adjust the pH to 5.8 using 1 M KOH, and make up the volume to 1 L with distilled water. Dispense the media into bottle, add Bacto agar (10 g/L), and autoclave. Upon sterilization, the media is cooled to about 50 °C and is poured into sterile petri plates (for a standard 15 cm square petri plate, 50 mL is sufficient). Once set, plates can be stored at 4 °C for a couple of weeks (or even longer).

3.2 Seed Sterilization

Aliquot *Arabidopsis* seeds in an Eppendorf tube (1.5 mL), and add approximately 1 mL of bleach solution. Mix well and incubate at room temperature for 7–8 min. Fill the tube with approximately ½ mL of water containing 0.15% Triton X-100 (or similar surfactant), and mix well. Allow the seeds to settle to the bottom of the tube, and then decant the solution in a waste bottle carefully. Fill the tube again with water containing Triton X-100, mix well, and then allow the seeds to settle. Decant the solution in a waste bottle. Repeat the last step one more time with water containing Triton X-100 and then two to three more times with pure distilled water (no Triton X-100). Squirt the seeds on a sterile Whatman number 1 filter paper using 70% ethanol. Allow the seeds to air dry for about 30–40 min in a laminar flow hood (*see* **Note 2**).

3.3 Plating

Use a sterile scalpel to unstuck the seeds from the filter paper, and plate them in a row on MS agar medium with gentle tapping. A micropipette can also be used to place the water-suspended sterile seeds (no ethanol wash) in a row, on MS plates (*see* **Note 3**). Vernalize the seeds in dark at 4 °C for 2 days.

3.4 Root Gravitropism Assay

Transfer the plates to a suitable growth room (typically for *Arabidopsis* we use 20 ± 2 °C with 16 h photoperiod; 100 μE light intensity), and place them vertically.

Seedlings are grown vertically for 4 days on MS plates, and the plates are then turned at an angle of 90°. Photographs are taken every 10 min for up to 24 h (*see* **Note 4**).

3.5 Image Analysis

The images can be analyzed for root gravitropic curvature using either RootTrace software or using manual image analysis.

Root trace can be used for a series of images taken from a fixed device or from an automated image acquisition system (e.g., Wells et al. [19]) with a defined camera and plate position(s). Images taken manually can also be analyzed if the camera position and the plate position are fixed for the duration of the experiment. However, if the camera position changes between different time points, RootTrace will struggle to locate the same root over time. In those situations, manual analysis is required.

3.5.1 Using RootTrace Software

RootTrace software is available from https://sourceforge.net/projects/roottrace. Installation instructions are included.

1. Open RootTrace software.
2. Click Advanced.
3. Click Invert box if necessary (for images taken using a gray scale infrared camera, i.e., dark roots on a light background).
4. Click Rotation if necessary (clockwise or anticlockwise). The roots should be predominantly growing down the image plane (from top to bottom).
5. Click Tip angle box.
6. Import images. (Normally no more than 20 images are required. A gravitropic response will be finished in 6 h. We routinely select three images every 30 min, up to 6 h, or two images every hour up to 8–10 h.)
7. Identify a region with up to ten roots per analysis (avoid roots that are too close to each other). Crop the image to include the selection.
8. Mark start points for the tracing. To do this, right-click somewhere further up the root from the tip. RootTrace will trace downward from this point. Do this for every root. If you do not

need a full root length measure, you can start tracing from nearer the tip to still capture rate of growth.

9. The next step is to define the background near each root; this allows the software to learn what the background appears like near each root. To do this, left-drag and make a square next to the root tip (it is best to drag the mouse top left to bottom right). This step needs to be done for all the selected roots (Order matters—select the background in the same order as you selected the root tips). You can do all the tips and then all the background patches, or you can interleave them (make sure the background numbers correspond to the root numbers in the image in the software).
10. Click the Misc tab, and change the tip point to fit setting to 6. This affects the distance back from the tip that RootTrace will fit a line through, in order to measure an angle for the tip. Smaller distances mean a quicker response to root angle changes, at the cost of more susceptibility to noise.
11. Make a new folder for data. (Create a folder in the same folder where the images are; calling the folder "output" is good practice, and if it exists before loading the images, RootTrace will automatically set it as a folder to store output files.)
12. Go to the main menu, and click GO to run the software.

    Once it is complete, check each image for tip angle tracking. Angles will be shown as orange lines offset from the root and represent the detected angle of the tip, and if it looks correct, you can optionally delete all the tracked photos from this newly created folder (not the originals) as well as all curvature files. The aim is to ensure tracking has been successful, to ensure the correct points are measured for the tip angles.

    Aim 1. RootTrace will terminate tracking when it detects the end of a root. It has different processes to determine when a root tip is located. If the default does not work, selecting Advanced → "End Detection version #2" may improve results.

    Aim 2. At places the root becomes almost transparent and thus causes tracking to fail (e.g., *see* Fig. 1). To overcome this, consider if you can start tracing nearer the tip or try choosing a different, more representative background sampling region (**step 9**).

    Aim 3. Do you need to track all the way to the tip? Tracking just the points where the bend is evolving may be sufficient to capture the change in angle. You can identify this point on the output images or in the Excel data.

    Aim 4. Rotating the images by 45° can improve tracking success, as RootTrace will not want to follow a root up toward the top of an image (e.g., which can occur with extreme bending responses). However, final angles in the output files must be adjusted accordingly (Fig. 2).

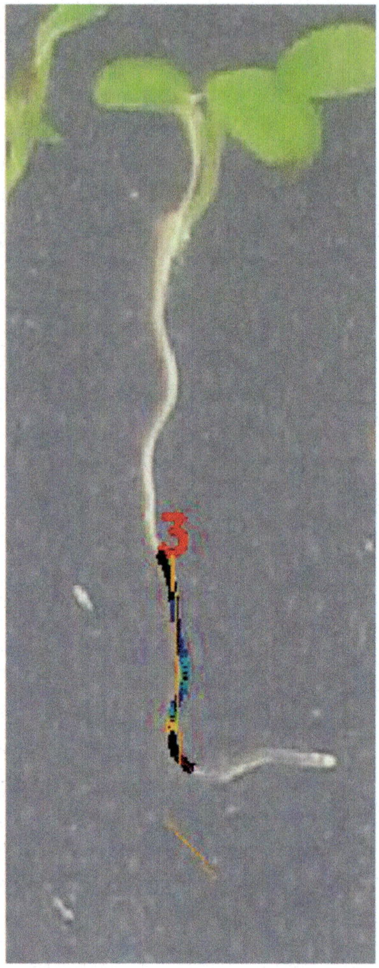

Fig. 1 RootTrace troubleshooting. Output of RootTrace overlaid (trace line and root and tip angles) on an *Arabidopsis* root image. In this example, two of RootTrace's assumptions have not been met here, causing tracking to fail. The root starts to grow upward, and the lighting of the root is such that the root becomes almost transparent (indicated by the arrow)

13. Multiple output files are produced. The data of interest for this work will be saved in an Excel CSV file called "tip angle to vertical."
14. You should delete all the data for all the roots that have not been correctly tracked (as judged above).
15. The statistical analysis can now be done on this data.

3.5.2 Manual Image Analysis

1. Take images every 30 min. (We typically take images every 30 min up to 8–10 h and additional image at 24 h.)
2. Measure the root angle using a protractor.

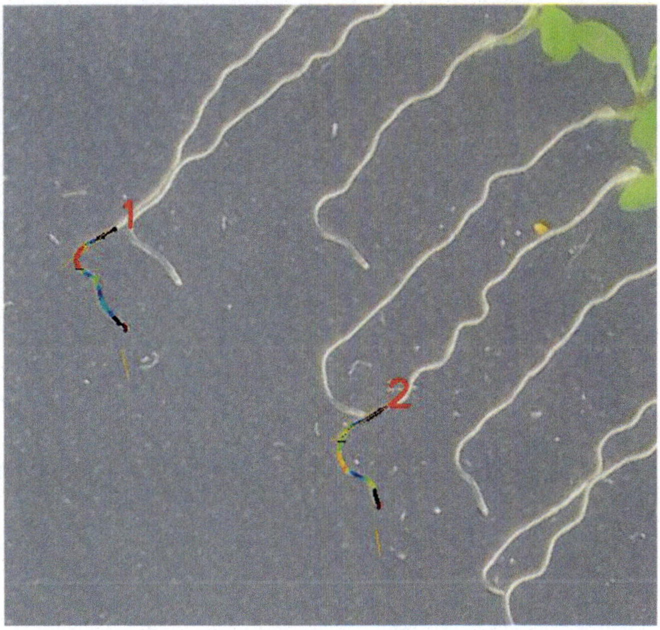

Fig. 2 Improving RootTrace output. Output of RootTrace overlaid (trace line and root and tip angles) on an *Arabidopsis* root image. This time, plate images have been rotated 45° before loading in RootTrace using image editing software. Note the tracking around the corners is now much more successful, as the assumptions of RootTrace have now been met; namely, the root appears to RootTrace to grow down the image plane—it does not curve upward

3. Do this for every root in all the images for all the samples/treatments.
4. The data can be presented as individual wheel diagrams [20] or as a time series [7].
5. For wheel diagram we routinely group the roots in 40° angles (0–40, 40–80, 80–120, etc.) and express them as percentage of roots bending at a given angle group. The size and the angle of the bar represent percentage of roots and angle, respectively (Fig. 3).
6. The data can also be expressed as a function of time (time series). This is done by calculating number of roots responding to the gravity stimulus at any given time (Fig. 4).

4 Notes

1. RootTrace requires a personal computer running Microsoft Windows (XP onward). Hardware requirements are modest; a PC with 2GB or more of RAM will be sufficient.

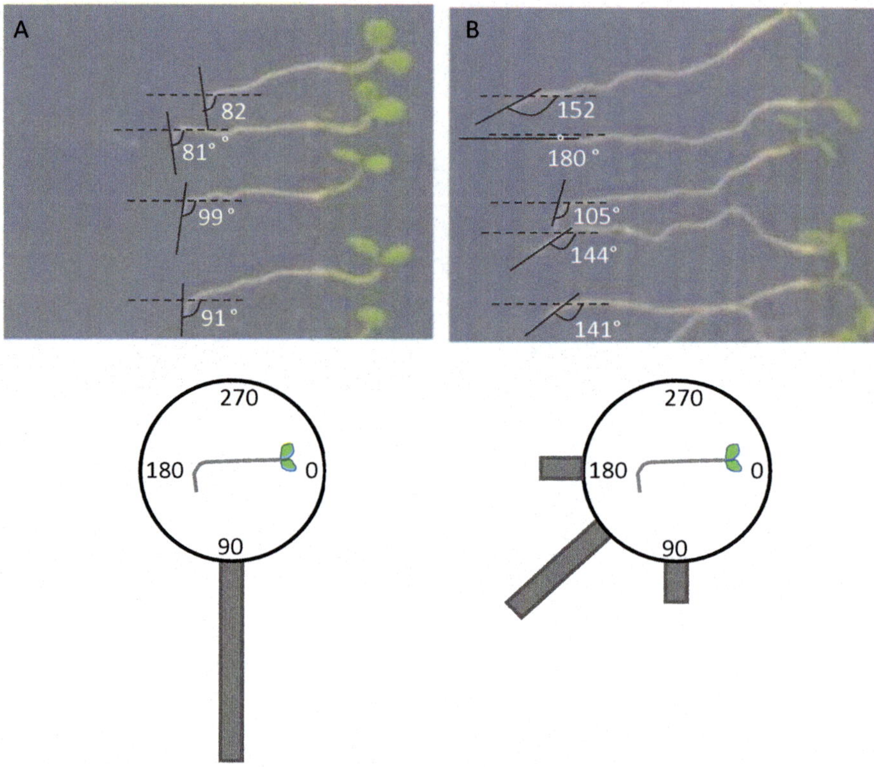

Fig. 3 Manual quantification of root gravitropic curvature. Seedlings were grown for 4 days, and then the plates were turned at 90°. Images were taken at specified time points (7 h after the bending in this example). A tangent line (solid line) was drawn tracing the root bend, and the angle of the root curvature from the vertical (dotted line) was measured. The angles were grouped as described in section B (manual image analysis) and were plotted on the wheel diagram as percentage of roots bending at a given angle group. For example, in **a**, all roots are gravitropic, whereas in **b**, the gravitropic response is compromised and is reflected in the wheel diagram by different angle groups and the size of the bars

2. If seeds are air dried, consider placing the sterile filter paper on a paper towel. This speeds up the drying as excess ethanol is soaked up by the paper towel.
3. Instead of drying the seeds on a filter paper, it is possible to use a micropipette for plating. This may provide more control when plating the seeds in a line. In that case there is no need for the last ethanol wash. After the water washes, seeds can directly be plated using a micropipette. Cutting the tip end slightly may help. It may be easier to suck only a few seeds in the cut tip and take more once they are plated. It is also possible to vernalize the seeds in the water prior to plating if required.
4. If photographs are taken manually, it is possible to take photographs at larger intervals, for example, every 20 or 30 min. In manual photography, it is better to start the experiment early in the day so that images can be taken for at least 7–8 h after the

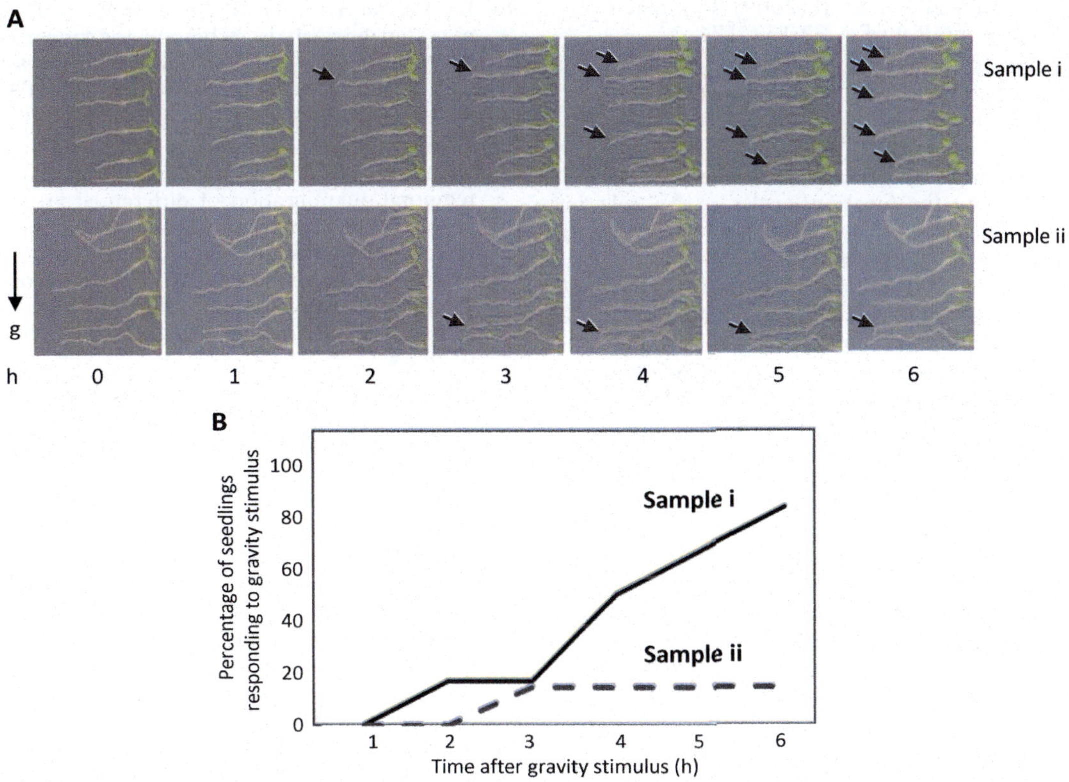

Fig. 4 Manual quantification of root bending kinetics. Root bending kinetics can be calculated from different time point images. In this example, images were taken every hour after the gravity stimulus (**a**) and the bending of root kinetics shown as percentage of root responding to the gravity stimulus (**b**). Arrowhead indicates the roots responding to the gravity vector

gravity stimulus. In most cases a visual gravitropic response can be seen within 4–6 h. A 24-h image can then be taken on day 2 if required.

Acknowledgment

We acknowledge support from the Biological and Biotechnological Science Research Council (BBSRC), European Research Council (ERC), Royal Society, and Newton International and Advanced Fellowship awards.

References

1. Lynch JP (2011) Root phenes for enhanced soil exploration and phosphorus acquisition: tools for future crops. Plant Physiol 156 (3):1041–1049. https://doi.org/10.1104/pp.111.175414
2. Sato EM, Hijazi H, Bennett MJ, Vissenberg K, Swarup R (2015) New insights into root gravitropic signalling. J Exp Bot 66(8):2155–2165. https://doi.org/10.1093/jxb/eru515

3. Swarup R, Bennett MJ (2009) Root gravitropism. Annu Plant Rev 37:157–174
4. Swarup R, Wells D, Bennett MJ (2013) Root gravitropism. In: Eshel A, Beeckman T (eds) The hidden half. CRC Press, Boca Raton, pp 19–34. IV edition:19–34
5. Blancaflor EB, Fasano JM, Gilroy S (1998) Mapping the functional roles of cap cells in the response of Arabidopsis primary roots to gravity. Plant Physiol 116(1):213–222
6. Swarup R, Bennett MJ (2014) Auxin transport: providing plants with a new sense of direction. Biochemist 36:12–16
7. Swarup R, Kramer EM, Perry P, Knox K, Leyser HM, Haseloff J, Beemster GT, Bhalerao R, Bennett MJ (2005) Root gravitropism requires lateral root cap and epidermal cells for transport and response to a mobile auxin signal. Nat Cell Biol 7(11):1057–1065. https://doi.org/10.1038/ncb1316
8. Boonsirichai K, Sedbrook JC, Chen R, Gilroy S, Masson PH (2003) ALTERED RESPONSE TO GRAVITY is a peripheral membrane protein that modulates gravity-induced cytoplasmic alkalinization and lateral auxin transport in plant statocytes. Plant Cell 15(11):2612–2625
9. Ottenschlager I, Wolff P, Wolverton C, Bhalerao RP, Sandberg G, Ishikawa H, Evans M, Palme K (2003) Gravity-regulated differential auxin transport from columella to lateral root cap cells. Proc Natl Acad Sci U S A 100 (5):2987–2991. https://doi.org/10.1073/pnas.0437936100
10. Rashotte AM, DeLong A, Muday GK (2001) Genetic and chemical reductions in protein phosphatase activity alter auxin transport, gravity response, and lateral root growth. Plant Cell 13(7):1683–1697
11. Knight TA (1806) On the direction of the radicle and germen during the vegetation of seeds. Philos Trans R Soc Lond 96:99–108
12. Swarup R, Peret B (2012) AUX/LAX family of auxin influx carriers-an overview. Front Plant Sci 3:225. https://doi.org/10.3389/fpls.2012.00225
13. Basu P, Pal A, Lynch JP, Brown KM (2007) A novel image-analysis technique for kinematic study of growth and curvature. Plant Physiol 145(2):305–316. https://doi.org/10.1104/pp.107.103226
14. Chavarria-Krauser A, Nagel KA, Palme K, Schurr U, Walter A, Scharr H (2008) Spatio-temporal quantification of differential growth processes in root growth zones based on a novel combination of image sequence processing and refined concepts describing curvature production. New Phytol 177(3):811–821. https://doi.org/10.1111/j.1469-8137.2007.02299.x
15. Qi X, Qi J, Wu Y (2007) RootLM: a simple color image analysis program for length measurement of primary roots in Arabidopsis. Plant Root 1:10–16
16. French A, Ubeda-Tomás S, Holman TJ, Bennett MJ, Pridmore T (2009) High-throughput quantification of root growth using a novel image-analysis tool. Plant Physiol 150 (4):1784–1795
17. Pound MP, French AP, Atkinson JA, Wells DM, Bennett MJ, Pridmore T (2013) RootNav: navigating images of complex root architectures. Plant Physiol 162(4):1802–1814
18. Murashige T, Skoog F (1962) A revised medium for rapid growth and bio assays with tobacco tissue cultures. Physiol Plant 15 (3):473–497
19. Wells DM, French AP, Naeem A, Ishaq O, Traini R, Hijazi H, Bennett MJ, Pridmore TP (2012) Recovering the dynamics of root growth and development using novel image acquisition and analysis methods. Philos Trans R Soc Lond B Biol Sci 367(1595):1517–1524
20. Swarup R, Kargul J, Marchant A, Zadik D, Rahman A, Mills R, Yemm A, May S, Williams L, Millner P, Tsurumi S, Moore I, Napier R, Kerr ID, Bennett MJ (2004) Structure-function analysis of the presumptive Arabidopsis auxin permease AUX1. Plant Cell 16(11):3069–3083. https://doi.org/10.1105/tpc.104.024737

Part II

Microscopy and Cell Biology

Check for updates

Chapter 9

Calcium Ion Dynamics in Roots: Imaging and Analysis

Ellie Himschoot, Melanie Krebs, Alex Costa, Tom Beeckman, and Steffen Vanneste

Abstract

Calcium sensors are indispensable tools to study the role of Ca^{2+} and visualize Ca^{2+} dynamics during biological processes. Over the past years, the field of Ca^{2+} imaging has strongly expanded by the development of a wide palette of sensors and optimization of sample handling. Here, we provide guidelines for imaging of the Ca^{2+} sensor R-GECO1 in *Arabidopsis thaliana* roots which can be interpolated to other intensiometric Ca^{2+} sensors. Furthermore, we demonstrate a procedure for image analysis of the acquired time-lapse recordings.

Key words R-GECO1, Ca^{2+} imaging, Confocal laser scanning microscopy, Spinning disk microscopy, Fiji, MATLAB, *Arabidopsis thaliana*

1 Introduction

Calcium is an important second messenger regulating a wide range of developmental processes and environmental responses in plants and animals [1–6]. Typically, the resting Ca^{2+} concentration in the cytoplasm is kept low, around 100–200 nM, while the Ca^{2+} concentration in the organelles and extracellular space is several orders of magnitude higher. This steep concentration gradient allows to generate rapid cytoplasmic Ca^{2+} signals by opening only a few Ca^{2+} channels. After a stimulus-induced cytosolic Ca^{2+} increase, the cytoplasmic Ca^{2+} levels are restored to the resting status. Based on the amplitude and duration of the Ca^{2+} signals, specific Ca^{2+} signatures can be defined for distinct cellular processes. Therefore, to investigate the role of Ca^{2+} in biological processes, it is critical to be able to capture accurately and with high resolution the distinct Ca^{2+} signatures. Given that Ca^{2+} is a bivalent ion, rather than a protein, its dynamics cannot be visualized using standard molecular biology approaches such as translational reporters. Instead, the visualization of Ca^{2+} is commonly done indirectly, based on the change in physical properties of dyes or genetically encoded Ca^{2+}

Daniela Ristova and Elke Barbez (eds.), *Root Development: Methods and Protocols*, Methods in Molecular Biology, vol. 1761, https://doi.org/10.1007/978-1-4939-7747-5_9,

indicators (GECI) upon binding of Ca^{2+}. The rapid loading, sensitivity, and broad dynamic range make fluorescent dyes highly attractive to study Ca^{2+} dynamics. This explains why, to date, multiple Ca^{2+} sensitive dyes, such as Fura-2, are still commonly used to capture dynamic Ca^{2+} signatures in animal cells [7–9]. However, such dyes have never been commonly used in plants, mainly due to technical difficulties to load the dyes uniformly into the plant tissues and the requirement of specimen manipulations which can introduce artifacts. In contrast, the plant field has benefited a lot from a wide range of genetically encoded Ca^{2+} indicators (GECI) which have the key advantage that they can be expressed both in specific tissues and in the entire plant. Moreover, they can be targeted to distinct subcellular compartments to dissect the mechanisms of Ca^{2+} signaling at the organelle level [10–13].

The GECIs can be subdivided in the chemiluminescence- and fluorescence-based indicators. Chemiluminescent indicators like aequorin have a large dynamic range; however, they suffer from rather low light emission levels (quantum yield), need coelenterazine loading into the cells and tissues, and do not allow imaging at cellular resolution [14]. Fluorescence-based sensors have a high spatial and temporal resolution and do not need loading of any cofactors. However, they have a limited dynamic range and usually do not have a linear Ca^{2+}-dependent readout. Among the fluorescence-based Ca^{2+} sensors, ratiometric and intensiometric sensors can be distinguished. The ratiometric sensors are usually based on Ca^{2+}-dependent Förster resonance energy transfer (FRET), in which a donor and acceptor fluorophore (e.g., CFP-YFP) are interconnected by a Ca^{2+}-controlled interaction module (e.g., calmodulin (CaM) and CaM-binding peptide M13). Upon Ca^{2+} sensing, the interaction module induces a conformational change thereby bringing both fluorophores in closer proximity allowing FRET. This FRET causes a simultaneous increase in acceptor fluorescence and decreased donor fluorescence resulting in an increased ratio of acceptor emission/donor emission which can be imaged using confocal microscopy. A great advantage of the ratiometric character of such sensor is that it allows to assess Ca^{2+} dynamics independently of the expression levels and correct for changes of focus. However, ratiometric sensors have a limited dynamic range. Intensiometric sensors are based on enhancement or decrement in fluorescence of a single fluorophore depending on the Ca^{2+}-controlled interaction module. In this case, Ca^{2+} binding leads to a (de)stabilization of the fluorophore and thus correspondingly an increase or decrease of brightness. These sensors generally have a relatively high dynamic range, and have fast on-off kinetics, but do not allow to correct for differences in expression levels. Multiple intensiometric probes such as the green fluorescent GCaMP series [15, 16] and GECOs with different spectral properties and sensitivities [17–19] have been developed over the past years and are continuously being improved. Recently the red fluorescent R-GECO1 ($K_d = 480$ nM, [19]) and the green fluorescent GCaMP6s ($K_d = 144$ nM, [20]) have been described as very sensitive probes for monitoring Ca^{2+} dynamics in *A. thaliana* [21, 22].

Here, we provide guidelines for image acquisition of the intensiometric R-GECO1 sensor, which also apply to other intensiometric Ca^{2+} sensors. In addition, we provide a 2D and 3D design for an imaging chamber prototype which facilitates image acquisition for real-time recording of Ca^{2+} dynamics during treatment while minimizing sample drift. Finally, we present a method for image analysis using Fiji (manual) and MATLAB (script).

2 Materials

2.1 Seed Sterilization

1. Eppendorf tubes (2 mL).
2. Desiccator in fume hood.
3. Sterilization solution: 5 mL concentrated HCl (37%), 100 mL sodium hypochlorite (NaClO, 13%).

2.2 Growth Medium (0.5× MS)

1. For 1 L growth medium: 10 g sucrose, 2.3 g MS salts, 0.5 g MES.
2. Dissolve in ±950 mL MilliQ.
3. Adjust pH to 5.7 using KOH.
4. Add MilliQ up to 1 L.
5. Add 8 g of agar, and mix well before autoclaving.

2.3 Plant Growth and Sample Preparation

1. Growth medium (*see* Subheading 2.2.)
2. Square sterile plates (12 cm × 12 cm).
3. Sterile tooth picks.
4. Epifluorescence binocular with a suitable filter.

2.4 Imaging Medium (Minimal Medium)

1. For 1 L growth medium: 2.132 g MES, 0.372 g KCl, 1.11 g $CaCl_2$.
2. Dissolve in ±950 mL MilliQ.
3. Adjust pH to 5.7 using KOH.
4. Add MilliQ up to 1 L.

2.5 Imaging Chamber

We use a customized imaging chamber (based on the imaging incubation chamber shown in ref. 23) suited for imaging on an inverted microscope. The design of the chamber allows to apply treatments during imaging while minimizing sample drift. Here we provide a 2D and 3D design for fabrication of the chamber by milling/cutting or 3D printing depending on the available equipment (*see* **Note 1**). The designs are made available through the following link: https://www.psb.ugent.be/docs/himschoot-et-al-2017-calcium-ion-dynamics-in-roots-chamber-designs-and-matlab-script. The required components for assembly of the chamber are listed below.

1. Customized imaging chamber with lid (*see* 2D and 3D design).
2. Small round petri dish, diameter 35 mm.
3. Low viscous silicon paste and swab.
4. Round cover glasses, diameter 24 mm thickness 1.
5. Chemically pure glass wool.
6. 2 M3 x 5 mm screws +2 washers with outer diameter 15 mm and inner diameter 4 mm.

2.6 A Widefield Microscope (WM), Confocal Laser Scanning Microscope (CLSM), or Spinning Disk Microscope (SDM)

The procedures described can be applied to any inverted WM, CLSM, or SDM.

2.7 Fiji

Fiji (*F*iji *I*s *J*ust *I*mageJ) is an open-source image processing package based on ImageJ [24]. The described procedures for image analysis with Fiji can be implemented using any version of Fiji. It is freely available from https://imagej.net/Fiji/Downloads.

2.8 MATLAB

MATLAB (*mat*rix *lab*oratory) is a multi-paradigm numerical computing environment which is popular among scientists for image processing. A script for image analysis *intensiometric_Ca_imaging.m* is provided under the following link: https://www.psb.ugent.be/docs/himschoot-et-al-2017-calcium-ion-dynamics-in-roots-chamber-designs-and-matlab-script. It was developed in MATLAB 2013a and is compatible with this and more recent MATLAB releases. In contrast to Fiji, it is not available for free, and a license is required. In case no MATLAB license is available, a free product trial can be requested on www.mathworks.com.

3 Methods

3.1 Seed Sterilization Using Chlorine Gas

1. Put the desired seeds into 2 mL Eppendorf tubes (the recommended amount of seeds is approx. 50 μL), and place the seeds into the desiccator in the fume hood.

2. Add 5 mL of concentrated HCl (37%) and 100 mL 13% sodium hypochlorite (NaClO) into the desiccator, close the desiccator immediately, and incubate the seeds with developed chloral gas for at least 3 h or overnight.
3. Let the majority of the toxic chloral gas escape from the desiccator in the fume hood. Close the tubes, and transfer them to the laminar flow to air out the remnant chloral gas for 15 min.

3.2 Plant Growth

1. Square sterile plates for in vitro growth are filled with 50 mL growth medium (0.5× MS) with agar.
2. Place the seeds using a sterile toothpick on the solidified medium. Ensure that the seeds are not too close to one another (±0.5 cm). This will facilitate screening for seedlings with good expression levels later on.
3. After sowing, the plates are transferred to the cold room (4 °C, dark) for 1 day vernalization.
4. Transfer the plates to the growth chamber (21 °C, continuous light) for 4–5 days.

3.3 Imaging Chamber Assembly and Sample Preparation (Fig. 1)

1. Before preparing the samples, it is advised to check the seedlings for good expression levels of the sensor. This can be done using an epifluorescence binocular with a suitable filter.
2. Apply a small amount of silicon paste on the rim of the base of the chamber using a swab and mount a round cover glass. The silicon paste will seal the chamber and prevent leakage.
3. Put a seedling in a small petri dish filled with liquid minimal medium for a few seconds to hydrate the seedling and prevent it from drying out. Transfer the seedling to the center of the cover glass, and cover it with a thin layer of glass wool. The glass wool will reduce seedling drift during treatment. Add 200 μL minimal medium to the chamber. It is crucial to perform these operations quickly to avoid the seedling would dry out.

Fig. 1 Imaging chamber assembly. (**a**) The imaging chamber components. (**b**) Assembled imaging chamber

4. Mount the top lid, with the opening parallel to the seedling, and mount the M3 screws with washers. The washer will hold down the top lid.
5. Let the mounted seedling recover for about 30 min before initiating the Ca^{2+} imaging. The mechanostimulation of the seedlings during sample preparation may render the seedlings less responsive to subsequent stimuli (*see* **Note 2**).
6. Prepare the solution for treatment in 100 μL minimal medium. Ensure the compound concentration is three times higher than the desired end concentration (*see* **Note 3**).

3.4 Ca^{2+} Imaging

The data processing procedure is compatible with widefield microscope (WM), confocal laser scanning microscope (CLSM), and spinning disk microscope (SDM) image acquisition. The main difference between WM and CLSM or SDM is the lack of optical sectioning thereby not only capturing fluorescence from the focal plane but also out-of-focus light which affects the resolution of the image. A WM typically uses a fluorescent lamp as an excitation source, while a CLSM and SDM are equipped with lasers. The main difference between CLSM and SDM is the method of scanning the sample. In case of CLSM, a point-by-point illumination is used, while in SDM multiple points are illuminated simultaneously, thereby speeding up image acquisition, which can be important to capture very fast Ca^{2+} kinetics. The WM also allows for fast image acquisition. Most steps described below apply to WM, CLSM, and SDM setups; where differences occur, the guidelines will be described for both devices separately.

R-GECO1 is an intensiometric Ca^{2+} sensor, meaning that an increase in fluorescence intensity reflects an increase in Ca^{2+} levels in a single emission channel [19]. As it is a red-shifted Ca^{2+} indicator, it can be excited by 561 nm wavelength, and emission light can be captured in a range of 620–650 nm wavelength.

1. Select an objective (*see* **Note 4**) and set the desired image resolution (e.g., 512 × 512) and bit depth (e.g., 16 bit).
2. Mount the imaging chamber on the stage, position the root tip in the center of the field of view (*see* **Note 5**), and focus on the root tip.
3. Set the percentage of laser, the pinhole size in case of CLSM, detector gain, and offset to minimize photobleaching during a time-lapse recording, and use the full dynamic range of the detector. The maximal Ca^{2+} signal should not result in pixel saturation, and the background should be close to the detector noise levels for optimal semiquantitative assessment of Ca^{2+} dynamics in the sample. Therefore, it is advised to perform a test experiment to estimate the extent of fluorescence intensity

increase which can be expected and adjust the settings accordingly.

4. For detection of the signal, microscopes can be equipped with photomultiplier tubes (e.g., CLSM) or (CCD) cameras (e.g., SDM). In case of a camera detector, it is required to set the exposure time during which the sample is illuminated and fluorescence will be collected by the camera. High laser power combined with long exposure times can cause photobleaching.
5. Set up the time-lapse recording (*see* **Note 6**). Time series are recorded by collecting a scan every 3–5 s. If higher temporal resolution is required, the frame rate can be increased. For very fast data acquisition, the SDM is preferred. However, increasing the frame rate will result in more frequent scanning of the sample and possible photobleaching.
6. Start the time-lapse recording. Allow the fluorescence level to stabilize (3–5 min) before applying a treatment.
7. Pause the time lapse, write down the frame number, apply the treatment, and resume the recording. Use 100 μL minimal medium supplemented with the 3× concentrated treatment (after addition, the end volume in the imaging chamber will be 300 μL, thereby diluting the treatment to 1×). The frame number will be important for normalization during image analysis of the data (*see* **Note 7**). In the case of a compound-induced Ca^{2+} increase, the fluorescence levels will increase.
8. Save the time-lapse recording. For image analysis the procedures for Fiji and MATLAB are described. It is therefore important that the file format used for saving the time-lapse recording is compatible with either program. A wide range of file formats can be opened and processed in Fiji (.lsm, .tiff, . ometiff, etc.). In case of MATLAB, the number of file formats is more limited (e.g., .lsm cannot be processed by the provided MATLAB script). It is advised to use .tiff or .ometiff files (*see* **Note 8**).

3.5 Image Analysis

Here, the fluorescence intensity dynamics will be analyzed for multiple regions of interest. Analysis of these plots will allow to identify the maximal amplitude, duration, onset, etc. of a Ca^{2+} response upon treatment. The procedure consists of five major steps: (1) image registration, (2) measurement of the background levels, (3) measurement of the fluorescence intensity in a selected ROI, (4) background subtraction, and (5) normalization of the background-corrected fluorescence intensities to the initial fluorescence intensity (*see* **Note 7**). These operations can be performed using different image processing software packages. Here, we provide the guidelines for manual image analysis in Fiji and a MATLAB script to automate part of the image analysis.

3.5.1 Fiji (Fig. 2)

1. Open the time-lapse recording by selecting "File" > "Open..."
2. Perform image registration. In case of movement/drift of the root during the time-lapse recording, it is essential to perform an image registration step to align the different frames. This is to avoid that the area of interest would move over time and would no longer be contained within the marked ROI.

   To perform intensity-based image registration, use the StackReg plugin in Fiji [25]. Go to "Plugins" > "Registration" > "StackReg". In case the plugin is not available, it can be downloaded here: http://bigwww.epfl.ch/thevenaz/stackreg/. Select a transformation method depending on the type of movement the root makes over time. The transformation methods are Translation, Rigid Body (translation + rotation), Scaled Rotation (translation + rotation + scaling), and Affine (more complex movement). In case of root movement due to growth or drift, translation is usually sufficient for proper image alignment.
3. Define a region for background measurement and regions of interest (ROIs). To mark a region, use a selection tool of choice (e.g., a polygon) in the toolbar. After drawing the region, it can be saved by going to "Edit" > "Selection" > "Add to Manager". The region for background measurement should be drawn outside of the root (Fig. 2). Multiple ROIs can be drawn at positions of choice. The number of ROIs is referred to as n.
4. Measure the mean fluorescence intensity in the background region and n ROIs. First, define the parameter that has to be measured. Go to "Analyze" > "Set Measurements..." > select "Mean gray value" > "OK." Next, measure the mean gray value in the different regions. In the ROI Manager window, select all regions, and click on "More" > select "Multi Measure" > enable "One row per slice..." > "OK." The Results window shows the mean fluorescence intensity for the background region (I_{BG}) and n ROI (I_{ROIn}) over time (Fig. 2). The obtained results are further processed in Excel.
5. Perform a background subtraction by calculating $(I_{\mathrm{ROIn}}-I_{\mathrm{BG}})(t)$ (Fig. 3). Subtract, per frame, the mean background signal I_{BG} from the mean signals in each of the n ROI, I_{ROIn}.
6. Normalize the background-corrected values by calculating $[(I_{\mathrm{ROIn}}-I_{\mathrm{BG}})(t)-(I_{\mathrm{ROIn}}-I_{\mathrm{BG}})(t_{ref})]/(I_{\mathrm{ROIn}}-I_{\mathrm{BG}})(t_{ref})$ (Fig. 4). The background-corrected mean intensities $(I_{\mathrm{ROIn}}-I_{\mathrm{BG}})$ for ROI n over time t are normalized to the background-corrected mean intensity $(I_{\mathrm{ROIn}}-I_{\mathrm{BG}})$ at a reference time point t_{ref}. The reference time point is usually a time point prior to the treatment, when fluorescence levels have

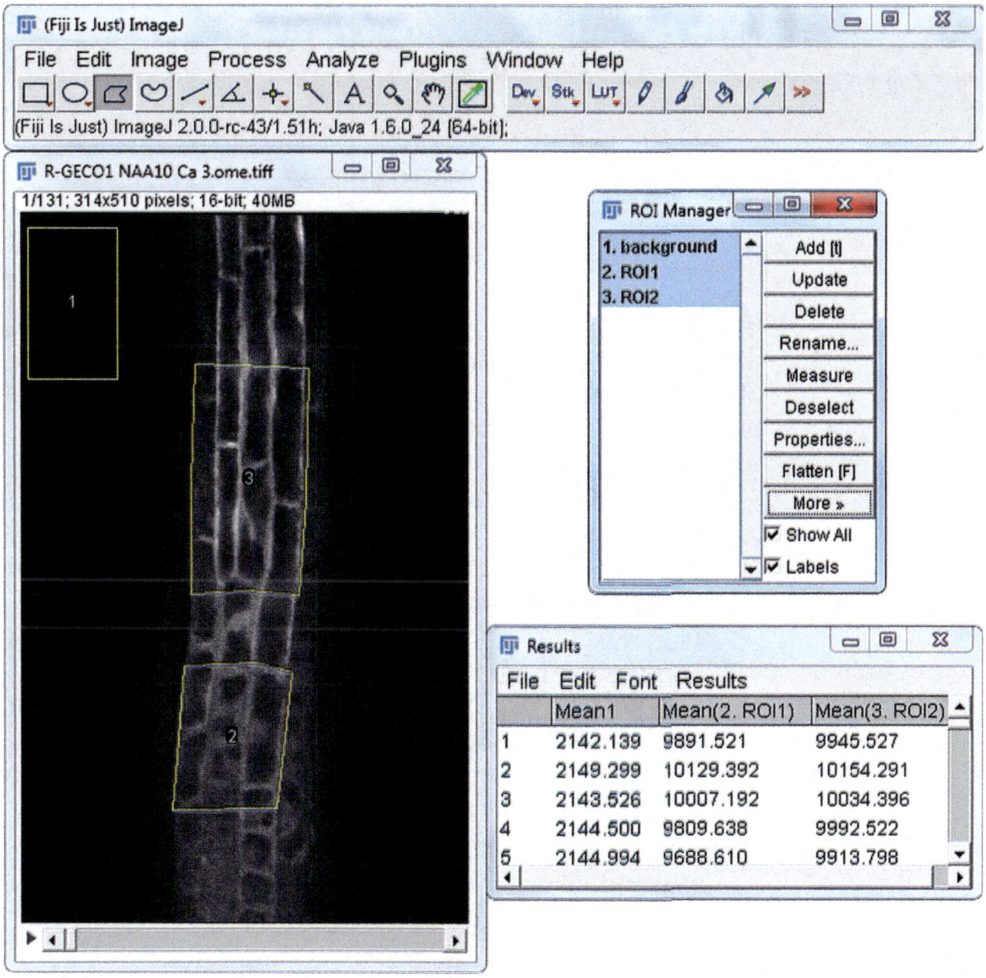

Fig. 2 Screenshot of Fiji interface. The ROI Manager window stores the background region and ROIs. The corresponding regions are displayed on the image at the left. The Results window shows the mean gray values for the different regions over time

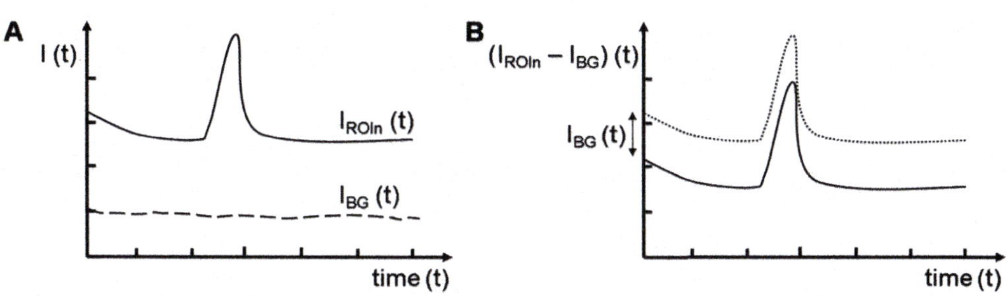

Fig. 3 Visualization of background subtraction procedure. (**a**) The measured mean fluorescence intensity values for ROI *n* (I_{ROIn} (*t*)) and the background region (I_{BG} (*t*)) over time. Note the initial drop in mean fluorescence intensity for ROI *n* due to photobleaching at the start of the time-lapse recording. (**b**) The mean fluorescence intensity values of ROI *n* before (dotted line) and after (full line) background subtraction

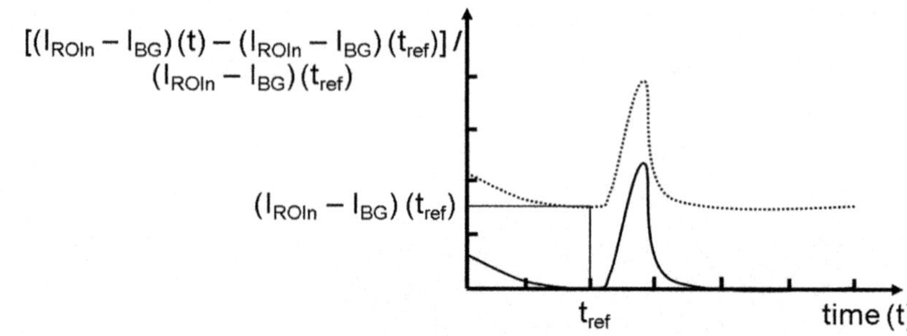

Fig. 4 Visualization of normalization procedure. The graph shows the background-corrected mean intensities over time before (dotted line) and after (full line) normalization. The background-corrected mean intensity at time point t_{ref} (= (I_{ROIn}−I_{BG}) (t_{ref})) is used for normalization

stabilized (*see* **Note 7**). For normalization, subtract and subsequently divide the background-corrected mean intensities for a ROI ($I_{ROIn}-I_{BG}$) (t) by the background-corrected mean intensity at the reference time point ($I_{ROIn}-I_{BG}$) (t_{ref}).

7. Plot the data in a line chart. The normalized background-corrected mean intensities [($I_{ROIn}-I_{BG}$) (t)-($I_{ROIn}-I_{BG}$) (t_{ref})]/($I_{ROIn}-I_{BG}$) (t_{ref}) are used for the y-axis values. Recalculate the number of frames to the corresponding time in seconds by taking into account the time interval between subsequent scans (x-axis values).
8. Plot analysis. Different parameters can be extracted from this plot and compared among treatments/genotypes such as the peak amplitude, response time, steepness of the peak, peak duration, etc.

3.5.2 MATLAB

The workflow of the provided MATLAB script *intensiometric_Ca_imaging.m* is schematically illustrated in Fig. 5 and is described in more detail below. The script does not include an image registration step. Therefore, it should be first performed in Fiji as described above (Subheading 3.5.1, **step 2**).

1. Getting started with MATLAB. Copy the provided script *intensiometric_Ca_imaging.m* to the directory used by MATLAB, and add it to the path (right-click > "Add to path") (Fig. 6). Ensure that all the movies are in the same directory and added to the path. The plots that will be generated by the script will be saved to this directory. In order to run the script, right-click on *intensiometric_Ca_imaging.m* in the "Current Folder" panel, and select "Run."
2. A "Select File to Open" window will pop up. Select the time-lapse recording of interest, and click "Open." If the time-lapse

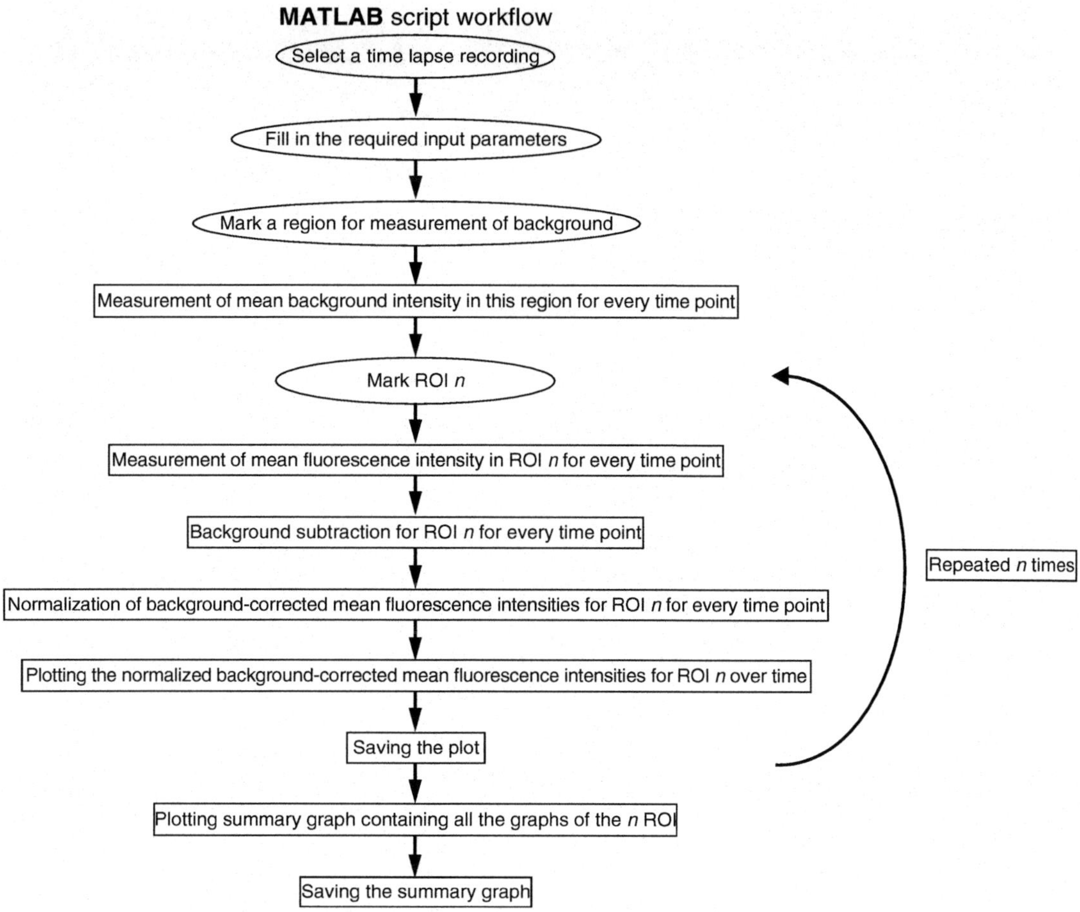

Fig. 5 Workflow of MATLAB script *intensiometric_Ca_imaging.m*. The oval shapes indicate steps where user input is required. The rectangular shapes mark the steps which are performed automatically while running the script

file is not visible in the opened folder, ensure that "All Files" is selected and that the time-lapse file is stored in this folder.

3. An "Input" window will be shown. Provide the required input parameters: the number of the frame of which the fluorescence intensity has to be used for normalization (t_{ref}), the time interval between scans in sec (here: 5 s), and the number of ROI ($= n$).
4. The first frame of the time-lapse recording will appear, together with a window containing the instructions to set a (polygon) region for measurement of background fluorescence. To proceed and start drawing the region, click "OK." Specify the region by selecting vertices of the polygon. To close the polygon, double-click or position the mouse over the first vertex and single-click. The polygon can be moved or resized using the mouse. When the polygon has the desired size and position, create a mask by right-clicking inside the region and selecting

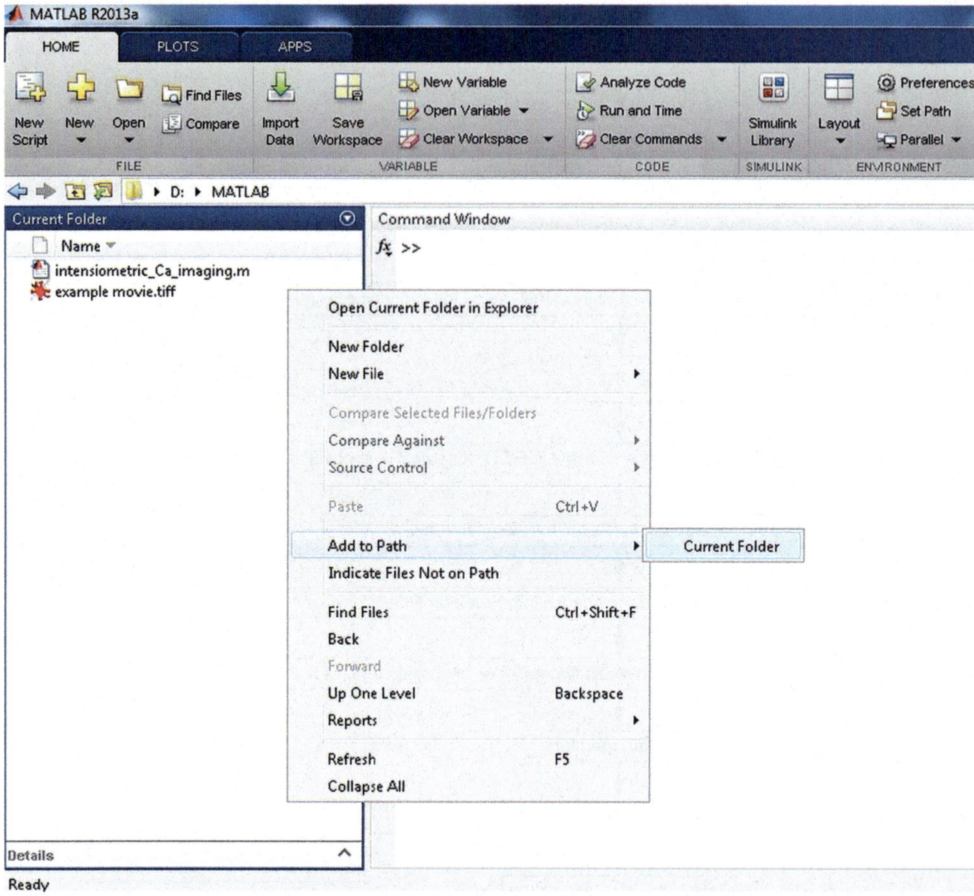

Fig. 6 Getting started with MATLAB. The current folder used in this case is D:\MATLAB. Copy the *intensiometric_Ca_imaging.m* script and the movies of interest to this folder, and add them to the active path as shown. The plots that will be generated by the script will be saved in this current folder

"Create mask" from the menu. Once this is done, the script will measure the mean background intensity I_{BG} in the defined region for every frame and store the values in the I_BG vector.

5. The next step is to mark a first ROI. As in **step 4**, the first frame of the time-lapse recording and a window containing the instructions for drawing the ROI will appear. To start drawing the ROI, click "OK." Draw the ROI as described in **step 4**. When the polygon is drawn, create a mask by right-clicking inside the region and selecting "Create mask" from the menu.

   The following tasks are performed automatically by the commands in the script. The mean fluorescence intensity I_{ROI1} in the defined region is measured for every frame, and the values are stored in the I_ROIn vector. A background subtraction is performed ($I_{ROI1}-I_{BG}$), and the background-corrected mean fluorescence intensities are saved in the I_corr

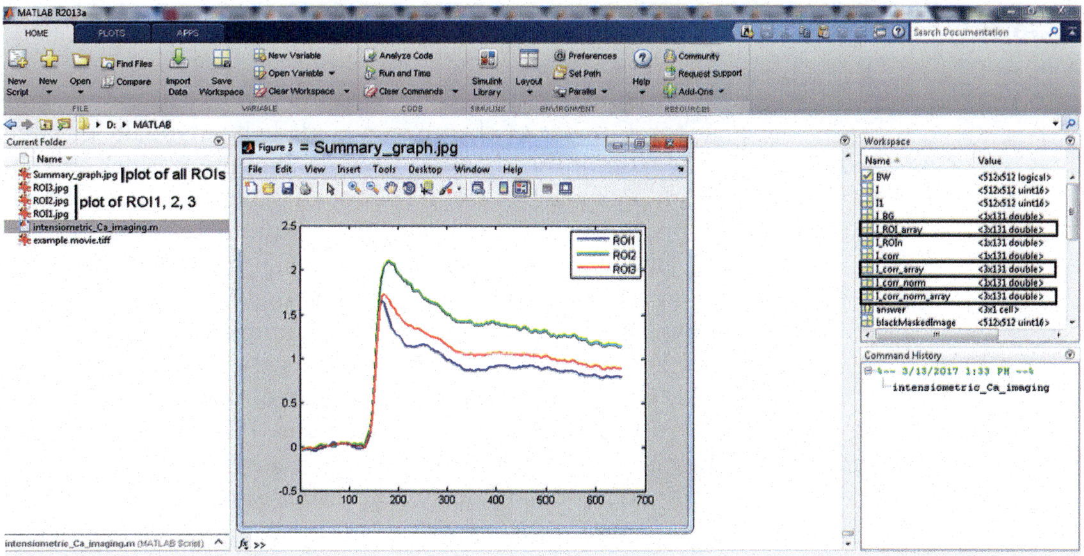

Fig. 7 MATLAB output. The script generates a plot for every individual ROI (ROI1.jpg, ROI2.jpg, etc.) and a summary plot (Summary_graph.jpg) which will be saved to the current folder. The workspace lists all the vectors containing the fluorescence intensities which have been measured/calculated while running the script. To access the values, *see* **Note 9**

vector. Based on the number of the frame for normalization defined in **step 3** (t_{ref}), the background-corrected mean fluorescence intensities for the ROI over time t $(I_{\text{ROI1}}-I_{\text{BG}})\,(t)$ are normalized to the background-corrected mean fluorescence intensity of the ROI at t_{ref} by subtracting and subsequently dividing by this value: $[(I_{\text{ROI1}}-I_{\text{BG}})\,(t)-(I_{\text{ROI1}}-I_{\text{BG}})\,(t_{ref})]/(I_{\text{ROI1}}-I_{\text{BG}})\,(t_{ref})$. The obtained values are stored in the I_corr_norm vector and plotted over time. The graph is saved in the MATLAB directory as JPEG file and is named "ROI1.jpg" (Fig. 7).

6. Depending on the number of ROIs (n), **step 5** is repeated n times.
7. In the case of multiple ROIs ($n > 1$), the script generates a single graph containing all the graphs of the n ROI (Fig. 7). The graph is saved as a JPEG file named "Summary_graph.jpg" in the MATLAB directory.
8. Analyze the graphs. Different parameters can be extracted and compared among treatments/genotypes such as the peak amplitude, response time, steepness of the peak, peak duration, etc. (*see* **Notes 9** and **10**).
9. Before proceeding to analysis of another time-lapse recording, ensure to rename the "Summary_graph" and "ROIn" plots. Otherwise they will be overwritten during a subsequent run of the script.

4 Notes

1. Two different methods were used for producing the imaging chamber. One set of chambers was made by manually machining the base of the chamber from an aluminum block and the lid from plexiglass. A second set of chambers was made by 3D printing from PETG plastic. Both methods resulted in functional imaging chambers. The manually machined chambers were made based on the provided 2D drawing. The 3D design is provided as a STL file which is the universal file format for most 3D printers. For 3D printing a fused deposition modeling (FDM) printer was used. Both the base of the chamber and the lid were printed from PETG filament. PETG has a melting temperature of 220–250 °C, and it is advised to print it on a heated bed of around 80 °C to prevent warping.
2. To increase the time efficiency during imaging, it is convenient to have several imaging chambers to allow imaging a sample while preparing another sample and let it recover for 30 min.
3. The total volume of the chamber is about 300 μL. Because 200 μL minimal medium is added during sample preparation, only an additional 100 μL can be applied for treatment, resulting in a threefold dilution of the treatment solution upon mixing it to the sample. Therefore, the concentration of any drug in the treatment solution should be 3× more concentrated than the desired end concentration.
4. It is recommended to use an objective with high numerical aperture (NA). For imaging of root cells, it is advised to use a 40× water immersion objective.
5. Keep in mind that root tips might grow out of the field of view during long time-lapse recordings. To increase the field of view, tile scanning can be performed at every time point. The frame rate might have to be decreased to have enough time to execute this operation. A SDM might be more suitable for this purpose as it allows for faster data acquisition.
6. Depending on the stimulus, the time of Ca^{2+} response can vary. Set the duration of the time-lapse recording according to the purpose of the experiment.
7. By normalization of the fluorescence intensities, the initial fluorescence intensities are brought to a similar level making comparison among different ROIs in a root or among different roots more convenient. Typically, at the start of the time-lapse recording, some photobleaching can occur. Therefore, it is advised to wait until the fluorescence levels have stabilized. For normalization, use the fluorescence intensity after fluorescence stabilization and prior to treatment (t_{ref}).

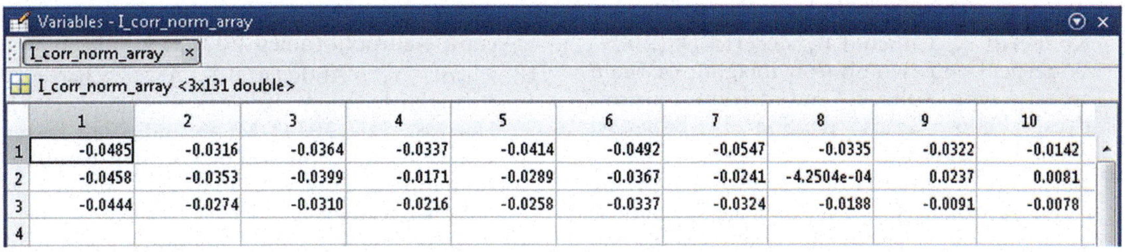

Variables - I_corr_norm_array

I_corr_norm_array

I_corr_norm_array <3x131 double>

	1	2	3	4	5	6	7	8	9	10
1	-0.0485	-0.0316	-0.0364	-0.0337	-0.0414	-0.0492	-0.0547	-0.0335	-0.0322	-0.0142
2	-0.0458	-0.0353	-0.0399	-0.0171	-0.0289	-0.0367	-0.0241	-4.2504e-04	0.0237	0.0081
3	-0.0444	-0.0274	-0.0310	-0.0216	-0.0258	-0.0337	-0.0324	-0.0188	-0.0091	-0.0078
4										

Fig. 8 Screenshot of the "Variables" window containing the data of the desired variable

8. A full overview of the file formats which are supported by Fiji can be found here: http://imagejdocu.tudor.lu/doku.php?id=faq:general:which_file_formats_are_supported_by_imagej. A list of MATLAB-compatible file formats is listed here: https://nl.mathworks.com/help/matlab/import_export/supported-file-formats.html.
9. To access the measured fluorescence intensity values and the ones calculated by the MATLAB script, go in the "Home" tab to "Open Variable," and select the variable of choice. A "Variables" window will be opened containing the data which can be selected and copied to, e.g., Excel for further analysis. An example for the I_corr_norm_array is shown below. A row corresponds to the normalized background-corrected mean fluorescence intensities of a ROI over time. In the example (Fig. 8), the data for three ROIs are shown.
10. The absence of changes in fluorescence intensity can be explained by the absence of Ca^{2+} signaling or by the lack of responsiveness of the seedling, e.g., when it is damaged. In order to distinguish between both scenarios, a positive control treatment can be applied at the end of each time-lapse recording (e.g., 0.1 or 1 mM ATP). If the response to the positive control occurs as expected, this suggests the seedling is intact, and the test treatment does not elicit a Ca^{2+} increase.

References

1. Clapham DE (2007) Calcium signaling. Cell 131:1047–1058
2. Knight H, Knight MR (2001) Abiotic stress signalling pathways: specificity and cross-talk. Trends Plant Sci 6:262–267
3. Rudd JJ, Franklin-Tong VE (1999) Calcium signaling in plants. Cell Mol Life Sci 55:214–232
4. Sanders D, Brownlee C, Harper JF (1999) Communicating with calcium. Plant Cell 11:691–706
5. Sanders D, Pelloux J, Brownlee C, Harper JF (2002) Calcium at the crossroads of signaling. Plant Cell 14:s401–s417
6. Scrase-Field SAMG, Knight MR (2003) Calcium: just a chemical switch? Curr Opin Plant Biol 6:500–506
7. Dolenšek J, Špelič D, Skelin Klemen M, Žalik B, Gosak M, Slak Rupnik M, Stožer A (2015) Membrane potential and calcium dynamics in beta cells from mouse pancreas tissue slices: theory, experimentation, and analysis. Sensors (Basel, Switzerland) 15:27393–27419

8. Homma R, Baker BJ, Jin L, Garaschuk O, Konnerth A, Cohen LB, Zecevic D (2009) Wide-field and two-photon imaging of brain activity with voltage- and calcium-sensitive dyes. Philos Trans R Soc B Biol Sci 364:2453–2467
9. Kettunen P (2012) Calcium imaging in the Zebrafish. In: Islam MS (ed) Calcium Signaling. Springer, Dordrecht, Netherlands, pp 1039–1071
10. Bonza MC, Loro G, Behera S, Wong A, Kudla J, Costa A (2013) Analyses of Ca(2+) accumulation and dynamics in the endoplasmic reticulum of Arabidopsis root cells using a genetically encoded Cameleon sensor. Plant Physiol 163:1230–1241
11. Krebs M, Held K, Binder A, Hashimoto K, Den Herder G, Parniske M, Kudla J, Schumacher K (2012) FRET-based genetically encoded sensors allow high-resolution live cell imaging of Ca2+ dynamics. Plant J 69:181–192
12. Loro G, Costa A (2013) Imaging of mitochondrial and nuclear Ca2+ dynamics in Arabidopsis roots. Cold Spring Harb Protoc 2013:pdb.prot073049
13. Mehlmer N, Parvin N, Hurst CH, Knight MR, Teige M, Vothknecht UC (2012) A toolset of aequorin expression vectors for in planta studies of subcellular calcium concentrations in Arabidopsis thaliana. J Exp Bot 63:1751–1761
14. Creton R, Kreiling JA, Jaffe LF (1999) Calcium imaging with chemiluminescence. Microsc Res Tech 46:390–397
15. Allen GJ, Kwak JM, Chu SP, Llopis J, Tsien RY, Harper JF, Schroeder JI (1999) Cameleon calcium indicator reports cytoplasmic calcium dynamics in Arabidopsis guard cells. Plant J 19:735–747
16. Denninger P, Bleckmann A, Lausser A, Vogler F, Ott T, Ehrhardt DW, Frommer WB, Sprunck S, Dresselhaus T, Grossmann G (2014) Male–female communication triggers calcium signatures during fertilization in Arabidopsis. Nat Commun 5:4645
17. Podor B, Y-l H, Ohkura M, Nakai J, Croll R, Fine A (2015) Comparison of genetically encoded calcium indicators for monitoring action potentials in mammalian brain by two-photon excitation fluorescence microscopy. Neurophotonics 2:021014
18. Zhao Y, Abdelfattah AS, Zhao Y, Ruangkittisakul A, Ballanyi K, Campbell RE, Harrison DJ (2014) Microfluidic cell sorter-aided directed evolution of a protein-based calcium ion indicator with an inverted fluorescent response. Integr Biol 6:714–725
19. Zhao Y, Araki S, Wu J, Teramoto T, Chang Y-F, Nakano M, Abdelfattah AS, Fujiwara M, Ishihara T, Nagai T, Campbell RE (2011) An expanded palette of genetically encoded Ca(2+) indicators. Science (New York, N.Y.) 333:1888–1891
20. Chen T-W, Wardill TJ, Sun Y, Pulver SR, Renninger SL, Baohan A, Schreiter ER, Kerr RA, Orger MB, Jayaraman V, Looger LL, Svoboda K, Kim DS (2013) Ultrasensitive fluorescent proteins for imaging neuronal activity. Nature 499:295–300
21. Keinath NF, Waadt R, Brugman R, Schroeder Julian I, Grossmann G, Schumacher K, Krebs M (2015) Live cell imaging with R-GECO1 sheds light on flg22- and chitin-induced transient $[Ca^{2+}]_{cyt}$ patterns in *Arabidopsis*. Mol Plant 8:1188–1200
22. Liu K-h, Niu Y, Konishi M, Wu Y, Du H, Sun Chung H, Li L, Boudsocq M, McCormack M, Maekawa S, Ishida T, Zhang C, Shokat K, Yanagisawa S, Sheen J (2017) Discovery of nitrate–CPK–NLP signalling in central nutrient–growth networks. Nature 545:311–316
23. Krebs M, Schumacher K (2013) Live cell imaging of cytoplasmic and nuclear Ca2+ dynamics in arabidopsis roots. Cold Spring Harb Protoc 2013:pdb.prot073031
24. Schindelin J, Arganda-Carreras I, Frise E, Kaynig V, Longair M, Pietzsch T, Preibisch S, Rueden C, Saalfeld S, Schmid B, Tinevez J-Y, White DJ, Hartenstein V, Eliceiri K, Tomancak P, Cardona A (2012) Fiji - an open source platform for biological image analysis. Nat Methods 9:676–682. https://doi.org/10.1038/nmeth.2019
25. Thevenaz P, Ruttimann UE, Unser M (1998) A pyramid approach to subpixel registration based on intensity. IEEE Trans Image Process 7:27–41

Check for updates

Chapter 10

Optimized Whole-Mount In Situ Immunolocalization for *Arabidopsis thaliana* Root Meristems and Lateral Root Primordia

Michael Karampelias, Ricardo Tejos, Jiří Friml, and Steffen Vanneste

Abstract

Immunolocalization is a valuable tool for cell biology research that allows to rapidly determine the localization and expression levels of endogenous proteins. In plants, whole-mount in situ immunolocalization remains a challenging method, especially in tissues protected by waxy layers and complex cell wall carbohydrates. Here, we present a robust method for whole-mount in situ immunolocalization in primary root meristems and lateral root primordia in *Arabidopsis thaliana*. For good epitope preservation, fixation is done in an alkaline paraformaldehyde/glutaraldehyde mixture. This fixative is suitable for detecting a wide range of proteins, including integral transmembrane proteins and proteins peripherally attached to the plasma membrane. From initiation until emergence from the primary root, lateral root primordia are surrounded by several layers of differentiated tissues with a complex cell wall composition that interferes with the efficient penetration of all buffers. Therefore, immunolocalization in early lateral root primordia requires a modified method, including a strong solvent treatment for removal of hydrophobic barriers and a specific cocktail of cell wall-degrading enzymes. The presented method allows for easy, reliable, and high-quality in situ detection of the subcellular localization of endogenous proteins in primary and lateral root meristems without the need of time-consuming crosses or making translational fusions to fluorescent proteins.

Key words *Arabidopsis*, In situ immunolocalization, Lateral root primordia

1 Introduction

Any living cell displays a high degree of subcellular complexity as illustrated by the existence of multiple complex organelles, a dynamic cytoskeleton, heterogeneities in membranes, and many more. Each of these subcellular structures is a place to which specific proteins become recruited to execute a specialized function. This illustrates the value of visualizing the subcellular localization of a protein of interest as part of the process to unravel its function.

Commonly used methods to visualize proteins within plant cells include the expression of translational fusion with fluorescent

Daniela Ristova and Elke Barbez (eds.), *Root Development: Methods and Protocols*, Methods in Molecular Biology, vol. 1761, https://doi.org/10.1007/978-1-4939-7747-5_10,

proteins [1, 2] and protein immunolocalization using specific antibodies [3, 4]. The greatest advantage of translational fusion to fluorescent proteins is that they allow for visualization in vivo the subcellular localization, turnover, and dynamics of the chimeric protein but often requires time-consuming stable transformation, and the fluorescent tag can interfere with the function of the protein resulting in mislocalization, altered turnover, and dynamics [5, 6]. Protein immunolocalization allows for the in situ detection of endogenous epitopes and is thus applicable on any genetic background but involves fixation of the tissue and is thus not amenable for in vivo studies [3, 7–9].

Current immunolocalization methods perform well for a variety of tissues and species [7, 8, 10–13]. In particular, the primary root meristem of *Arabidopsis thaliana* allows for easy whole-mount in situ immunolocalization given its small dimensions, conserved tissue organization, and relatively weak diffusion barriers [3, 12, 14]. The success of in situ fluorescence immunolabeling depends largely on the properties of the antigen, the antibody, and the target tissue. The antigenicity of the epitope sequence needs to be considered in order to raise efficient antibodies. However, the affinity and specificity of the antibody to the targeted antigen cannot be actually predicted as they depend to a large extend on the immune response of the host animal in which the antibody is raised [15]. Besides evaluating the size, complexity, predicted hydrophobicity, topology, and subcellular localization of the protein for selecting an epitope as antigen [15], the epitope sequence must be as unique as possible to minimize nonspecific binding. One possibility to improve the efficiency of the antibody is to purify and concentrate the antibody from the serum [16]. It should also be noted that a good performance of an antibody on Western blots is no guarantee for a good performance for whole-mount in situ immunolocalization. Therefore, every antibody should be carefully tested for specificity and its dilutions experimentally optimized.

Given the diversity of both the plant tissues and the proteins, a universal method to facilitate in situ immunolabeling of any protein of interest in any tissue of the plant body does not exist. Many variations in immunolocalization procedures have been developed for specific applications in moss [17, 18] and maize [19], but also for different *Arabidopsis* tissues, such as ovules [8], differentiated root or leaf epidermal cells [20], and root hairs [18, 21], and even for specific subcellular structures, such as microtubules [9, 11], cell wall carbohydrates [22], and plant hormones [23–26]. In each of these protocols, the choices of fixation and tissue permeabilization represent a critical determinant of success, reflecting a trade-off between epitope and tissue conservation versus tissue permeability.

In this chapter we present a robust method that is well suited for whole-mount in situ immunolocalization of many proteins in the endomembrane system or even peripherally attached to these membranes. We suggest fixation in a mannitol-containing solution with an alkaline pH and permeabilization using a mixture of cell wall-degrading enzymes. The method also works well for lateral root primordia when an additional step to remove hydrophobic barriers is included.

1.1 Fixation

The fixation aims to the formation of stable chemical bonds between the cell components, immobilizing organelles and stabilizing the epitopes in situ. In doing so, fixation also reduces tissue permeability, a feature that is required for whole-mount immunolocalization. By manipulating the fixing agent, its concentration, and the duration, it is possible to control the strength of the fixation.

Commonly used fixing agents are the inorganic osmium tetroxide, potassium dichromate and mercury chloride, the organic picric acid and acetic acid, acrolein, glyoxal, periodate lysine paraformaldehyde (PFA), formaldehyde (FA), and glutaraldehyde (GA). The latter two are the most frequently used for immunolocalization applications in order to form covalent bonds of methylene bridges between amino groups of proteins or nucleic acids [27, 28], while they have minor reactivity against lipids [29, 30]. FA is a small molecule, which diffuses easily through the tissues. From the other side, GA is much bigger with two aldehyde groups, which, although its diffusion is more slow, it forms stronger irreversible bonds. In mixtures they combine the simplest quickly diffusing aldehyde, FA, and the strong chemical fixation of a dialdehyde, GA [28, 31]. Besides, alkaline pH increases the cross-linking rate of FA [28, 32, 33]. Moreover, given that most protein isoelectric points are between 5.5 and 8, proteins in alkaline pH are charged, soluble, and, thus, more reactive to cross-linking with aldehydes. However, residual GA in the sample generates significant amounts of undesired autofluorescence, especially in alkaline conditions [34]. Thus, it is important to eliminate residual GA after fixation by neutralization with a strong reducing agent [33, 34].

As an alternative to the commonly used PFA in PBS fixatives [3, 7], we propose the use of a fixative based on FA and GA, in a mannitol-containing alkaline buffer to result in a good trade-off between epitope preservation and tissue permeability. We found this fixative, termed MFG (from mannitol, formaldehyde, glutaraldehyde) is particularly successful in protein immunolocalization for proteins resident at the PM/GA/TGN, such as clathrin heavy chain (Fig. 1a, b), GNOM (Fig. 1c), and PINs (Fig. 1d–i).

1.2 Permeabilization of the Cell Wall

The success of immunolocalization is further limited by the accessibility of the epitope. Therefore, it is crucial that diffusion barriers are taken down to allow interaction of the antibodies with the antigen in the fixed tissue. Diffusion in plant tissues is further compromised by

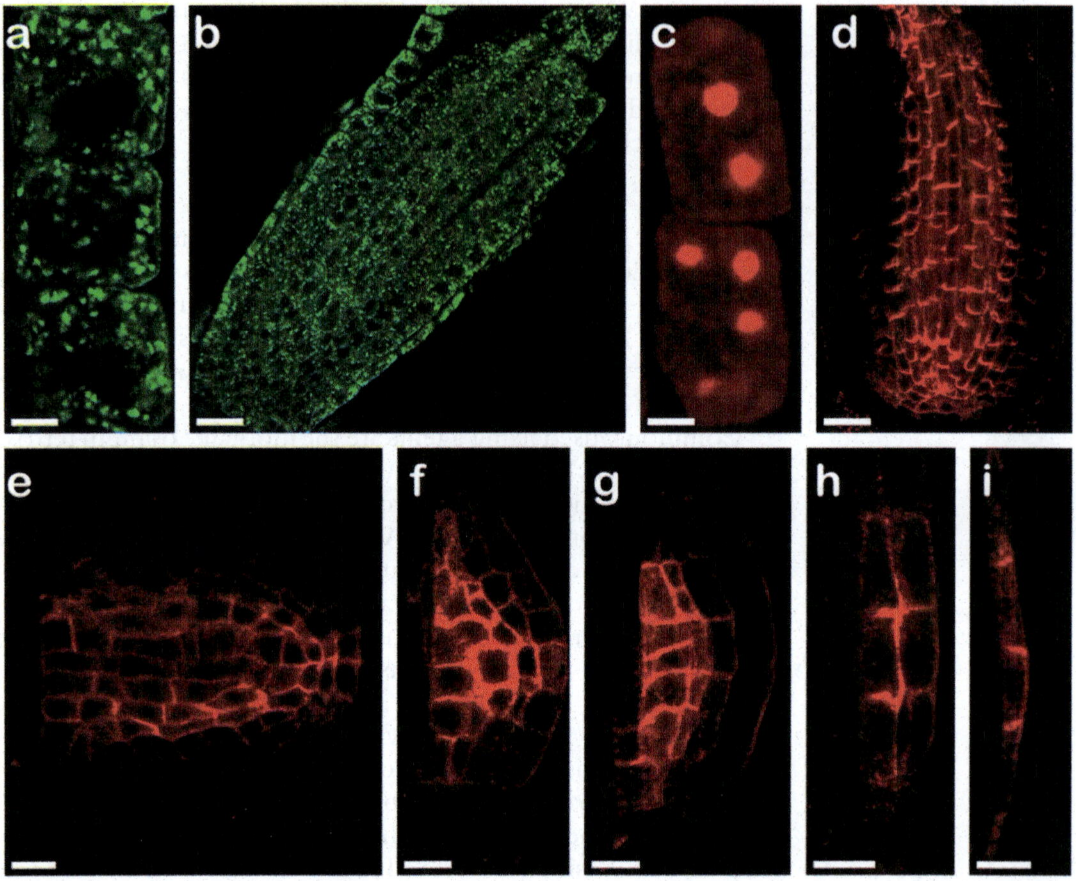

Fig. 1 Examples of immunolocalizations in *Arabidopsis* seedlings. (**a**) Immunolocalization of clathrin in root epidermal cells and (**b**) root tip of wild-type (ecotype *Columbia*) seedlings with anti-CHC (rabbit) and secondary antibody (donkey) labeled with AlexaFluor®488. (**c**, **d**) Immunolocalization of GNOM-3xMYC in root epidermal cells of pGNOM::GNOM:3xMYC seedlings treated with 50 μM BFA for 30 min with anti-MYC (mouse) and secondary antibody (goat) labeled with Cy3® (red). (**e–i**) PIN1 immunolocalization with anti-PIN1 (goat) and stained with AlexaFluor®555 (donkey) in (**d**) primary root, (**e**) emerged, (**f**, **g**, **h**) intermediate, and (**i**) first stage of lateral root primordium. Bars: 10 μm, except for **h** and **i** that is 5 μm

the cuticle [35–37] and the cell wall [17, 38, 39]. Therefore it is important to include steps for permeabilization of the cell wall.

The plant cell wall consists of complex mixtures of carbohydrate polymers, including cellulose, hemicelluloses, and pectin [39]. The permeabilization of the cell wall matrix can be achieved by cocktails of cell wall-degrading enzymes. Depending on the species and tissue, the cell wall matrix composition may differ, requiring alternative enzyme mixtures and conditions (concentrations, temperatures, buffer, and pH). Whereas good immunolocalization in the root apical meristem can be achieved via cell wall enzyme mixtures, such as Driselase, this enzyme mixture is not very efficient for immunolocalization in young lateral root primordia.

Instead we propose an enzyme mix containing cellulase, pectinase, and pectolyase in a simple PBS buffer (pH 7.4), and it is applied at 37 °C. This cocktail (termed CePePe, after its containing enzymes) allows for efficient immunolocalization of the auxin transporter PIN-FORMED1 (PIN1) in stage I LRP of *Arabidopsis* seedlings (Fig. 1e–i).

1.3 Reduction of Hydrophobicity

Immunolocalization of antigens in undifferentiated cells, such as the root apical meristem of *Arabidopsis*, is relatively straightforward compared to other tissues. Aboveground tissues and mature roots are covered with the cuticle that protects them from excessive water loss, from pests and pathogens, and partially from UV light and repels water and debris [40, 41]. The cuticle consists of layers of hydrophobic organic molecules, such as fatty acids, alcohols, ketones, aldehydes, esters, and their complex polymers [35–37, 42], which interfere with penetration of the water-soluble assay solutions. In morphological studies, solubilization and removal of organic compounds is applied via organic solvents [43, 44]. Usually, these organic solvents are heptane, xylene, or xylene alternatives, such as the commercially available d-limonene (Histo-Clear®, National Diagnostics). In the following protocol, we suggest an optional step with xylene treatment in order to remove the cuticular layers, ensuring successful immunolocalization in the shoot apex, cotyledons, and hypocotyl of membrane proteins, like PIN1.

In total the combined application of MFG fixative, CePePe enzyme mix, and the optional xylene treatment are integrated in a whole-mount in situ immunolocalization method for increased efficiency localization of difficult antigens in recalcitrant tissues in *Arabidopsis thaliana*.

2 Materials

2.1 Plant Material and Growth Conditions

We used *Arabidopsis thaliana* seedlings of ecotype *Columbia* and plants expressing the GNOM coding sequence fused to a MYC tag, pGNOM::GNOM-MYC [45]. Seeds were sterilized with 70% ethanol for 3 min, dried in the sterile flow bench, and then sown and germinated on half-strength Murashige and Skoog supplemented with 0.8% (w/v) plant agar and grown vertically at 21 °C, in continuous 100 μmol/m^2/s light for 4 days.

2.2 Buffers and Solutions

1. MFG fixative solution: final concentrations 25 mM K_2HPO_4, 0.4 M mannitol, 1 mM $CaCl_2$, 4% PFA, 0.025% GA, pH 10. Optionally supplemented with 0.1% Triton™ X-100, termed MFG-T. Store at 4 °C. To prepare 100 mL, add 2.5 mL of 1 M K_2HPO_4 (pH 10.0) to 35 mL of distilled water in a beaker with a magnetic stirrer. In a separate beaker, add 1 mL of 0.1 M $CaCl_2$ to 35 mL of distilled water. Mix the diluted $CaCl_2$

solution with the diluted K_2HPO_4 (*see* **Note 1**). The solution should be clear. Add 4 g of PFA. Adjust pH to 12.0 with KOH (start with one pellet) to catalyze the hydrolysis of PFA (*see* **Note 2**). Wait until the solution is clear (10–15 min). Adjust pH to 10.0 with HCl (37 or 10%) (*see* **Note 3**), and add 50 μL of 50% GA (*see* **Notes 4** and **5**) and 7.27 g of mannitol. Check pH to 10.0 and wait until the solution is clear (>5 min). Adjust the total volume to 100 mL with distilled water. Store at 4 °C for 3–4 weeks maximum (*see* **Notes 6–8**).

2. Phosphate-buffered saline (1× PBS): 137 mM NaCl, 2.7 mM KCl, 1.42 mM Na_2HPO_4, 0.24 mM KH_2PO_4, adjust pH to 7.4 with KOH. Prepare 10× concentrated solution for your convenience. Store at room temperature.
3. CePePe enzyme solution: 1% cellulase, 1% pectinase, and 0.1% pectolyase in PBS buffer, pH 7.4. Store as a 10× concentrated stock at −20 °C (*see* **Notes 9–12**).
4. Permeabilization solution: 3% IGEPAL® CA-630, 10% DMSO in PBS, pH 7.4. Prepare fresh.
5. Blocking solution: 3% bovine serum albumin (BSA) in PBS, pH 7.4. Prepare fresh.
6. Antibody solutions: primary and secondary antibodies are diluted in blocking solution.
7. Antifade mounting solution: 25% DAPCO, 90% glycerol, in PBS, pH 7.4 with KOH or H_2SO_4. Store at −20 °C.

2.3 Antibodies

1. Polyclonal rabbit anti-CHC in 1/800 dilution (AS10 690; Agrisera).
2. Monoclonal mouse anti-MYC in 1/1000 dilution (R951-25; Thermo Fisher).
3. Polyclonal goat anti-PIN1 in 1/900 dilution (sc-27163; Santa Cruz Biotechnologies).
4. Polyclonal donkey anti-rabbit-AlexaFluor®488 in 1/600 dilution (A21206; Thermo Fisher).
5. Polyclonal goat anti-mouse-Cy3 in 1/600 dilution (ab97035 Abcam).
6. Polyclonal donkey anti-goat-AlexaFluor®555 in 1/600 dilution (A21432; Thermo Fisher).

2.4 Reagents: Chemicals

Plant growth medium: half-strength Murashige and Skoog basal salts, supplemented with 1.0% (w/v) sucrose, 0.05% (w/v) MES (pH 5.7), and 0.8% (w/v) plant agar.

2.5 Equipment

1. Laminar flow, culture plates, breathable tape, soft tweezers, pipettes, and tips.

2. Refrigerator and culture champers with controllable temperature and light regime.
3. Tissue cultures plates with 6, 12, or 24 wells.
4. Snap cap glass vials.
5. Mini vacuum gas pump with vacuum desiccator.
6. Fume hood.
7. Rocking shaker and oven at 37 °C.
8. Glass slides.
9. Confocal microscope, equipped with appropriate lasers and filter.
10. Optional: automatic programmable liquid handling robot (*see* **Note 26**).

3 Methods

A workflow for the immunolocalization method can be installed in a lab so that it can routinely provide reliable information about the protein localization. The complete method includes several protocols with many steps and critical points; therefore the installation of a workflow and regular application requires careful planning and maintenance. Chemicals and solutions should be dedicated for immunolocalization and where mentioned must be prepared fresh (*see* below). Success can be tested with immunolocalization of antigen with antibodies of confirmed efficiency. This can be the reference to any problems or reduced performance in the future.

Once successful, immunolocalization can include new antibodies with unknown efficiency against antigens of known or unknown localization. The efficiency and specificity of new antibodies can be essentially evaluated with basic biochemical experiments, like SDS-PAGE with protein samples from knockout mutants versus samples from plants expressing tagged or truncated versions of the protein of interest. Protein sample volume comparison and antibody dilutions may provide information about the antigen detection sensitivity and useful dilutions. The specificity can be tested using knockout mutants and seedlings expressing tagged protein of interest.

For a regular immunolocalization workflow, the availability of confocal imaging platform is critical. Samples of immunolocalization can be stored in darkness at 4 °C for several days, but we suggest that imaging of all samples should be performed within a week. Needless exposure to light or higher temperatures must be avoided.

3.1 Immuno-localization Procedure

1. Add 1 mL of MFG or MFG-T fixative in well of tissue culture well plates.
2. Fix plant tissues (*see* **Notes 19–22**) by dipping them in the MFG or MFG-T fixative solution (*see* **Note 23**).
3. Apply vacuum for 1 h in room temperature.
4. Optional: protocol for the *Reduction of hydrophobicity* for LRPs (*see* Subheading 3.2).
5. Wash three times in PBS-T for 5 min (*see* **Note 25**).
6. Wash three times in 0.1% Triton ™ X-100 in distilled water for 5 min.
7. Replace with CePePe enzyme solution.
8. Incubate for 30 min at 37 °C.
9. Wash three times in PBS-T for 5 min.
10. Replace with permeabilization solution.
11. Incubate for 30 min at room temperature.
12. Repeat once **steps 8** and **9**.
13. Wash three times in PBS-T for 5 min.
14. Replace with blocking solution.
15. Incubate for 1 h at 37 °C.
16. Replace with primary antibody solution.
17. Incubate for 4 h at 37 °C.
18. Wash five times in PBS-T for 5 min.
19. Replace with secondary antibody solution.
20. Incubate for 3 h at 37 °C.
21. Wash five times in PBS-T for 5 min.
22. Wash five times in distilled water for 5 min.
23. Optional for nuclear staining: replace with DAPI solution and incubate for 10 min at room temperature. Wash twice in distilled water for 5 min.
24. Mount in antifadant solution on slides.
25. Carefully spread all tissue to avoid overlap using a stereomicroscope.
26. Proceed to microscopy.

3.2 Reduction of Hydrophobicity

This is an optional protocol for immunolocalization in LRP (and green tissues). Proceed in glass vials (*see* **Note 13**).

1. Wash the fixed plant tissues twice in distilled water for 5 min (*see* **Note 24**).
2. Dehydrate in 100% methanol for 30 min.
3. Replace methanol for 100% ethanol and incubate for 5 min.

4. Replace ethanol for 100% xylene and incubate for 15–30 min.
5. Wash twice in 100% ethanol at least for 5 min. (Repeat until the green tissues are totally cleared.)
6. Optional break point. Samples may be stored in sealed vials with 100% ethanol at 4 °C for 1–3 days.
7. Rehydrate in 80-60-30-15% ethanol and twice in PBS buffer for minimum 5–10 min each step.
8. Proceed with **step 3** of the *Immunolocalization* procedure.

3.3 Microscopy

1. Use 10× or 20× objective lenses to identify and mark the position of root tips and/or LRPs and image acquisition of organs in broader plane. Such images may indicate the spatial protein expression.
2. Use higher magnification objective lenses (40× or 63× water-corrected lens) for detailed imaging of the marked positions.
3. For simultaneous labeling of different targets with specific primary antibodies, care should be taken to avoid cross-reactivity among the secondary antibodies to prevent false positive labeling (*see* **Note 4**).
4. The samples can be stored up to 1 week at 4 °C, but it is recommended to image them as soon as possible to avoid sample deterioration.
5. Properties of the confocal imaging setup (laser strength, pin-hole, etc.) should be preserved in comparative studies between different samples.
6. Ensure that you image tissues in which the immunolabeled protein is expressed.
7. Signal specificity should be always evaluated by including positive and negative controls, especially at the first use of the antibody. The best negative controls are full knockout mutants, and positive controls can be seedling expressing verified tagged versions of the protein of study.

4 Notes

1. Strictly follow the order of mixing components as presented. Avoid mixing concentrated $CaCl_2$ and phosphate buffer, as this will result in calcium phosphate precipitation.
2. While addition of KOH pellets helps to dissolve the PFA, we found that the quality of immunolocalization is negatively affected when the pH goes above 12. Do not overuse the KOH, i.e., add small amounts at a time and wait until the KOH pellets dissolve before adding more. Be patient in dissolving the PFA.

3. The MFG fixative may be titrated to pH 12 to improve its performance, but the strong alkaline solution is very detrimental for the integrity of the plant tissues. In this case, tissues must be treated with extra care.
4. The use of GA in the strongly alkaline fixative destroys the tertiary structure of proteins, thereby destroying the fluorophore of expressed fluorescent proteins such as GFP. In case the fluorescent tag needs to be visualized, a specific antibody against the fluorescent protein needs to be included.
5. The GA stock can be stored in aliquots (10 mL) at −20 °C to avoid too many freeze-thaw cycles.
6. The MFG at pH 10 can be stored at 4 °C for several weeks, whereas MGF at pH 12.0 can only be stored for 3 days. Freezing must be avoided.
7. Triton X-100™ in the fixative is added prior to use and it is only required for immunolocalization in lateral root primordia and green tissues.
8. PFA is a presumed carcinogen and should be treated with caution. Prepare the fixative in the fume hood; wear appropriate protective clothing and gloves. All waste streams should be treated accordingly. The fixation should be performed in the fume hood.
9. It is preferable to prepare the CePePe mix from aliquots of tenfold concentrated stocks for each enzyme (10% cellulase, 10% pectinase, and 1% pectolyase in PBS pH 7.4), which can be stored in aliquots at −20 °C. In 7 mL of PBS (pH 7.4), add 1 mL of 10% cellulase, 1 mL of 10% pectinase, and 1 mL of 1% pectolyase. Mix by inversion. Aliquot in 2 mL tubes. Freeze in −20 °C. Thaw on ice before use.
10. We suggest a short centrifugation step of the 10× CePePe to remove insoluble debris prior to aliquoting.
11. Enzyme activity may differ between different batches and may therefore require slight adjustments in the concentration of the CePePe.
12. When the protein immunolocalization fails, one can consider evaluating alternatives to the CePePe enzyme solution such as 1.5% Driselase in PBS pH 7.4 [3] or 0.2% Driselase + 0.15% pectolyase in 2 mM MES pH 5.0 [7].
13. Do not use polystyrene or other xylene-soluble plasticware. Glass vials/tubes (5–8 mL) or Eppendorf tubes are recommended.
14. Plant tissues start bleaching after **step 6**, and, when totally cleared, they must be white or semitransparent. Ethanol steps can be repeated to achieve effective clearing.

15. Treatments with alcohols and xylene make the plant tissues very brittle. Thus handling, especially in the first steps of rehydration, must be done with care.
16. All steps are performed at room temperature. If needed, 37 °C can be applied to facilitate clearing with xylene. If doing so, use glass vials with hermetic caps.
17. In order to avoid cellular collapse, the steps of dehydration, application of the organic solubilizer, but especially rehydration of the tissue must be done as gentle and slow as possible.
18. Ensure a thorough ethanol wash to remove all xylene before the rehydration steps, as remnant xylene will render aqueous solutions milky. In case of milky appearance, dehydrate the samples in ethanol until clear and rehydrate again.
19. The method returns about 60–80% of intact root meristems when the MFG fixative has pH 10 and about 15–20% when pH is 12.0. Therefore depending on the targeted tissue (roots, LRP, cotyledons, etc.), the amount of input tissue should be considered.
20. All steps are performed at room temperature ($\approx$21 °C), unless mentioned otherwise.
21. 6-, 12-, or 24-well plates can be used to perform the method.
22. Fixation is typically carried out with 0.5–1 mL fixative solution in 24-well tissue culture plate. Ensure that all tissues are covered with fixative throughout the total time of fixation.
23. Fixative solution can be supplemented with Triton™ X-100 for immunolocalization in LRPs.
24. Fixative may be mixed with an equal volume of heptane to facilitate removal of hydrophobic layers of green tissues. Such fixation should be performed in glass vials.
25. Neutralization of GA by strong reducing agents like $NaBH_4$ is not required, because of the low GA concentration.
26. The method can easily be automated using a microfluidics robot, for all the steps after fixation or after the optional hydrophobicity treatment (plastic tubing is usually not compatible with the solvents).

References

1. Crivat G, Taraska JW (2012) Imaging proteins inside cells with fluorescent tags. Trends Biotechnol 30:8–16
2. Cutler SR, Ehrhardt DW, Griffitts JS, Somerville CR (2000) Random GFP::cDNA fusions enable visualization of subcellular structures in cells of Arabidopsis at a high frequency. Proc Natl Acad Sci U S A 97:3718–3723
3. Sauer M, Paciorek T, Benková E, Friml J (2006) Immunocytochemical techniques for whole-mount in situ protein localization in plants. Nat Protoc 1:98–103
4. Paciorek T, Sauer M, Balla J et al (2006) Immunocytochemical technique for protein localization in sections of plant tissues. Nat Protoc 1:104–107

5. Moore I, Murphy A (2009) Validating the location of fluorescent protein fusions in the endomembrane system. Plant Cell Online 21:1632–1636
6. Margolin W (2012) The price of tags in protein localization studies. J Bacteriol 194:6369–6371
7. Pasternak T, Tietz O, Rapp K et al (2015) Protocol: an improved and universal procedure for whole-mount immunolocalization in plants. Plant Methods 11:50
8. Escobar-Guzmán R, Rodríguez-Leal D, Vielle-Calzada J-P, Ronceret A (2015) Whole-mount immunolocalization to study female meiosis in Arabidopsis. Nat Protoc 10:1535–1542
9. Šamajová O, Komis G, Šamaj J (2014) In: Komis G, Šamaj J (eds) Immunofluorescent localization of MAPKs and colocalization with microtubules in Arabidopsis seedling whole-mount probes, Methods in molecular biology. Springer, New York, NY, pp 107–115
10. Chen W, Baldwin TC (2007) An improved method for the fixation, embedding and immunofluorescence labeling of resin-embedded plant tissue. Plant Mol Biol Report 25:27–35
11. Szechyńska-Hebda M, Wędzony M, Dubas E et al (2006) Visualisation of microtubules and actin filaments in fixed BY-2 suspension cells using an optimised whole mount immunolabeling protocol. Plant Cell Rep 25:758–766
12. Boutté Y, Grebe M (2014) Immunocytochemical fluorescent in situ visualization of proteins in Arabidopsis. In: Sanchez-Serrano JJ, Salinas J (eds) Arabidopsis protocols. Humana Press, Totowa, NJ, pp 453–472
13. Pawlowski W, Grelon M, Armstrong S (2013) Plant meiosis, Methods and protocols. life sciences. Springer, New York
14. Omelyanchuk NA, Kovrizhnykh VV, Oshchepkova EA et al (2016) A detailed expression map of the PIN1 auxin transporter in Arabidopsis thaliana root. BMC Plant Biol 16:5
15. Wang Y, Wu W, Negre NN et al (2011) Determinants of antigenicity and specificity in immune response for protein sequences. BMC Bioinformatics 12:251
16. O'Kennedy R, Murphy C, Devine T (2016) Technology advancements in antibody purification. Antibody Technol J 6:17–32
17. Knox JP (2008) Revealing the structural and functional diversity of plant cell walls. Curr Opin Plant Biol 11:308–313
18. Willats WGT, McCartney L, Knox JP (2001) In-situ analysis of pectic polysaccharides in seed mucilage and at the root surface of Arabidopsis thaliana. Planta 213:37–44
19. Forestan C, Carraro N, Varotto S (2013) Protein immunolocalization in maize tissues, Methods in molecular biology. Springer, New York, pp 207–222
20. Wasteneys GO, Willingale-Theune J, Menzel D (1997) Freeze shattering: a simple and effective method for permeabilizing higher plant cell walls. J Microsc 188:51–61
21. Braun M, Baluška F, von Witsch M, Menzel D (1999) Redistribution of actin, profilin and phosphatidylinositol-4,5-bisphosphate in growing and maturing root hairs. Planta 209:435–443
22. Larson ER, Tierney ML, Tinaz B, Domozych DS (2014) Using monoclonal antibodies to label living root hairs: a novel tool for studying cell wall microarchitecture and dynamics in Arabidopsis. Plant Methods 10:30
23. Avsian-Kretchmer O (2002) Indole acetic acid distribution coincides with vascular differentiation pattern during Arabidopsis leaf ontogeny. Plant Physiol 130:199–209
24. Benková E, Michniewicz M, Sauer M et al (2003) Local, efflux-dependent Auxin gradients as a common module for plant organ formation. Cell 115:591–602
25. De Diego N, Rodríguez JL, Dodd IC et al (2013) Immunolocalization of IAA and ABA in roots and needles of radiata pine (Pinus radiata) during drought and rewatering. Tree Physiol 33:537–549
26. Forestan C, Varotto S (2013) Auxin immunolocalization in plant tissues, Methods in molecular biology. Springer, New York, pp 223–233
27. Grizzle WE (2009) Special symposium: fixation and tissue processing models. Biotech Histochem 84:185–193
28. Thavarajah R, Mudimbaimannar VK, Rao UK et al (2012) Chemical and physical basics of routine formaldehyde fixation. J Oral Maxillofac Pathol 16:400
29. Jones D (1969) The reaction of formaldehyde with unsaturated fatty acids during histological fixation. Histochem J 1:459–491
30. Gigg R, Payne S (1969) The reaction of glutaraldehyde with tissue lipids. Chem Phys Lipids 3:292–295
31. Eltoum I, Fredenburgh J, Grizzle WE (2001) Advanced concepts in fixation: 1. Effects of fixation on immunohistochemistry, reversibility of fixation and recovery of proteins, nucleic acids, and other molecules from fixed and processed tissues. 2. Developmental methods of fixation. J Histotechnol 24:201–210
32. Bacallao R, Sohrab S, Phillips C (2006) Guiding principles of specimen preservation for confocal fluorescence microscopy. In: Pawley JB

(ed) Handbook of biological confocal microscope. Springer, Boston, MA, pp 368–380

33. Migneault I, Dartiguenave C, Bertrand MJ, Waldron KC (2004) Glutaraldehyde: behavior in aqueous solution, reaction with proteins, and application to enzyme crosslinking. BioTechniques 37(790–6):798–802
34. Lee K, Choi S, Yang C et al (2013) Autofluorescence generation and elimination: a lesson from glutaraldehyde. Chem Commun (Camb) 49:3028–3030
35. Nawrath C (2006) Unraveling the complex network of cuticular structure and function. Curr Opin Plant Biol 9:281–287
36. Jenks MA, Eigenbrode SD, Lemieux B (2002) Cuticular waxes of Arabidopsis. Arabidopsis Book 1:e0016
37. Kunst L, Samuels AL (2003) Biosynthesis and secretion of plant cuticular wax. Prog Lipid Res 42:51–80
38. Gilbert HJ (2010) The biochemistry and structural biology of plant cell wall deconstruction. Plant Physiol 153:444–455
39. O'Neill MA, York WS (2003) The composition and structure of plant primary cell walls In: The plant cell wall. Rose JKC (ed.) Vol. 8, Blackwell Publishing Ltd, Oxford, UK
40. Yeats TH, Rose JKC (2013) The formation and function of plant cuticles. Plant Physiol 163:5–20
41. Kolattukudy PE (2001) Polyester in higher plants. In: Babel W, Steinbuchel A (ed.) Advances in Biochemical Engineering/Biotechnology, Vol. 71. Springer Berlin Heidelberg, Berlin, Heidelberg
42. Kolattukudy PE (2001) Advances in Biochemical Engineering Biotechnology, Vol. 71. Springer Berlin Heidelberg, Berlin, Heidelberg
43. de Almeida Engler J, Van Montagu M, Engler G (1998) Whole-mount in situ hybridization in plants. In: Martinez-Zapater JM, Salinas J (eds) Arabidopsis protocols. Humana Press, Totowa, NJ, pp 373–384
44. Tainaka K, Kuno A, Kubota SI et al (2016) Chemical principles in tissue clearing and staining protocols for whole-body cell profiling. Annu Rev Cell Dev Biol 32:713–741
45. Geldner N, Anders N, Wolters H et al (2003) The Arabidopsis GNOM ARF-GEF mediates endosomal recycling, auxin transport, and auxin-dependent plant growth. Cell 112:219–230

Chapter 11

Light Sheet Fluorescence Microscopy Optimized for Long-Term Imaging of Arabidopsis Root Development

Paolo Baesso, Ricardo S. Randall, and Giovanni Sena

Abstract

Light sheet fluorescence microscopy (LSFM) allows sustained and repeated optical sectioning of living specimens at high spatial and temporal resolution, with minimal photodamage. Here, we describe in detail both the hardware and the software elements of a live imaging method based on LSFM and optimized for tracking and 3D scanning of *Arabidopsis* root tips grown vertically in physiological conditions. The system is relatively inexpensive and with minimal footprint; hence it is well suited for laboratories of any size.

Key words Microscopy, Live imaging, Morphodynamics, *Arabidopsis* root, Tracking, Control software

1 Introduction

Live imaging is a necessary tool to characterize the dynamics in developmental processes. While the details of the phenomenon under investigation will dictate the optimal magnification, spatial resolution, frequency of observations (temporal resolution), and duration of the observation, it is also crucial that the sample is kept alive, in the field of view and in focus during the entire session.

Fluorescence microscopy offers many advantages, especially when internal tissues are to be imaged. If the sample is relatively transparent, a common solution is optical sectioning, usually achieved through one of the existing realizations of confocal microscopy [1]. Unfortunately, such methods rely on scanning the sample with a laser beam focused on a single point, resulting in high energy levels absorbed locally by the tissue. Therefore, frequent and prolonged time-lapse imaging through confocal microscopy can easily lead to severe damage and possibly death of the specimen [2].

Light sheet fluorescence microscopy (LSFM), or selective plane of illumination microscopy (SPIM), is an alternative to standard confocal microscopy that in the past few years has been successfully

Daniela Ristova and Elke Barbez (eds.), *Root Development: Methods and Protocols*, Methods in Molecular Biology, vol. 1761, https://doi.org/10.1007/978-1-4939-7747-5_11,

applied to developmental biology [3, 4]. A confocal microscope illuminates a large portion of the sample but collects emitted light only from a thin focal plane (thus effectively realizing optical sections in the emission step). LSFM instead focuses the laser beam into a thin "sheet" of light to illuminate at once only a small portion of the sample at any given time (thus achieving actual optical sectioning), while collecting all light emitted in the direction perpendicular to the sheet [5]. LSFM results in significantly less energy being absorbed by the tissue, allowing prolonged and frequent observations [6].

The root of *Arabidopsis thaliana* is a popular model system for studying dynamic processes related to plant development and physiology. Its diameter is on average between 100 and 150 μm, and it is almost completely optically transparent, making it an ideal tissue for fluorescence live imaging. Many key developmental and physiological processes (e.g., lateral root formation, tissue repair and regeneration, hormone and ion fluxes, etc.) exhibit rapid dynamics over relatively long periods. For the reasons described above, LSFM offers unique advantages for studying these processes at high spatial and temporal resolutions, by generating a long time-lapse series of frequent optical sections of growing *Arabidopsis* roots.

Since pioneering works originally demonstrated the successful application of LSFM to *Arabidopsis* roots [7, 8], other setups have been developed to expand the potentials of the system [9]. To date, LSFM has been used with *Arabidopsis* roots to study cell division dynamics [7, 10], lateral root formation [8, 11–14], calcium dynamics [15, 16], cell plate formation, root hair growth [17], and developmental zonation [18].

Full protocols have been published, developed around custom-made setups based on the design originally used in animal systems [19], a compact version inspired by the open-source hardware vision [20] and a commercial platform sold by Zeiss [21]. More recently, innovative hardware add-ons have been developed independently [10, 22].

Here, we provide a comprehensive description of the protocols for the development of both hardware and software for one of the simplest versions of LSFM adapted for *Arabidopsis* roots. The setup is optimized for sustaining long-term (days) vertical growth of the primary root, whose tip is automatically tracked and scanned in 3D every few minutes. All the optical and motorized elements can be purchased off-the-shelf; a couple of custom-made parts can be easily 3D–printed, while the control software is written in MATLAB and freely available on a dedicated repository on GitHub.

2 Materials

STL files required for 3D printing and electric wiring schematics can be found under: https://github.com/GiovanniSena/LSFM/tree/master/Hardware.

2.1 Optics and Motors

All the optical elements of the system are installed on a pneumatic optical table to reduce vibrations. Figure 1 shows a schematic of the setup.

1. Laser. The excitation beam is produced by a free space laser diode (Fig. 1a). The laser wavelength is chosen based on the fluorophore excitation spectrum; the power requirement depends on the abundance and properties of the fluorophore, the sensitivity of the CCD camera, and the quality of the optics: in our setup, a power between 10 and 30 mW is used.
2. Illumination optics group. The laser beam passes through a series of optical elements to generate the thin "sheet of light"

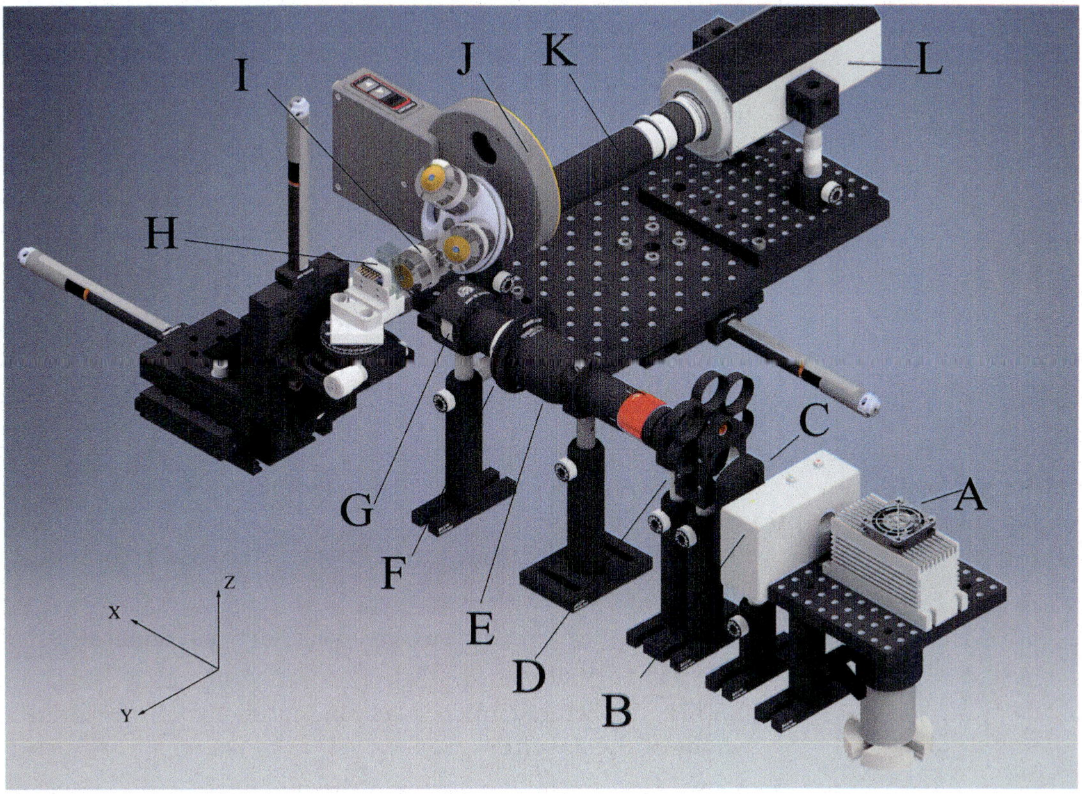

Fig. 1 Schematic of the main optical components of the light sheet microscope setup. The elements are mounted on a pneumatic optical table (not shown in the figure) and controlled remotely via MATLAB software. (**a**) laser module; (**b**) safety shutter; (**c**) shutter; (**d**) neutral density filter wheel; (**e**) beam expander; (**f**) iris; (**g**) cylindrical lens; (**h**) sample chamber (GIC); (**i**) objective; (**j**) emission filter wheel; (**k**) tube lens; (**l**) CCD camera

used for the excitation of the sample, specifically an interlock safety shutter (Fig. 1b), a remotely controlled shutter (Fig. 1c), a filter wheel (Fig. 1d) carrying neutral density filters to adjust the laser power, a Galilean expander (Fig. 1e) to increase the laser beam diameter, an iris aperture (Fig. 1f) to shape the beam diameter, and a cylindrical lens (Fig. 1g) (*see* **Note 1**).

3. Stage and sample chamber. The sample is located in a custom-made growth and imaging cuvette (GIC, described below) (Fig. 1h) positioned to locate the sample at the focal point of the cylindrical lens (*see* **Note 2**). The GIC is mounted on a three-axis motorized stage that allows translating the sample in a 25.4 mm range with 0.05 μm precision in all three dimensions.
4. Imaging optics group. The light emitted by the sample is collected by a series of optical elements, mounted perpendicularly to the illumination axis, and focused on the CCD detector of a commercial microscope camera (Fig. 1l). This is achieved by coupling an infinity-corrected objective (Fig. 1i) with a tube lens (Fig. 1k). A motorized filter wheel carries band-pass filters (Fig. 1j) that can be inserted between objective and tube lenses. The band-pass range must be selected so that it encompasses the excitation wavelength. A manual turret can be used to change magnification by switching between different objectives. The distance between the objective and the GIC can be adjusted using a secondary motorized stage located under the imaging optics.
5. Enclosure. To maintain the system in darkness and obtain optimal imaging conditions, the illumination and imaging groups, together with the stage and the GIC, are enclosed in a box equipped with movable windows to access the system. A custom-made LED cluster is mounted on the roof of the box, above the sample holder, to illuminate the seedling between imaging events.

2.2 Growth and Imaging Cuvette (GIC)

All of the below, other than the plastic cuvette, must be autoclaved prior to use (*see* **Note 3**).

Figure 2 shows the main parts of the GIC and the steps used to mount the specimen in it.

1. 3.5 mL optical quartz cuvette with internal dimensions of $10 \times 10 \times 40$ mm.
2. 0.5 mm diameter stainless steel pin, cut to match the internal height of the cuvette.
3. A soft sleeve to be used around the pin, which will be pushed into the corner of the cuvette to create the channel for the root (*see* **Note 4**). Crucially, the sleeve is shorter than the pin, so that the liquid medium can diffuse in and out of the channel from the top and the bottom of it.

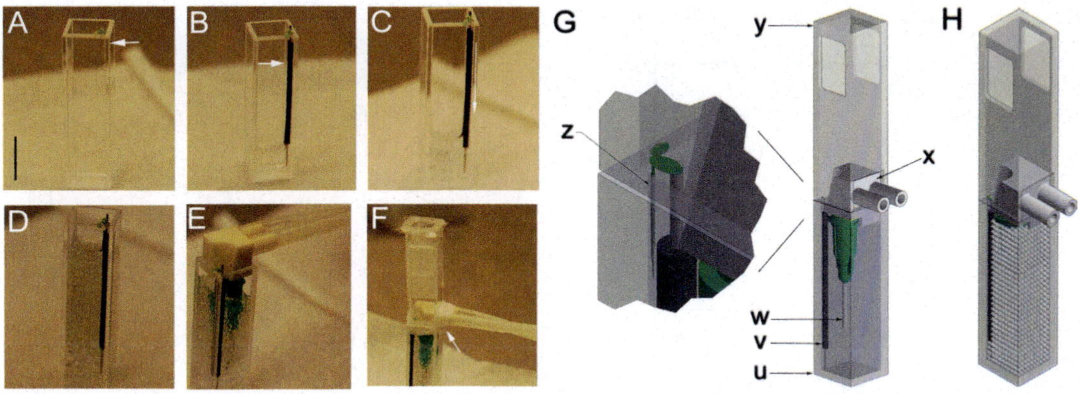

Fig. 2 Mounting the seedling within the GIC. (**a**) the seedling is positioned in the glass cuvette, with the root laying in one of the corners (arrow); (**b**) the pin with its black sleeve (arrow) constrains the root in the corner; (**c**) large beads (Ø = 2 mm) on the bottom of to the cuvette, to hold the pin in position (arrow); (**d**) small beads (Ø = 1 mm) filling the rest of the cuvette; (**e**) the nylon joint inserted in the cuvette; (**f**) the complete GIC, connected to the perfusion tubings and sealed with micropore tape (arrow); (**g**) schematic of the assembled GIC (beads not shown); (**h**) schematic of GIC with the beads. Details: u, glass cuvette; v, pin with sleeve; w, dispensing needle; x, nylon joint; y, top cuvette; z, root in the corner of the cuvette. Bar, 10 mm

4. 1 and 2 mm diameter glass beads.
5. 3D–printed nylon joint (Fig. 3), to connect the GIC to the perfusion system (*see* **Note 5**). The joint has an inlet path bringing fresh medium to the cuvette and an outlet path removing spent medium from the top of the cuvette.
6. Dispensing needle, to be connected to the nylon joint to inject the fresh medium at the bottom of the cuvette.
7. Plastic "top" cuvette, with cutout windows covered by breathable tape on the side and closed at the top with a glass coverslip.
8. The GIC is positioned in the ad hoc, 3D–printed, holder sitting on the motorized stage, as shown in Fig. 1h and Fig. 5.

2.3 Perfusion System

Nutrients and oxygen are provided to the plant by means of a continuous perfusion system. Its main components are shown in Fig. 4 (*see* **Note 6**).

1. Air pump (Fig. 4a) connected to a porous "air" stone, to oxygenate the medium in the reservoir bottle (Fig. 4c).
2. Air filters (Fig. 4b and e).
3. Reservoir bottle (Fig. 4c), containing sterile medium (*see* **Note 7**).
4. Cold trap (Fig. 4d), to condensate and capture water vapor in the vented air escaping the reservoir (*see* **Note 8**).
5. Peristaltic pump (Fig. 4f), to drive the feeding line.
6. Peltier cell (Fig. 4g), to control the temperature of the feeding line.

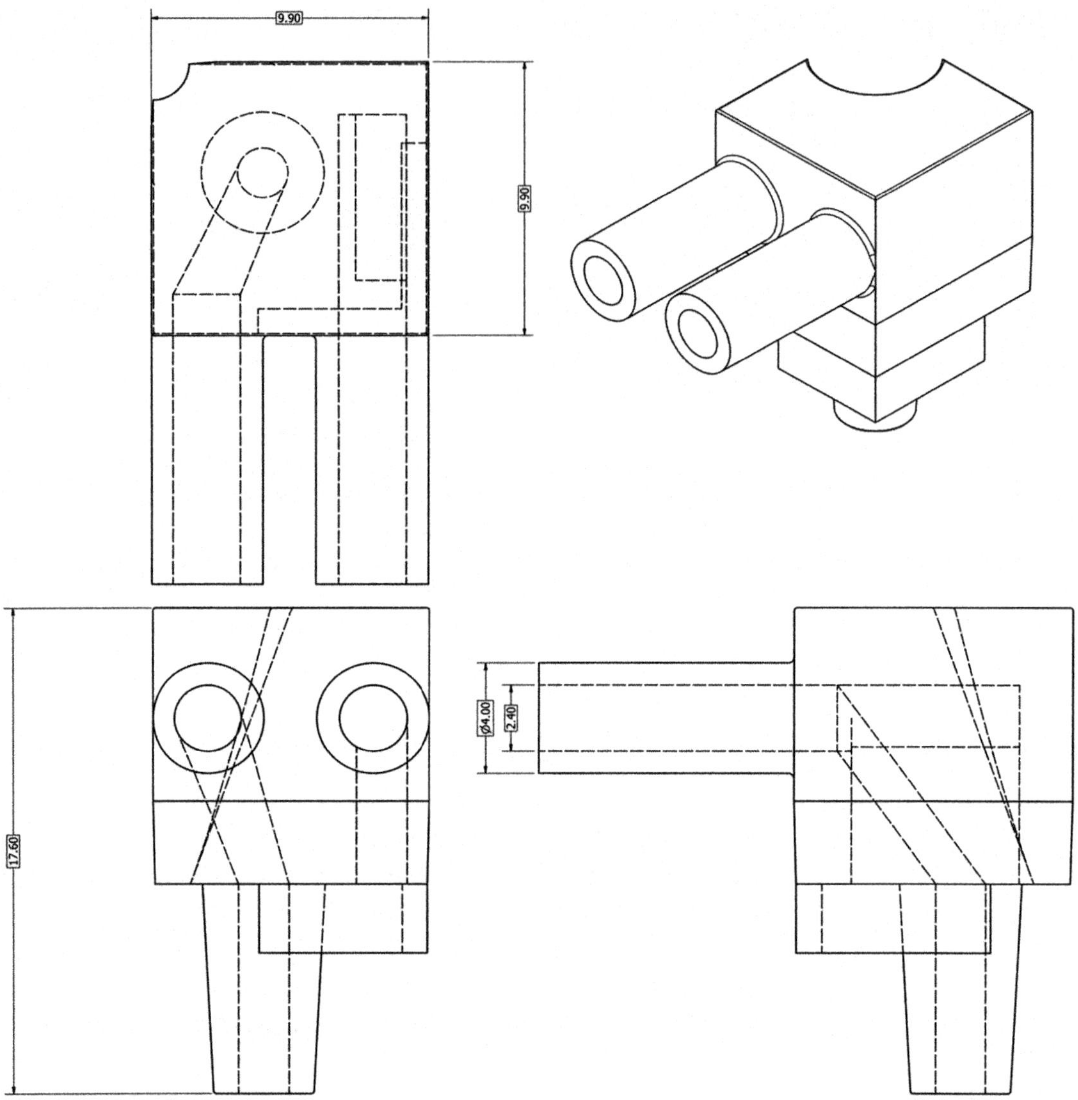

Fig. 3 Drawings of the nylon joint. Measurements in mm

7. GIC (Fig. 4h), as described above.
8. Waste bottle (Fig. 4i), to collect the spent medium from the waste line.
9. Vacuum pump (Fig. 4j), to drive the waste line.
10. Inflow (blue) and outflow (red) silicone tubing.

2.4 Temperature and Light Control Systems

The temperature of the medium in the inflow line is continuously monitored and automatically adjusted with a Peltier cell via feedback control. The seedling is illuminated by an LED cluster only between imaging scans.

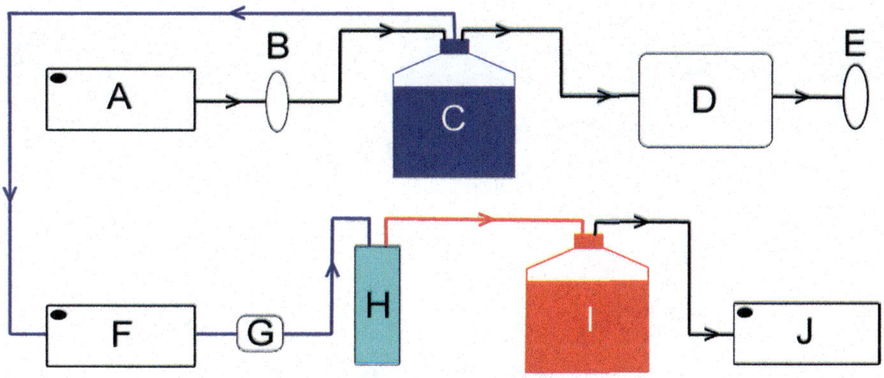

Fig. 4 Schematic of the perfusion system with its main components. The feeding line is shown in blue, the waste line in red. Air lines are shown in black. Arrows indicate the direction of flow. (**a**) air pump; (**b**) air filter; (**c**) media bottle; (**d**) cold trap; (**e**) air filter; (**f**) peristaltic pump; (**g**) Peltier cell; (**h**) GIC; (**i**) waste media bottle; (**j**) vacuum pump

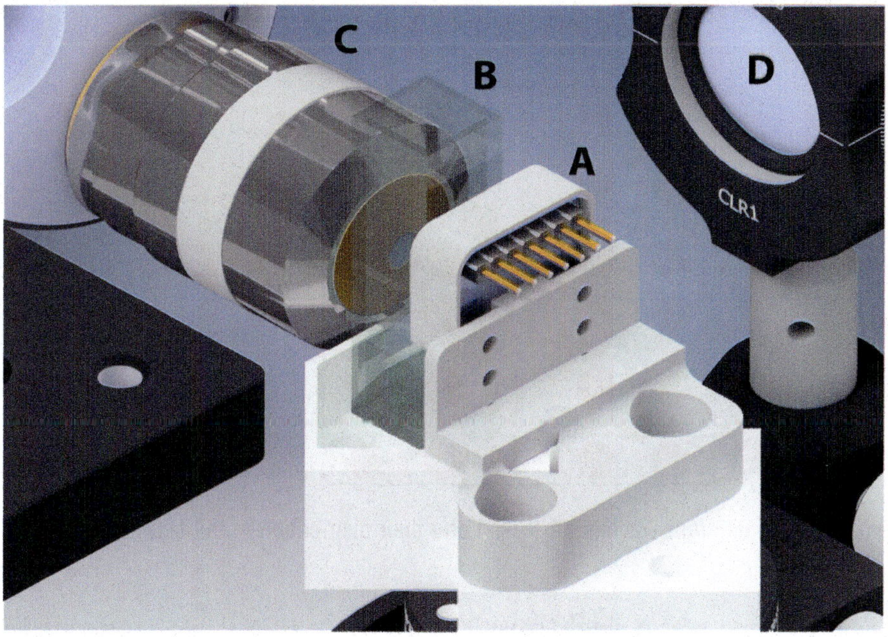

Fig. 5 Schematic close-up of the cuvette holder. (**a**) IR temperature sensor; (**b**) GIC; (**c**) objective; (**d**) cylindrical lens

1. Infrared sensor, controlled via an Arduino board (Genuino Uno Rev. 3) and mounted in front of the GIC (Fig. 5) to remotely monitor the temperature of the liquid medium.
2. Heat exchanger element (Fig. 6), constituted by a Peltier cell held in contact with a serpentine by a 3D–printed clamp. The Peltier cell is controlled through a separate Arduino board controlled via MATLAB.

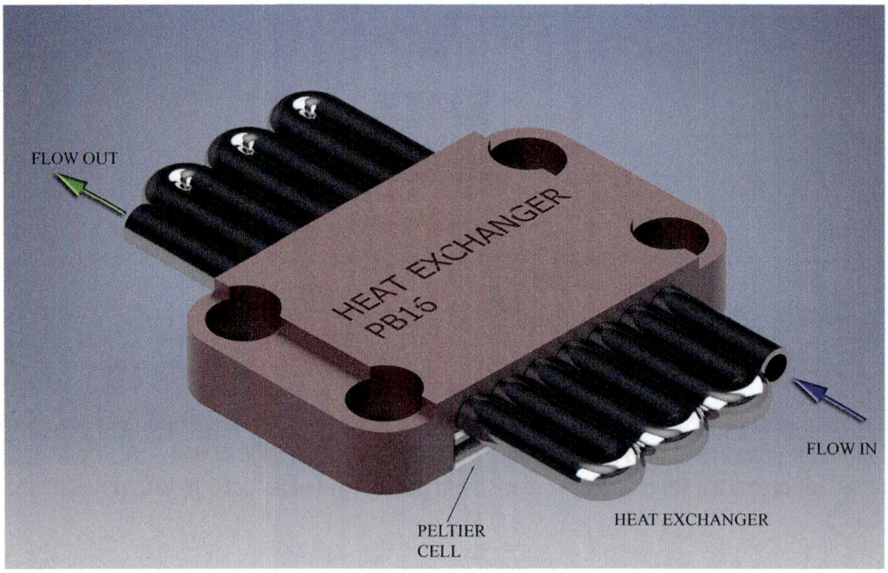

Fig. 6 Schematic of the heat exchanger element used in the perfusion system

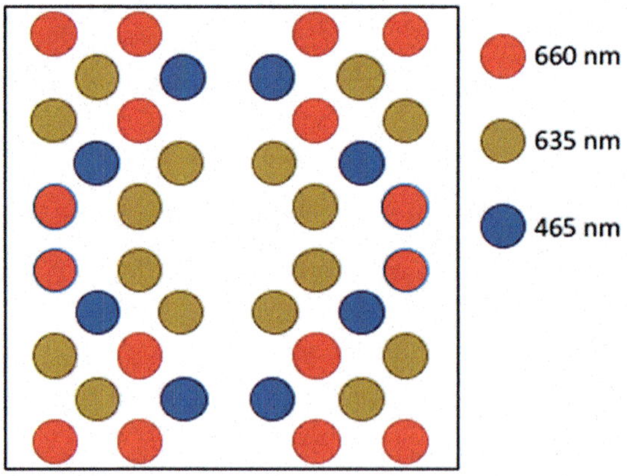

Fig. 7 Schematic of the different wavelengths used and their distribution in the LED cluster

3. A single cluster of 40 LEDs is installed approximately 10 cm above the cuvette. The cluster has been designed to provide uniform illumination at wavelengths 465, 635, and 650 nm (Fig. 7). The cluster switch is controlled through a LabJack board (TRD U3 with LJTick-RelayDriver) using MATLAB.

2.5 Control Software

The following system components are under the control of a single program developed in MATLAB and run on standard desktop computer (quad-core, 16GB RAM, mid-range graphic card): laser shutters, GIC stage motors (x, y, z axes), emission filter wheel, camera stage motors (x, y axes), LED cluster, perfusion inflow peristaltic pump, Peltier cell, and infrared temperature sensor.

A graphical user interface (GUI) allows the operator to control the system and to run an automated imaging routine. The automated routine is composed by a sequence of steps repeated at a fixed interval of time. *See* Subheading 3 for a detailed description of each step.

The MATLAB software to control the system and provide the GUI is stored in a dedicated GitHub repository: *https://github.com/GiovanniSena/LSFM*. All the used drivers are off-the-shelf from manufacturers.

2.6 Plant Material and Solutions

1. As presented, the protocol is designed to be used with *Arabidopsis thaliana*, but not limited to any specific genetic background. Since the system is based on fluorescence microscopy, a fluorescent reporter or dye must be present.
2. Sterilization solution: 50% v/v house bleach, 0.05% v/v Triton X-100.
3. Solid medium (agar plates): 1× Murashige and Skoog (MS) Basal Medium, 0.5% w/v sucrose, 0.05% w/v MES hydrate, 0.8% w/v agar, pH 5.7 with KOH.
4. Liquid medium (perfusion): ¼× Murashige and Skoog (MS) Basal Medium with Gamborg's vitamins, 0.5% w/v sucrose, 0.05% w/v MES hydrate, pH 5.7 with KOH.

3 Methods

3.1 Agar Plates for Seed Germination

1. Fill a 1 L beaker containing a magnetic stir bar with 900 mL water, and add 5 g sucrose, 0.5 g MES hydrate, and 1.75 g MS. Use a magnetic stirring plate to dissolve the reagents. Once dissolved, add KOH until the solution reaches pH 5.7, and then make the solution up to 1 L with distilled water. Add 8 g agar.
2. Autoclave. The media may be kept for several hours in an oven with a temperature of 55 °C.
3. Pour the medium into sterile petri dishes under sterile conditions.

3.2 Seed Sterilization and Germination

Perform the following under sterile conditions:

1. Resuspend seeds in sterilization solution for 3 min. Immediately rinse the seeds with sterile water. Repeat the rinsing six more times to eliminate bleach residues.
2. To synchronize seed germination, stratification is applied by keeping the sterile seeds in the dark and at 4 °C for 2 days before transferring onto a solid agar plate.
3. Using a sterile plastic transfer pipette (*see* **Note 9**), place the seeds onto an MS agar plate. The seedlings will need up to 3 cm

of space underneath the seed for growth of the root. Seal each plate with either Parafilm or breathable tape, to minimize the risk of contamination.

4. Place the plates in standard germinating conditions: 23 °C with 120 μmol/m^2/s light using a light/dark cycle of 16/8 h (*see* **Notes 10** and **11**).

3.3 Mounting the Seedlings into the GIC

Perform the following under sterile conditions:

1. The seedlings should be mounted when the root is between 15 and 25 mm long, typically 3 or 4 days post-germination.
2. Before mounting, confirm the signal intensity from the root under a fluorescence stereo microscope.
3. Find the "optical surface" of the cuvette, as defined by the manufacturer, to identify the corner where the root will be located.
4. Use a small amount of MS liquid medium to wet the corner of the cuvette. The surface tension of the liquid facilitates the correct positioning of the root.
5. Remove the seedling from the plate using sterile forceps and carefully position the root in the wet corner (*see* **Note 12)** (Fig. 2a). Ideally, the entire length of the root should be in the cuvette, while the hypocotyl and cotyledons should protrude from its top. The correct position of the root should be checked after each of the following steps.
6. Using sterile forceps, gently position the pin in the cuvette, to trap the root in the corner (*see* **Note 13**) (Fig. 2b and g).
7. Slowly fill the cuvette with some of the perfusion medium and the 2 mm glass beads (*see* **Note 14**) (Fig. 2c).
8. Fill the rest of the cuvette with 1 mm glass beads, up to about 10 mm from the top of the cuvette (Fig. 2d).
9. Attach the nylon joint to the dispensing needle via the Luer lock, and gently push the needle through the beads, until the joint sits tightly in the cuvette, effectively capping it (Fig. 2e).
10. Slide the plastic "top" cuvette over the insert, taking care not to pinch the cotyledons. Seal the cuvettes together using breathable tape (Fig. 2f).

3.4 Starting the Perfusion

Use Fig. 4 as reference for the parts mentioned here.

1. At the beginning of the experiment, the reservoir bottle is filled with sterile liquid medium.
2. Mount the feeding line into the peristaltic pump (*see* **Note 15**).
3. Under sterile conditions, insert the air stone connected to the air line in the reservoir bottle; ensure that this bottle contains a

magnetic stir bar. Do not yet turn on either the air pump or the peristaltic pump.

4. Start stirring the medium on the stirring plate.
5. Connect the feeding line to the serpentine connected to the Peltier cell.
6. Connect the other end of the serpentine to the waste line, thus bypassing the GIC and closing the circuit.
7. Connect the waste bottle to the vacuum pump.
8. Turn on the air pump (to bubble air into the reservoir), the peristaltic pump (to circulate the feeding line), and the vacuum pump (to start the suction in the waste line), in this order.
9. Let the system prime until the feeding line is completely filled with medium (*see* **Note 16**), and then stop the air, peristaltic, and vacuum pumps, in this order.
10. Prime the short tubing connected to the GIC's feeding inlet, using a manual syringe to inject liquid medium until it reaches the hypocotyl of the seedling in the GIC.
11. Position and secure the GIC on the microscope's holder (Fig. 1h).
12. Slightly loosen the cap of the waste bottle, to eliminate any residual negative pressure from the waste line.
13. Disconnect the waste line from the serpentine, and connect it to the outflow inlet of the GIC.
14. Connect the serpentine to the inflow inlet of the GIC. The circuit should now be closed again, with the GIC between the serpentine and the waste bottle (Fig. 4).
15. Tighten the cap of the waste bottle.
16. Turn on, in this order: the vacuum pump, the air pump, and finally the peristaltic pump (*see* **Note 17**).

3.5 Setting Up Initial Conditions and Starting the Automated Imaging

With the seedling mounted in the GIC (Subheading 3.3) and the perfusion system operative (Subheading 3.4), it is time to position the root in the field of view and to start the scanning.

1. Switch to brightfield conditions, by turning on the LEDs, closing the shutter, and removing the emission filter.
2. Use the motors controlling the camera's position to find the root tip and to focus onto it (*see* **Note 18**).
3. Switch to fluorescence conditions by turning off the LEDs, opening the shutter, and positioning the required emission filter.
4. The root is already in focus with the objective (**step 2** above), but it will only fluoresce when it is illuminated by the light sheet: use the dedicated command on the software GUI to lock together the camera and the stage movements in the dimension

perpendicular to the light sheet (to keep the root in focus), and move them until fluorescence emission from the root becomes visible (*see* **Note 19**).

5. Once fluorescence is observed, one should ensure that the root is illuminated by the thinnest part of the sheet (TPS), which is located at the focal point of the cylindrical lens. First, make sure that the camera and the stage movements are locked together in both dimensions (to keep the root in the field of view and in focus). Then apply small adjustments perpendicular to the light path, until the signal in emission reaches maximum intensity. Finally, apply small translations in the direction parallel to the light sheet and repeat the previous step. Since light is most concentrated at the TPS but it illuminates a smaller volume of the root, one can look for increased emission of the details in focus and decreased emission of those out of focus, as a proxy for being in the TPS.
6. Manually scan the root in the axial dimension (the direction perpendicular to the light sheet, used to generate the stack of optical sections during automated acquisition) to determine the (a) thickness of the root in microns and (b) position of the root's surface closest to the objective.
7. Unlock stage and camera movements.
8. Manually focus on the root's surface closest to the objective.
9. Use the parameters in the *Config* panel of the GUI to specify (a) scanning step length (usually 4 μm), (b) total number of steps (root's diameter measured in **step 6**, plus some buffer space on either side of the root), (c) temporal interval between scans (usually 10 min), and (d) temperature to maintain in the GIC liquid medium (usually 23 °C) (*see* **Note 20**).
10. In the GUI, start the automated scan.

3.6 Automated Z-Stack Acquisition

1. At each programmed time point, a 3D stack of optical sections is created by moving the root through the light sheet in steps and recording an image at each position.
2. During the scan, a standard autofocus routine [23] is executed at intervals defined by the user.
3. The raw data of each stack is saved to disk, and its maximum intensity projection is visualized in the software GUI.
4. A new stack is acquired at temporal intervals defined in the *Config* panel of the software (*see* **step 9** in Subheading 3.5).

3.7 Root Tracking

A tracking algorithm was developed to determine the movement of the root tip between two subsequent scans and to adjust the cuvette position to maintain the root tip in the field of view.

A root scan creates a stack of N grayscale images, which can be saved as a 3D matrix where each element is the intensity value of the pixel. In order to minimize the number of scans imposed on a root and therefore to minimize any photodamage, the tracking is solely based on the 3D stacks already acquired on the current and previous time points. In essence, it is assumed that in the previous time point, the root tip was in the center of the field of view and that in a single time interval, it grew only slightly out of it.

The actual steps involved are (Fig. 8):

1. Acquire 3D stack at time t: image I_t. If this is the first image of the scan, the system pauses until the required time has elapsed.

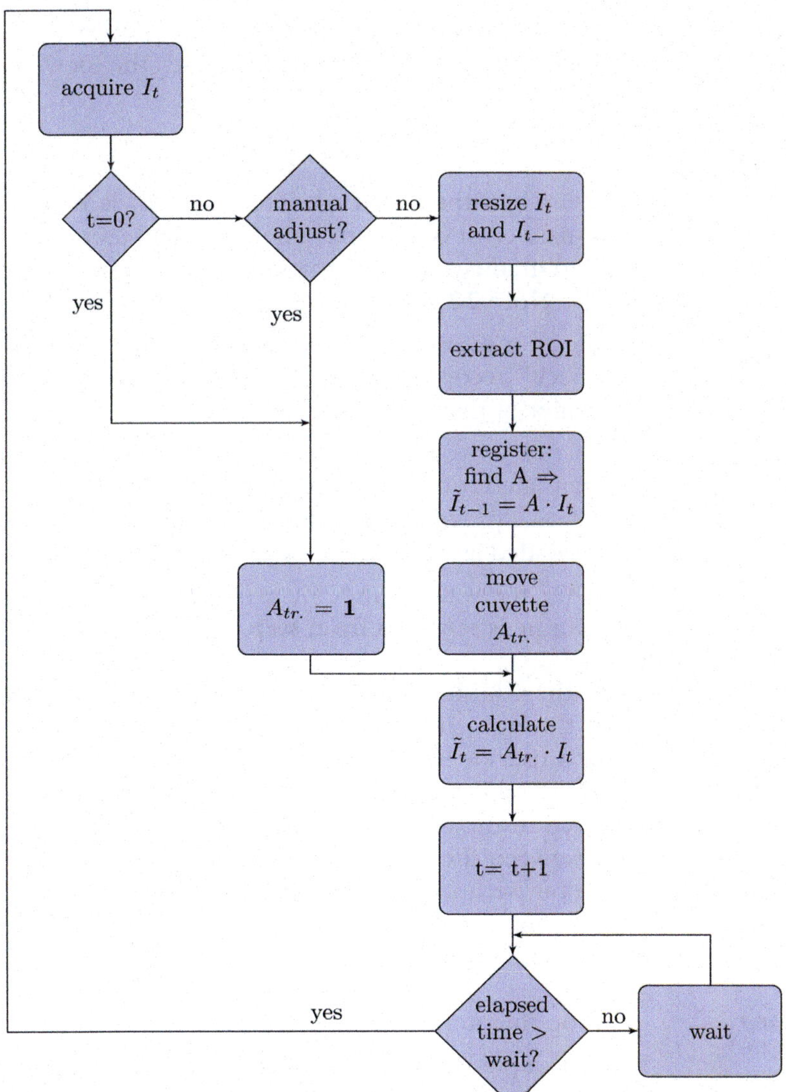

Fig. 8 Block diagram of the main operations performed by the software during automated data acquisition

Similarly, if the user has manually adjusted the motors, the system treats the stack as if it were the first of a new acquisition.

2. Retrieve the virtually transformed stack obtained at time *t-1* (*see below*): image $\tilde{I}_{t-1}$.
3. Rescale I_t and $\tilde{I}_{t-1}$, to reduce the computational time required for the tracking algorithm. The scaling factor is defined by the user and is typically equal to 2 in X and Y (i.e., the image size is reduced by a factor 4).
4. Extract a region of interest from the full stack, to further reduce computational time and to eliminate potential external objects (dead cells, optical artifacts, etc.), by removing a fixed {ΔX, ΔY, ΔZ} region on the perimeter of the stack. For simplicity, the rescaled and cropped stacks are still referred to as I_t and $\tilde{I}_{t-1}$.
5. Registration is performed to find the new transformation matrix A_{tr} (*see* **Note 21**), such that

$$I_t = A_{tr} \cdot \tilde{I}_{t-1}$$

The algorithm used to determine A_{tr} is based on the mean square error of the pixel intensities and uses a standard iterative step gradient descent to estimate the best transformation [24] (*see* **Note 22**).

6. Stage movement: the stages holding the GIC are moved "back" accordingly to A_{tr}^{-1}, to bring the root tip back in the center of the field of view.
7. Virtual movement: instead of taking a second scan of the now translated root (*see above*), the last image I_t is virtually (computationally) translated with A_{tr} to generate $\tilde{I}_t$, which will be uploaded as $\tilde{I}_{t-1}$ at the beginning of the next time point.
8. After the waiting time determined by the user is elapsed, the routine is repeated from **step 1**.

3.8 Temperature Control

While the GUI is running, the temperature of the cuvette is monitored through the IR sensor (Fig. 5a). If the temperature drops below a user-defined threshold (T_{target}), the software activates the Peltier cell and provides energy to the heat exchanger. The energy provided to the cell is increased linearly as a function of the difference between the current temperature T and T_{target}.

In our setup, the Peltier cell was configured to act as a heater, but an identical cell could easily be installed to provide cooling capabilities as well.

3.9 Cleaning and Maintenance

1. Stop the pumps.
2. Uncouple the GIC from the perfusion system and remove it from the stage.

3. Close the perfusion circuit by connecting together the feeding and the waste lines.
4. Remove the air stone and the feeding tube from the media bottle, and place them in a new bottle containing hot clean water, pumping air through the stone and circulating the hot water through the closed system for at least 45 min.
5. Clean the cuvette using optical glass cleaning solution and lens wipes; rinse it with 100% isopropanol and air dry it.
6. Use a syringe to flush the tubes attached to the cuvette, and the feeding needle still mounted on the insert, with hot water.
7. Wrap all items in individual autoclave bags and autoclave.

4 Notes

1. The shape of the illumination beam after the cylindrical lens is determined by two factors: the minimum width w_0 of the sheet of light (or thinnest part of the sheet, TPS, at the focal point of the cylindrical lens) and its extension, Δx (or for how long the sheet's minimum thickness is maintained in the dimension parallel to the laser beam).

   These parameters are approximated by:

$$w_0 \cong 1.22\,\lambda \frac{f}{d}$$

$$\Delta x \cong \frac{2\pi}{\lambda} {w_0}^2$$

   where λ is the laser wavelength, f is the focal length of the cylindrical lens, and d is the beam diameter hitting the cylindrical lens (i.e., the iris aperture). The laser wavelength will have to be determined on the basis of the fluorophore to be imaged. We currently use genetic reporters tagged with GFP or YFP, and we chose a laser with $\lambda = 488$ nm; we chose to use a lens with $f = 50$ mm and $d = 7.5$ mm to obtain nominal values $w_0 \cong 4\ \mu m$ and $\Delta x \cong 203$ μm, which are compatible with a desirable axial resolution (e.g., the average distance between cell nuclei is >4 μm) and a typical *Arabidopsis* root diameter. Notice that the optical expander must be chosen so that the diameter of the expanded beam is sufficient to cover the whole iris aperture. After the cylindrical lens, the laser beam travels through air, glass (cuvette wall), and aqueous medium before reaching the sample in the cuvette. The optical path is therefore composed by multiple indexes of refraction, and the exact position of the cylindrical lens focal point (the actual location of the TPS) will have to be determined empirically. We use fluorescent beads embedded in Phytagel to visualize the light sheet profile and precisely determine the position of the TPS inside the cuvette.

2. Maintaining the root in the objective's field of view and in focus while it is growing requires subsequent adjustments of the camera position. Our custom control software contains specific automated "tracking" and "focusing" routines designed to automatically move the camera and follow the root movements (see the software description). For this reason, the objective and the camera are rigidly connected and mounted on a second motorized stage able to translate in x and y.
3. It is important that all items that come into contact with the specimen and growth media are sterile and free of compounds that might be toxic to the seedling. Most components can be autoclaved, with the exception of the plastic cuvette and microporous tape. To reduce the likelihood of potentially toxic compounds accumulating on the surfaces of items during autoclaving, we use small autoclaving bags, commonly used for the sterilization of surgical tools.
4. We find that electrical heat-shrink tubing works well as a soft sleeve around the pin. Fit a length of this sleeve over the pin that leaves 3 mm of needle exposed at either end; use a source of heat, such as a heat gun, to shrink the sleeve onto the pin. The actual size of the channel in the cuvette's corner is determined by the total diameter of the pin + sleeve element pushed against the cuvette corner. Elementary geometry indicates that $d \cong 0.17\ D$, where d is the maximum diameter of a cylinder (i.e., the root) fitting in the channel and D is the diameter of the combined pin + sleeve element. We currently use a pin with diameter 500 μm and a sleeve with thickness 450 μm, resulting in $d = 240$ μm.
5. The insert should be 3D–printed in nylon or any other material capable of withstanding the autoclaving process.
6. We use autoclavable silicone tubing connected by means of quick-release shutoff couplings. This facilitates the setup of the perfusion system and allows us to quickly replace the reservoir and waste bottles during prolonged experiments, while maintaining sterile conditions.
7. The medium in the reservoir bottle should be maintained well mixed during the experiment. For this purpose, we keep a magnetic bar in it and place the bottle on a large stirring plate.
8. The relief line is needed to maintain the pressure within the bottle at equilibrium with the atmospheric pressure. The excess of air introduced by the air pump escapes through this route via a filter to reduce the risk of contamination of the medium. The cold trap reduces the risk of condensation within the filter, which would cause its occlusion.

9. We keep the sterile seeds resuspended in sterile water and use a plastic transfer pipette to dispense one seed at a time on the solid agar plate. Since the seedlings will not be grown for long on the plate, little space is needed between seeds, i.e., 5 mm horizontally and ca. 30 mm vertically.
10. To minimize the possibility of the cotyledons being submersed at any time during the experiment, a mild etiolation increases the distance between the cotyledons, and the root can be considered. To ensure proper growth of the cotyledons prior to mounting in the GIC, 2 days of normal growth should follow etiolation. Thus, one would grow the seedlings in the dark for 2 days following germination and a subsequent 2 days in standard light/dark cycle conditions.
11. The transition between the "germinating" light/dark regime and the "imaging" one might result in physiological stress to the plant. Depending on the scope of the experiment, it might be considered to germinate the seeds already under the "imaging" regime, simulating the same alternation of light and darkness that the seedlings will then experience in the GIC.
12. To ensure that the seedling does not dry out during the mounting procedure, one can use a nebulizer loaded with sterile water, to increase the relative humidity of the air in the working area.
13. It is crucial that the pin traps the root in the corner of the cuvette. To achieve this, position one end of the pin in the cuvette's bottom corner, and carefully pivot around it to ensure that the pin is positioned as in Fig. 2c. Using a magnifying lens, make sure that the root is indeed located within the channel formed between pin and cuvette walls. If not, repeat this step.
14. The beads will push on the pin and hold it in position.
15. It is critical that the feeding line is properly and fully mounted in the peristaltic pump. It might be easier to perform this step when the line is still empty of fluid.
16. This is critical to eliminate air bubbles from the feeding line. They would disturb the root during acquisition.
17. To avoid flooding, before turning on the peristaltic pump regulating the inflow, make sure that the vacuum pump is creating enough negative pressure in the waste bottle and that suction in the GIC is working adequately.
18. This is best achieved with a low magnification objective, such as 10× or 20×.
19. Once some signal is detected by the camera (i.e., the root is now illuminated by the light sheet), focusing might be required. Unlock the camera movement from the stage, and apply small steps on the camera to bring the root back in focus. Then lock the movements again.

20. The temporal interval between scans defines the temporal resolution of the dataset. A few considerations must be taken into account when choosing this parameter, to find the right balance between conflicting priorities: root tracking works better with small temporal intervals; during scanning the LEDs are off, therefore scanning too frequently limits the photosynthetic yield of the seedling; frequent imaging may also result in phototoxicity for the plant or photo-bleaching for the fluorophore.
21. Since the stage can only be translated linearly, we impose A_{tr} to represent a simple rigid translation in 3D, which will effectively represent the more complex roto-translation of the root tip. The MATLAB function used for this is *imregtform*.
22. Note that this tracking routine does not depend on the conical shape of the root tip, so it is equally amenable for uncut as well as regenerating tissue.

Acknowledgments

This work was supported by BBSRC grant BB/M002624/1.

References

1. Yuste R (2011) Imaging: a laboratory manual. CSH Press, New York
2. Dixit R, Cyr R (2003) Cell damage and reactive oxygen species production induced by fluorescence microscopy: effect on mitosis and guidelines for non-invasive fluorescence microscopy. Plant J 36:280–290
3. Huisken J, Swoger J, Del Bene F et al (2004) Optical sectioning deep inside live embryos by selective plane illumination microscopy. Science 305:1007–1009
4. Huisken J, Stainier DYR (2009) Selective plane illumination microscopy techniques in developmental biology. Development 136:1963–1975
5. Greger K, Swoger J, Stelzer EHK (2007) Basic building units and properties of a fluorescence single plane illumination microscope. Rev Sci Instrum 78:023705
6. Reynaud EG, Krzic U, Greger K, Stelzer EHK (2008) Light sheet-based fluorescence microscopy: more dimensions, more photons, and less photodamage. HFSP J 2:266–275
7. Sena G, Frentz Z, Birnbaum KD, Leibler S (2011) Quantitation of cellular dynamics in growing Arabidopsis roots with light sheet microscopy. PLoS One 6:e21303
8. Maizel A, von Wangenheim D, Federici F et al (2011) High-resolution live imaging of plant growth in near physiological bright conditions using light sheet fluorescence microscopy. Plant J 68:377–385
9. Berthet B, Maizel A (2016) Light sheet microscopy and live imaging of plants. J Microsc 263:158–164
10. de Luis Balaguer MA, Ramos-Pezzotti M, Rahhal MB et al (2016) Multi-sample Arabidopsis growth and imaging chamber (MAGIC) for long term imaging in the ZEISS Lightsheet Z.1. Dev Biol 419:19–25
11. Lucas M, Kenobi K, von Wangenheim D et al (2013) Lateral root morphogenesis is dependent on the mechanical properties of the overlaying tissues. Proc Natl Acad Sci U S A 110:5229–5234
12. Rosquete MR, von Wangenheim D, Marhavý P et al (2013) An Auxin transport mechanism restricts positive Orthogravitropism in lateral roots. Curr Biol 23:817–822
13. Vermeer JEM, von Wangenheim D, Barberon M et al (2014) A spatial accommodation by neighboring cells is required for organ initiation in Arabidopsis. Science 343:178–183

14. von Wangenheim D, Fangerau J, Schmitz A et al (2016) Rules and self-organizing properties of post-embryonic plant organ cell division patterns. Curr Biol 26:439–449
15. Costa A, Candeo A, Fieramonti L et al (2013) Calcium dynamics in root cells of Arabidopsis Thaliana visualized with selective plane illumination microscopy. PLoS One 8:e75646
16. Candeo A, Doccula FG, Valentini G et al (2017) Light sheet fluorescence microscopy quantifies calcium oscillations in root hairs of Arabidopsis thaliana. Plant Cell Physiol 58 (7):1161–1172
17. Berson T, von Wangenheim D, Takáč T et al (2014) Trans-Golgi network localized small GTPase RabA1d is involved in cell plate formation and oscillatory root hair growth. BMC Plant Biol 14(252)
18. Novák D, Kuchařová A, Ovečka M et al (2015) Developmental nuclear localization and quantification of GFP-tagged EB1c in Arabidopsis root using light-sheet microscopy. Front Plant Sci 6:1187
19. von Wangenheim D, Daum G, Lohmann JU et al (2014) Live imaging of Arabidopsis development. Methods Mol Biol 1062:539–550
20. von Wangenheim D, Hauschild R, Friml J (2017) Light sheet fluorescence microscopy of plant roots growing on the surface of a gel. J Vis Exp. https://doi.org/10.3791/55044
21. Ovečka M, Vaškebová L, Komis G et al (2015) Preparation of plants for developmental and cellular imaging by light-sheet microscopy. Nat Protoc 10:1234–1247
22. Jeandupeux E, Lobjois V, Ducommun B (2015) 3D print customized sample holders for live light sheet microscopy. Biochem Biophys Res Commun 463:1141–1143
23. Geusebroek JM, Cornelissen F, Smeulders A, Geerts H (2000) Robust autofocusing in microscopy. Cytometry 39:1–9
24. van der Bom IMJ, Klein S, Staring M et al (2011) Evaluation of optimization methods for intensity-based 2D-3D registration in x-ray guided interventions. In: Dawant BM, Haynor DR (eds). 8th IEEE International Symposium on Biomedical Imaging (ISBI 2011). International Society for Optics and Photonics, pp 796223–796223-15

Check for updates

Chapter 12

Histological Profiling Over Time to Optimize Root Cell Type-Specific Reporter Lines for Cell Sorting

Beatriz Lagunas, Liam Walker, Rana Muhammad Fraz Hussain, Ian Hands-Portman, Kate Woolley-Allen, and Miriam L. Gifford

Abstract

Cell type-specific marker lines expressing fluorophores such as GFP or GUS can be used as starting material from which single cell types can be isolated by fluorescence-activated cell sorting (FACS) and/or for the study of root development. Establishing the stability of these lines is an essential step prior to further study to ensure that marker expression and localization is stable over time and during environmental perturbations of interest to researchers applying these lines as treatments. Here, we detail the use of root cross sectioning to investigate marker expression throughout the length and width of the root using the model legume *Medicago truncatula* as an example. In order to deal with the fact that plant cell walls are highly autofluorescent, we also describe the usage of confocal microscopy to conduct a lambda scan to discriminate autofluorescence from marker molecule expression.

Key words Root development, FACS, Root cell types, Root GFP expression, Root cross sections

1 Introduction

A key aspect of multicellularity is that it permits an organism the ability to differentiate cells into specialized types. Different cell types have distinct fates and functions conferred by differential gene expression. In order to fully understand the contribution of different tissues to plant metabolic processes, such as growth and development, each tissue must be analyzed separately. Cell type-specific approaches to study developmental processes are able to unearth effects or responses that may be obscured at the whole organism level due to combined signals from mixed cell types and thus are highly informative.

Fluorescence-activated cell sorting (FACS) is a specialized form of flow cytometry that enables the isolation of subpopulations of cells from a liquid suspension, based on their fluorescence properties. When applied to study biological systems such as multicellular organisms, this enables the study of processes at the cell-type level.

Daniela Ristova and Elke Barbez (eds.), *Root Development: Methods and Protocols*, Methods in Molecular Biology, vol. 1761, https://doi.org/10.1007/978-1-4939-7747-5_12, © Springer Science+Business Media, LLC 2018

In plant research, FACS is most commonly used to segregate cell types or cells at particular response stages labeled with fluorescent markers, such as green fluorescent protein (GFP), from non-labeled cells within root or shoot tissue. In these approaches, root or leaf material is subject to an enzymatic treatment to break down connective cell walls and disaggregate individual cells (protoplasts) from each other into a liquid suspension. It has been shown that isolated protoplasts can retain their reporter expression and the analytical features of a FACS machine enable the user to discriminate between marked and unmarked cells and sort accordingly, resulting in enrichment of a specific subpopulation of cells [1]. Numerous approaches can be followed after FACS to investigate the isolated cell populations, including transcriptomics, proteomics, and metabolomics [2–4].

In order to generate marker lines that can be used for FACS, plants must be transformed to express a reporter gene in a cell type of interest or that is expressed in cells undergoing a response of interest. For cell types this requires the identification of suitable tissue-specific promoters from which to express the marker. Because FACS relies on the presence of this marker signal to determine which cells to sort, establishing that the localization of marker expression remains stable not only through developmental time but also in response to any intended treatments is a prerequisite of further study. Examples of previously well-characterized cell type-specific lines in *Arabidopsis thaliana* include the epidermal marker *GL2* [5], the cortical marker *Co2* [6], and the E3754 line marking the pericycle [7]. Generating stable cell type-specific lines in higher plants with more complex genomes than *Arabidopsis* or that are less amenable to transformation methods is more challenging. Furthermore, expression can be lost between generations as a consequence of silencing [8]. In this chapter, we include description of an approach based on microscopy of root cross sections to observe marker expression throughout the breadth of the root. We use the example of a transgenic *Medicago truncatula* line that expresses GFP in the epidermis [9], although the localization becomes more widespread as the root develops, likely due to altered promoter activity during development. In other reporter molecule cases, it is possible that markers could disperse throughout the root via plasmodesmata [10].

Although GFP is the most commonly used fluorescent reporter, it is possible to sort cells expressing other labels, including GUS. In this chapter, we detail a modified GUS staining protocol using a fluorescent substrate that can be used for FACS. Many plants have some level of endogenous GUS-like activity [10] which will lead to low levels of background fluorescence, but it is possible to circumvent this when choosing sort gates on a FACS machine.

Autofluorescence describes emission of light upon excitation from natural components of cells. Although root cells exhibit less autofluorescence than leaf cells (which contain highly autofluorescent chloroplasts), there are nevertheless numerous autofluorescent compounds in root cells. If these compounds exhibit autofluorescence in the same spectrum as a fluorescent marker, this can make screening for marker expression and cell sorting more challenging.

The major source of autofluorescence in root cells is from compounds present in cell walls (such as lignin, cutin, or suberin) [11], which might remain in the protoplast solution, either alone or associated with protoplasts even after disaggregation. Bona fide marker expression can usually be differentiated from autofluorescence by careful selection of optical filters to avoid signal overlap. In cases where it is suspected that autofluorescence overlaps with the signal from a fluorophore, lambda scanning can be a useful tool to separate the signals by a process termed linear unmixing. A lambda scan records a series of images from a specimen over a user-defined range of emission wavelengths. To use this technique for unmixing, it is essential that the collected images are free from saturated pixels, and regions can clearly be identified where fluorophores are spatially separated. In this chapter, we describe a procedure for lambda scanning for this purpose.

2 Materials

2.1 Root Tissue Cross-Section Preparation

1. Roots from plants expressing a fluorophore or GUS.
2. Conical flask.
3. Microwave.
4. Vibratome.
5. Light microscope.
6. Petri dishes of 6 cm diameter.
7. Pen-type thermometer.
8. Forceps.
9. Scalpel.
10. Superglue.
11. Stainless steel double-edge blades.
12. Microscope slides.

13. Microscope slide coverslips.
14. Distilled water.
15. Agar powder.
16. Propidium iodide solution to 100 μg/mL.

2.2 GUS Staining of Plant Tissue

1. 5 mL tubes with lid.
2. Sodium phosphate buffer 100 mM (pH 6.8).
3. ImaGene Green™ C12FDGlcU GUS Gene Expression Kit (Thermofisher catalog number 2908).
4. D-glucaric acid 1,4-lactone.
5. X-gluc solution: X-gluc (5-bromo-4-chloro-3-indolyl-beta-D-glucuronic acid, cyclohexylammonium salt) 0.5 mM, methanol (20%), NaEDTA pH 8 10 mM, sodium phosphate buffer 100 mM, $K_4Fe(CN)_6 \cdot 3H_2O$ 1.4 mM, $K_3Fe(CN)_6$ 1.4 mM, Triton X-100 0.1%.
6. Methanol.

2.3 Confocal Microscopy of Medicago Root Cross Sections

Confocal microscope.

3 Methods

Carry out all procedures at room temperature on a laboratory bench unless otherwise specified. Carefully read manufacturer's instructions before using the Imagene Green gene expression kit.

3.1 Root Tissue Cross-Section Preparation

1. For each sample, weigh 5 g of agar powder in to a conical flask, and add distilled water up to the 100 mL mark.
2. Using a steel blade, cut fresh roots (around 5 mm pieces) from seedlings of interest (*see* **Note 1**), and place them into a petri dish containing distilled water.
3. Heat the agar/water mixture in a microwave until it is fully dissolved (the solution should assume a honey-like texture). Be careful to avoid super-heating, and always wear full protective equipment. Allow to cool (with stirring) until <70 °C.
4. Carefully pour the agar solution into a fresh petri dish, and stir continuously using the pen-type thermometer until the mixture reaches 58 °C.
5. Working as quickly as possible, submerge the root sections vertically in the agar (*see* **Note 2**). Use tweezers to reorientate the root sections if necessary.

6. Keeping the petri dishes flat, transfer them to 4 °C to allow the agar to solidify more quickly.
7. Using a sharp scalpel, cut 1 cm^2 agar blocks around each suspended root section, ensuring that the root section is in the center of the block.
8. Using superglue, fix the agar block to the vibratome block (*see* **Note 3**), and proceed to cut cross sections. During cross sectioning, the razor blade vibrates from side to side as it moves toward the block and moves backward after each slice. Vibration amplitude and speed will need to be optimized for each plant species tissue. To generate Medicago cross sections that can easily be resolved using microscopy, cut the cross sections to 70–100 μm thick.
9. Place individual cross sections onto a microscope slide, and keep them moist using distilled water. Using a light microscope, confirm that the orientation of the root is correct (perpendicular to the cross-section plane). Discard cross sections that have root that is not correctly aligned.
10. Prepare cross sections for confocal microscopy by staining correct cross sections with (cell wall-staining dye) propidium iodide (100 μg/mL) and applying a cover slip (do not remove the dye).

Cross sections should be isolated from a number of different root sections, proportional to the seedling root length to ensure that the fluorophore expression is consistent along the root. Cross sections should also be isolated at a range of root development stages since promoter gene expression is not always stable over developmental time (*see* Fig. 3). In addition, silencing of fluorophore expression can occur (*see* Fig. 4); therefore, studies should be performed after generation propagation.

3.2 GUS Staining of Plant Tissue

1. Immerse *Medicago truncatula* roots in 100 mM sodium phosphate buffer (pH 6.8) containing Imagene Green substrate to a final concentration of 50 μM. Inhibit the reaction after 2 h using D-glucaric acid 1,4-lactone to a final concentration of 0.1 mM (in Imagene Green kit).
2. Carry out X-gluc activity staining of *Medicago truncatula* roots: submerge them in X-gluc solution and vacuum infiltrate the solution until maximum pressure then incubate the roots overnight at 37 °C, protected from light. Root cross sections of the *Medicago truncatula* roots stained with either X-gluc or Imagene Green are shown in Fig. 1.

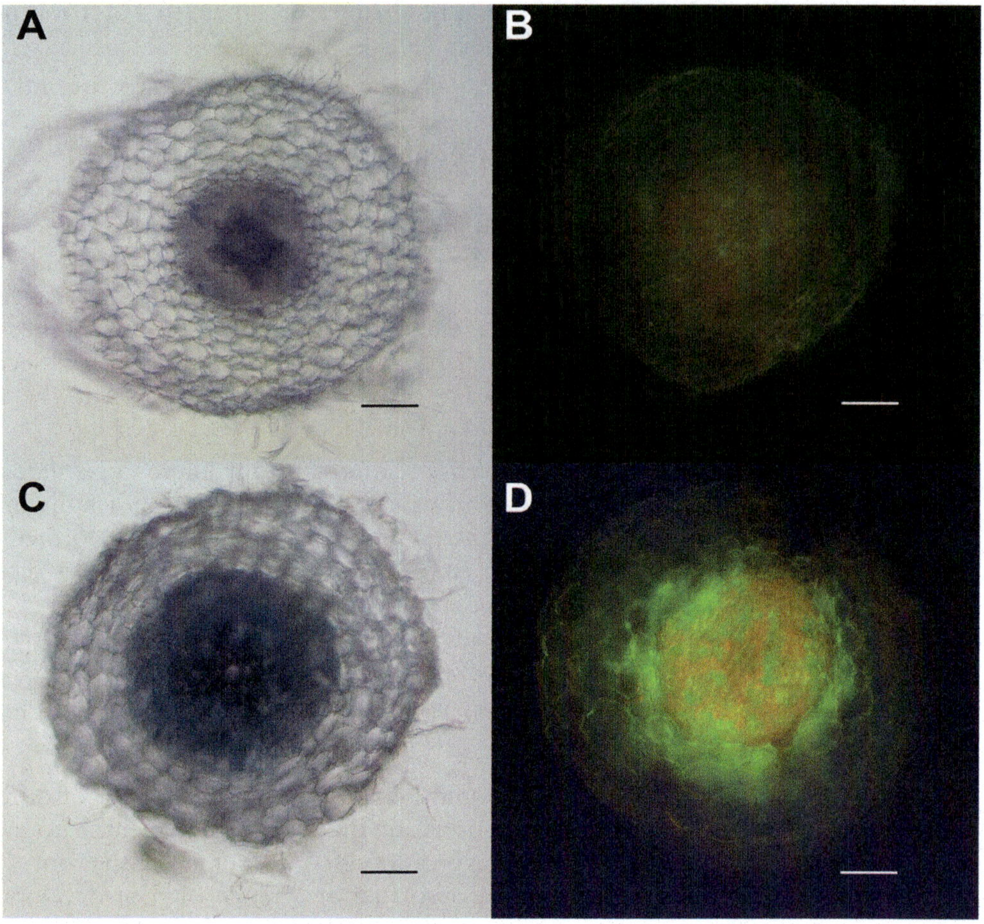

Fig. 1 Confocal and light microscopy *Medicago truncatula* root cross sections from A17 (**a**, **b**) and a GUS enhancer trap line (**c**, **d**). The enhancer trap line was kindly provided by Michael Schultze (University of York; unpublished). In (**a**) and (**c**), roots were incubated with X-gluc, leading to presence of a blue product where the x-glucuronidase (GUS) enzyme is expressed. In (**b**) and (**d**), roots were incubated with Imagene Green, leading to a green product in the same location. In A17 wild-type roots (**b**), basal fluorescence is visible due to the presence of endogenous GUS-like enzymes. Scale bars = 0.1 mm

3. Perform protoplast generation and FACS (fluorescence-activated cell sorting) as previously described in [1]. Representative FACS dot plots from analysis of an enhancer trap GUS line (in the A17 background), and the wild-type (non-transgenic) A17 root protoplasts are shown in Fig. 2.

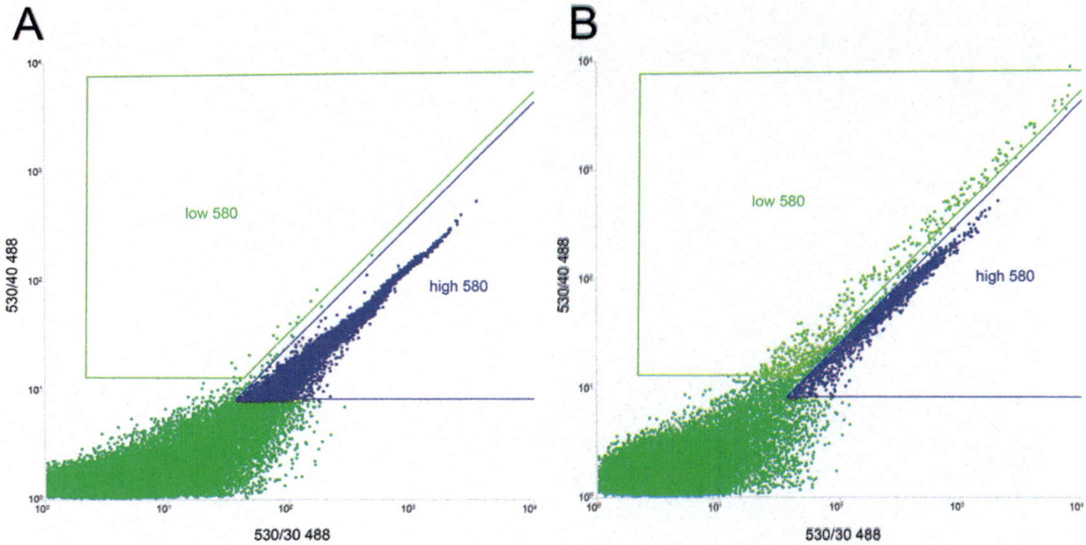

Fig. 2 Dot plots from *Medicago truncatula* root protoplast cell sorting. (**a**) Protoplasts from A17 roots incubated with Imagene Green. (**b**) Protoplasts from a GUS enhancer trap line incubated with Imagene Green. Roots were incubated with Imagene Green and the subjected to protoplast generation. To identify GFP-/Imagene Green-stained plant cells and gate them for sorting using FACS, it is necessary to plot the fluorescence on the 530 ± 20 nm filter (typical of GFP emission) versus the 580 ± 15 nm filter (associated with plant cell wall autofluorescence). Plant cell wall autofluorescence overlaps with the GFP-like emission spectra of the GUS-Imagene Green incubated cells. The GFP-like protoplast population is visible toward the 530 axis (low, bright green 580 nm gate) in comparison to the autofluorescence debris population that is visible toward to the 580 axis (high, blue 580 nm gate). In this case, selection of the low 580 nm gate in the graphs would enable the GUS-expressing (Imagene Green-stained) cells to be sorted

3.3 Confocal Microscopy of Medicago Root Cross Sections

1. To detect GFP or Imagene Green expression, use the 488 nm laser and visualize fluorescence emission in the 500–550 nm range.
2. To detect PI staining on a different channel, use the 488 nm laser and visualize fluorescence emission in the 600–700 nm range.
3. It is useful to also collect the transmitted light (PMT) on a third channel to be able to overlay channels afterward and confirm the spatial location of fluorophore expression.

   Medicago truncatula root cross sections from the cell type-specific line *pExpansin7::GFP* at 4 and 7 days after germination are shown in Fig. 3.

   Medicago truncatula root cross sections from the cell type specific line *pExpansin7::GFP* in the T2 and T3 generations are shown in Fig. 4.

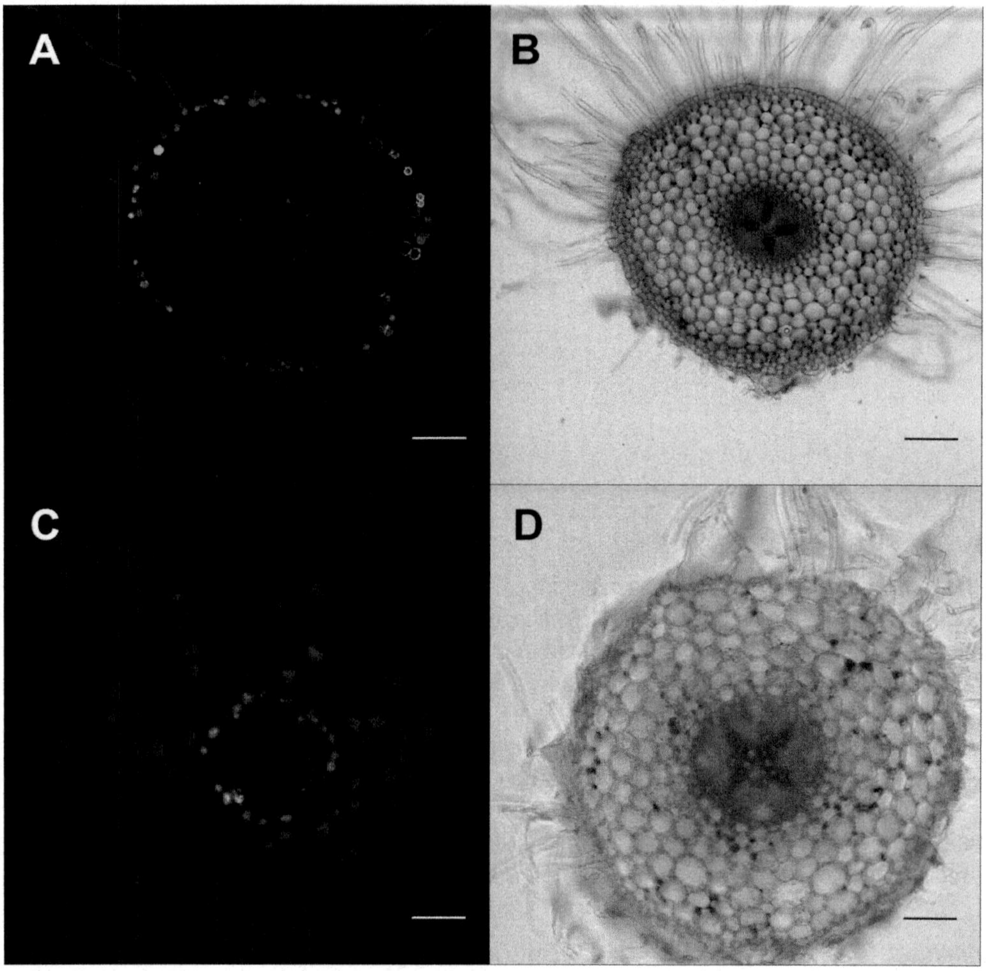

Fig. 3 Confocal microscopy of *Medicago truncatula* root cross sections from a *pExpansin7::*GFP line at 4 days (**a**, **b**) or 7 days (**c**, **d**) after germination. (**a**, **c**) Images of GFP expression. (**b**, **d**) Transmission light images of the same samples. At 4 days, GFP fluorescence is localized to the epidermis. However, at 7 days of growth, GFP fluorescence is visible in almost all cell types and much less visible in the epidermis. Promoter cell type stability during plant development is one of the most important issues to address when performing cell type-specific studies in plants. Scale bars = 0.1 mm

4. For a lambda scan analysis, carry out fluorescence scanning using 488 nm laser excitation, and visualize and record fluorescence emission from 490 nm to 600 nm in 3 nm steps. Define regions of interest (ROI) over the cross-section image designating background (no specimen), autofluorescence, PI, and GFP areas within the specimen.

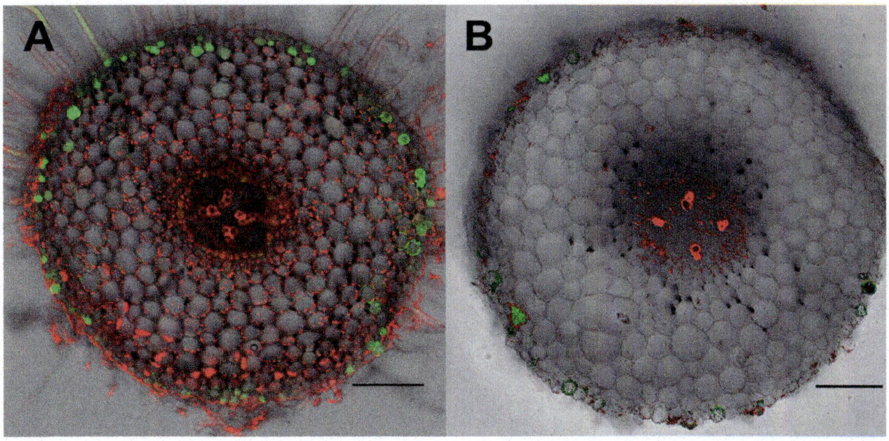

Fig. 4 Confocal microscopy of *Medicago truncatula* root cross sections from the *pExpansin7::*GFP line at 4 days of root development. (**a**, **b**) Overlay of transmission light, propidium iodide fluorescence (red) and GFP emission (green). (**a**) T2 generation. (**b**) T3 generation post-transformation. Silencing of GFP expression can be a problem when propagating GFP lines and multiple transformants should be generated to try to limit the impact of this on experiments. Scale bars = 0.1 mm

5. Plot mean fluorescence intensity vs wavelength for each ROI, and compare it to the GFP emission spectra. An example of a lambda scan analysis in different regions of interest (ROI) of a *Medicago truncatula* root cross section can be seen in Fig. 5.

4 Notes

Root tissue cross-section preparation

1. You can snap the double-edge blades in two and use them separately; if doing this, use protective gloves and take care. Using a fresh blade every time you generate cross sections is essential for success. However, as an alternative, they can be cleaned in ethanol then stored dry for reuse.
2. If you do not manage to orientate the root sections vertically, you can reshape the agar blocks to generate blocks in which the root section is perpendicular to the sectioning face. Alternatively, you can use a vibratome where the blade angle can be adjusted.
3. You can use a microtome for very thin or very thick roots, but a different mounting strategy to embed the tissue in agar must be used in each case.

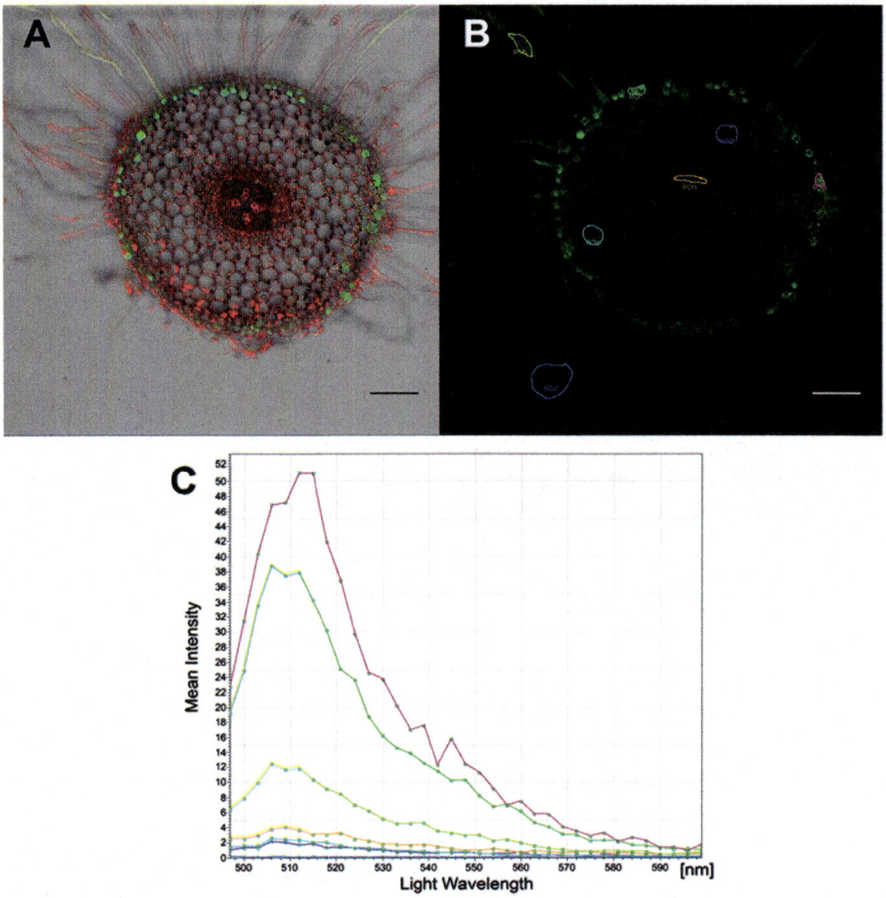

Fig. 5 Confocal microscopy of *Medicago truncatula* root cross section from a *pExpansin7::*GFP stable line at 4 days of root development. (**a**) Overlay of transmission light, propidium iodide fluorescence (red) and GFP emission (green). (**b**) Maximum projection of emitted GFP fluorescence after a lambda scan together with selection of regions of interest (ROI) for analysis. Pink and dark green ROIs are regions on epidermal cells; light green ROI is on a root hair; orange ROI is on pericycle cells; light blue and purple ROIs are on cortical cells, and dark blue ROI is the background non-sample signal. (**c**) Fluorescence emission spectrum of the ROI selected in (**b**). Scale bars = 0.1 mm

References

1. Carter AD, Bonyadi R, Gifford ML (2013) The use of fluorescence-activated cell sorting in studying plant development and environmental responses. Int J Dev Biol 57 (6–8):545–552
2. Birnbaum K, Shasha DE, Wang JY, Jung JW, Lambert GM, Galbraith DW, Benfey PN (2003) A gene expression map of the Arabidopsis root. Science 302(5652):1956–1960
3. Moussaieff A, Rogachev I, Brodsky L, Malitsky S, Toal TW, Belcher H, Yativ M, Brady SM, Benfey PN, Aharoni A (2013) High-resolution metabolic mapping of cell types in plant roots. Proc Natl Acad Sci U S A 110(13):E1232–E1241
4. Petricka JJ, Schauer MA, Megraw M, Breakfield NW, Thompson JW, Georgiev S, Soderblom EJ, Ohler U, Moseley MA, Grossniklaus U, Benfey PN (2012) The protein expression landscape of the Arabidopsis root. Proc Natl Acad Sci U S A 109 (18):6811–6818

5. Lin Y, Schiefelbein J (2001) Embryonic control of epidermal cell patterning in the root and hypocotyl of Arabidopsis. Development 128 (19):3697–3705
6. Heidstra R, Welch D, Scheres B (2004) Mosaic analyses using marked activation and deletion clones dissect Arabidopsis SCARECROW action in asymmetric cell division. Genes Dev 18(16):1964–1969
7. Gifford ML, Dean A, Gutierrez RA, Coruzzi GM, Birnbaum KD (2008) Cell-specific nitrogen responses mediate developmental plasticity. Proc Natl Acad Sci U S A 105(2):803–808
8. Rajeevkumar S, Anunanthini P, Sathishkumar R (2015) Epigenetic silencing in transgenic plants. Front Plant Sci 6:693
9. Vernie T, Kim J, Frances L, Ding Y, Sun J, Guan D, Niebel A, Gifford ML, de Carvalho-Niebel F, Oldroyd GE (2015) The NIN transcription factor coordinates diverse nodulation programs in different tissues of the Medicago truncatula root. Plant Cell 27 (12):3410–3424
10. Han X, Kim JY (2016) Integrating hormone- and micromolecule-mediated signaling with plasmodesmal communication. Mol Plant 9 (1):46–56
11. Goodwin R (1953) Fluorescent substances in plants. Annu Rev Plant Physiol 4:283–304

Check for updates

Chapter 13

Long-Term In Vivo Imaging of Luciferase-Based Reporter Gene Expression in Arabidopsis Roots

Wei Xuan, Davy Opdenacker, Steffen Vanneste, and Tom Beeckman

Abstract

Plants have an amazing capacity to adjust their growth to environmental limitations. This is particularly relevant for the root system that tunes its developmental pattern to mine the soil for water and nutrients while avoiding patches of soil that contain biotic and abiotic stress agents. Because most developmental processes are taking place gradually while roots are growing, it is often difficult to correlate gene expression events with specific developmental processes that are not necessarily coinciding in time and space. Therefore, in vivo detection and quantification of gene expression over a long period under gravitational conditions can be instrumental in dissecting complex processes in the root. Here, we describe a method for long-term imaging of luciferase dynamics in growing *Arabidopsis* roots that express a luciferase gene driven by the auxin-output reporter *DR5*, in the context of lateral root development.

Key words *DR5* oscillation, Pre-branch site, Lateral root, *Arabidopsis* root, Luciferase

1 Introduction

The root system is crucial for the uptake of water and nutrients from the soil while anchoring the plant. The complexity in root architecture is determined by a finely tuned regulation of lateral root (LR) spacing and by root growth that ensures optimal mining of the soil for heterogeneously distributed water and nutrients. This process involves a highly coordinated, flexible adaptation of the gene regulatory networks that underlie LR development in response to environmental opportunities or limitations. Due to the spatial separation of developmental stages along the primary root, it is often difficult to extract causal relationships between gene expression and the developmental progression. Therefore, imaging of the root growth pattern combined with the analysis of the spatiotemporal changes of gene expression patterns and/or levels has the potential to represent a major advance in our understanding of plant growth and development. However, live imaging of plant roots under natural conditions is more cumbersome due to the

Daniela Ristova and Elke Barbez (eds.), *Root Development: Methods and Protocols*, Methods in Molecular Biology, vol. 1761, https://doi.org/10.1007/978-1-4939-7747-5_13, © Springer Science+Business Media, LLC 2018

rapid elongation of roots and their propensity to grow along the gravitational vector.

Advanced technologies, such as RootChip, RootArray, RootScope, and GlO-Roots, have been developed for live imaging of plant root development and gene expression dynamics by quantifying fluorescence reporter genes that are fused to root specific genes or promoters [1–5]. Recently, bioluminescent (BL) imaging has become an indispensable tool for in vivo visualizing molecular events at cellular and tissue levels in plant research, such as gene expression analysis, protein-protein interaction, and protein-DNA binding [6–9]. There are several available BL reporter gene systems allowing in vivo imaging. For instance, firefly and click beetle luciferase (Luc) are ATP-dependent and require D-luciferin as a substrate, whereas Renilla and Gaussia Luc are ATP independent and use coelenterazine as a substrate. By oxidizing their substrates, these BL proteins generate light of different wavelengths (e.g., 578 nm for firefly luciferase, 537 nm for click beetle green luciferase, 615 nm for click beetle Red luciferase, and 465 nm for Renilla and Gaussia luciferases). By visualizing the expression of a firefly Luc reporter gene driven by the synthetic auxin signaling output promoter *DR5* (*DR5:Luc*), auxin signals were found to oscillate in a region proximal to the root meristem (oscillation zone—OZ), to install in a recurrent way discrete regions of high auxin signaling, known as pre-branch sites, in which cells acquire the competence to develop into a lateral root primordium [10–13].

The Luc reporter system is highly suited for dynamic expression analysis: Luc is a very sensitive reporter generating a strong signal output over a wide dynamic range; it is not susceptible to photobleaching and does not suffer from tissue-derived autofluorescence. Moreover, luciferase levels reflect very well the activity of the promoter that drives its transcription due to the absence of a protein maturation lag and a rapid turnover (half-life of ~3 h), making it a very good real-time reporter of transcriptional dynamics [14, 15]. This is particularly relevant for monitoring oscillatory activity of gene expression. Moreover, Luc imaging allows for simultaneous visualizing of gene expression in all developmental regions of growing roots over extended periods of time. In contrast, fluorescence-based imaging techniques have higher resolution but are seriously limited in the time frame over which the root can be imaged. Importantly, *DR5:Luc* expression can be visualized in different plant species even in thin layers of soil [4], thus closely approximating conditions that occur in the field. In conclusion, in vivo recording of Luc reporters in vertically growing roots represents a valuable tool to study the gene expression patterns that underlie root architecture.

Here, we describe a conventional imaging system which enables the real-time imaging of luciferase-based reporters over a 24-h period in *Arabidopsis* seedling roots. A detailed protocol, including the administration of substrates, the use of alternative luciferase proteins, steps in image processing, quantification of the luminescence, and general troubleshooting, is presented together with examples illustrating the power of this method. It is further explained how kymographs can be generated to assist the analysis of the spatiotemporal signal dynamics along the root.

2 Materials and Equipments

2.1 Materials

1. *Arabidopsis thaliana* transgenic lines *DR5:Luciferase* (*DR5:Luc*) [10].
2. 12 × 12 cm square petri dishes.
3. Tweezers.
4. Breathable tape.
5. Micropipettes.
6. 50 mL Falcon tubes.

2.2 Reagents

1. 10 mM D-luciferin stock solution in 0.01% Tween 80 in water and store at −20 °C. Before use, dilute it to 1 mM final concentration with 0.01% Tween 80.
2. *Half-strength Murashige and Skoog (½ MS) medium*:
   0.5× MS salts, 1% sucrose, 0.5 g/L 2-(N-morpholino) ethanesulfonic acid MES, pH 5.7, and 0.8% w/v agar. Autoclave before use.

2.3 Luciferase Imaging System

1. Bioluminescence recording device:
   A typical bioluminescence recording device carries a charge-coupled device (CCD) camera, a plate holder, light systems, and a XY mobile table, which are located inside a dark cabinet to avoid light interfering (Fig. 1a). The CCD can be adapted in the top or at the side for monitoring luminescence in specific plant tissues and whole plants. The plate holder is used to fix the plate vertically in the bioluminescence recording device, and XY mobile stage is used to precisely adjust the location of the plates to focus on the root tips (Fig. 1b).

   In this study, Lumazon carrying a CCD camera from Princeton Instruments Ltd. (Trenton, NJ, USA) or NightSHADE LB985 in vivo plant imaging system (BERTHOLD TECHNOLOGIES, https://www.berthold.com/en/bio/in_vivo_plant_imager_NightSHADE_LB985) carrying a deep-cooled slow scan CCD camera from Andor Instruments Ltd. (Belfast, UK) were used for luciferase imaging.

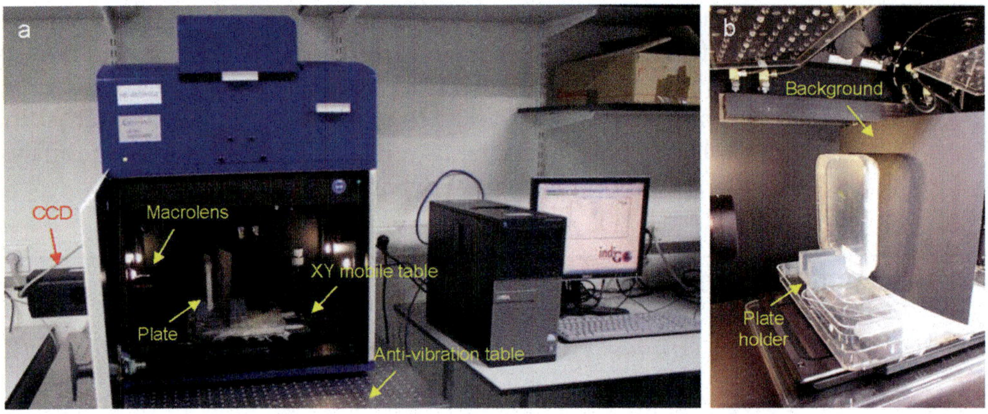

Fig. 1 Setup of the experiment. (**a**) Overview of the bioluminescence recording system (Nightshade LB985) for vertical long-term luciferase imaging. (**b**) Details of the plate setup inside the system. The bioluminescence recording system imaging system is placed on an anti-vibration Table. CCD camera is adapted at the sideway and connected with a macro lens. The agar plate is fixed opposite to the macro lens on the XY mobile table with a plate holder, and a black paper is put at the back of the plate as background

2. An appropriate macro lens (Nikon AF Micro-Nikkor 60 mm f/2.8D, aperture max f/2.8, min f/32) is used for long-term imaging of the luciferase signal in the root tip region (*see* **Note 1**), whereas the zoom lens from the respective imaging system is used for single imaging of pre-branch sites along the primary root in a bigger field of view. A zoom lens (Kenko PRO1D AC Close-up No.3 62 mm) can be adapted to the macro lens to increase the magnification.

2.4 Software for Image Analysis

1. Bioluminescence recording device control software (e.g., WinView/32 or indiGO™).
2. Fiji software packages (http://fiji.sc/).
3. Excel (Microsoft Office).

3 Methods

3.1 Plant Growth Condition

1. *Arabidopsis thaliana* seeds were surface sterilized with 70% (v/v) ethanol for 5 min and 20% (v/v) bleach for 12 min. The seeds were subsequently rinsed five times with sterile water and kept at 4 °C. After 3 days of imbibition, seeds were sown on petri dishes (12 × 12 cm) containing sterile ½ MS medium and grown under continuous light at the temperature of 21 °C (*see* **Note 2**).

3.2 Procedures for Long-Term Imaging of Luciferase Signal in the Root Tip

3.2.1 Preparation of the Plates for Long-Term Luciferase Imaging

1. To minimize variability, control the volume of ½ MS medium by adding each time 50 ml of growth medium per petri dishes (12 × 12 cm) (*see* **Note 3**). For analyzing the effects of hormones and or other compounds, add them to cooled (50 °C) ½ MS medium.
2. Leave the plates in the laminar flow bench until the medium has solidified.

3.2.2 Setup of the Imaging System

1. Place the CCD camera in the lateral position, to image the vertically growing root (Fig. 1a).
2. Turn on the machine and open its controlling software.
3. Install the macro lens on the CCD (A teleconverter is used to connect the macro lens and CCD). The image frame is around 2.6 × 2.6 cm at its maximum magnification.
4. Allow the CCD to cool to its optimal working temperature (e.g. −70 °C) to get a good signal-to-noise ratio. It will take a few minutes to reach the target temperature and become stable).

3.2.3 Spraying D-Luciferin Solution on the Plates

1. Take one Eppendorf tube of D-luciferin solution (1 mM) while keeping it shielded from light.
2. Mix D-luciferin solution by pipetting up and down, and then add 0.5 mL D-luciferin (1 mM) solution into the container of the airbrush (*see* **Note 4**).
3. Spray the D-luciferin solution evenly across the surface of agar (*see* **Note 5**).
4. Put the plates horizontally on the bench, in the dark with the lid open, until all liquid has been absorbed by the medium (| / – 15 min).

3.2.4 Transfer of Arabidopsis Seedlings to Petri Dishes and Focusing

1. Select ten 3-day-old *DR5:Luc* seedlings, and place them on a row about 2–3 cm from the top of the plate, and seal the plate (*see* **Note 6**).
2. Install the plate vertically in a plate holder in the recording unit, with the lid of the plate facing the lens.
3. Place a piece of black paper behind the plate to minimize background light (Fig. 1b).
4. Focus the camera on the root tips using visible light (adjust exposure time to 50–100 ms, depending on the light intensity).
5. Move the plate with XY table, and adjust the height of the plate holder to have the root tips in the center of the imaging field.
6. Focus on the root tips: First, move the plate toward the lens until it reaches the lens, and check whether the outer ring of the macro lens is in contact with the lid of the plate. Then,

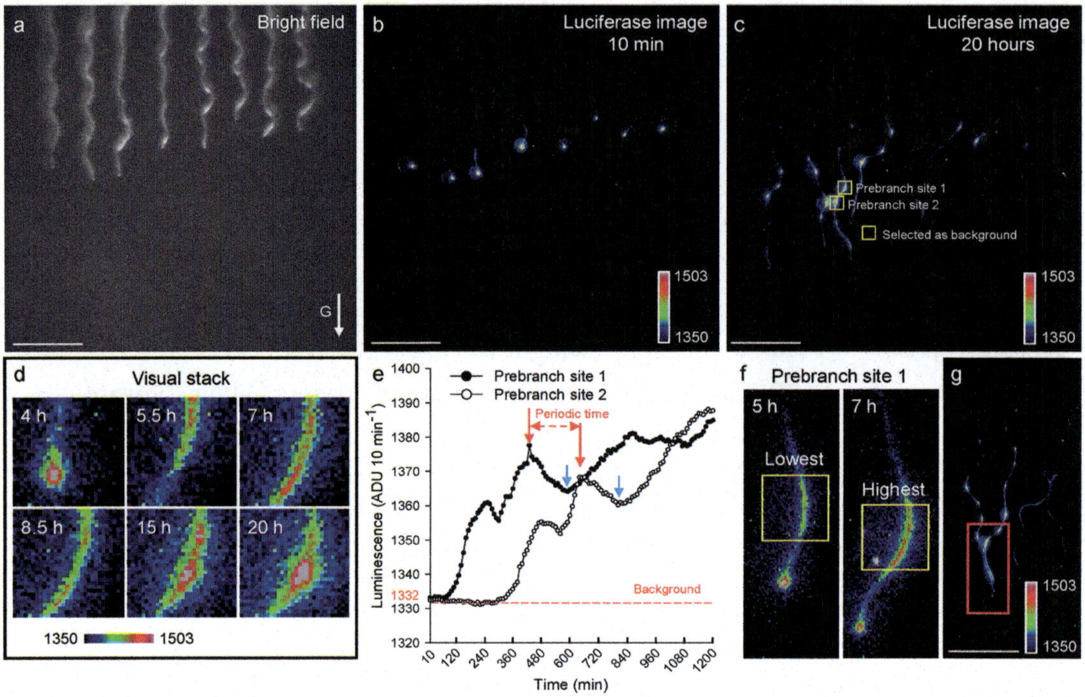

Fig. 2 Scheme of the quantification of *DR5:Luc* signals in the primary root over time. (**a**) A bright field image of the roots before starting long-term imaging. (**b**) First luminescence image obtained after starting the long-term imaging (10 min exposure). (**c**) Last luminescence image of the time series. Yellow boxed regions were selected for further luciferase signal quantification. (**d**) Visualization of *DR5:Luc* signal dynamics in yellow square blocked region in **c** (labeled as pre-branch site 1) with "Virtual Stack." (**e**) Quantification of *DR5:Luc* signal changes during the formation of two consecutive pre-branch sites as labeled in **c** by using "Virtual Stack." Red arrows indicate the highest *DR5:Luc* signal in the OZ, whereas blue arrows indicate the formation of pre-branch sites. (**f**) Images show lowest and highest signals of *DR5:Luc* in the OZ before the formation of pre-branch site 1. (**g**) An example of two twisted and interacting roots that are expressing *DR5:Luc* and are thus excluded from further analysis. Scale bars, 0.5 cm

move back the plate until the outline of the root can be visualized within the field, and make the fine adjustments of distance between plate and lens to focus on the OZ in root tip. Magnify the image field in the software to facilitate focusing, and zoom out for actual imaging (*see* **Note 7**).

3.2.5 Live Imaging

1. Open the aperture of the macro lens completely to allow the maximum detection of luciferase signal.
2. Take a bright field image on the focused root, and save the image in TIFF format (16-bit) (Fig. 2a).
3. Turn off the lights, and start live imaging. The sequential images are taken every 10 min with 10 min exposure time (*see* **Note 8**).
4. Stop the process or wait for the end of the process. Save the image series in TIFF format (16-bit) for further analysis.

3.3 Single Imaging of Luciferase Signal in the Whole Root

Transfer 3-day-old *DR5:Luc* seedlings to a plate containing 1/2 MS medium (with or without chemicals), and grow them for another 5 days before luciferase imaging (*see* **Note 9**).

3.3.1 Sample Preparation for Luciferase Imaging

3.3.2 Setup of the Imaging System

1. Place the CCD camera in a position for top-view imaging.
2. Replace the macro lens with a zoom lens to capture the whole plate.
3. Allow the CCD to cool to its optimal working temperature (e.g., −70 °C to get a good signal-to-noise ratio).

3.3.3 Spraying D-Luciferin Solution on the Plates

1. Open the plate with 8-day-old seedlings under low-light conditions.
2. Spray the D-luciferin solution (1 mM) gently on the seedlings. Avoid strong airflow from airbrush that may displace the seedlings from their original positions in the plates (*see* **Note 10**).
3. Wait for 10 min to allow a homogenous penetration of the D-luciferin in the roots.

3.3.4 Focusing and Single Imaging

1. Put the plate horizontally in a mobile black stage inside the recording unit with the plate open.
2. Turn on the lights in the luminescence device, and focus the camera on the plant root.
3. Move the plate with XY table, and adjust the height of the plate holder to locate the root tip at the center of the imaging field.
4. Zoom in on the root tip, and make the fine adjustments of distance between plate and lens to focus on the root tip.

3.3.5 Imaging

1. Open the aperture of the macro lens completely.
2. Take a bright field image on the focused root and save the image in TIFF format (16-bit)
3. Turn off the lights, and start the imaging with 10–20 min exposure time.
4. Save the image in TIFF format (16-bit).

3.4 Data Analysis

The luciferase signal can be quantified by measuring the analog-digital units per pixel using the Fiji software. In all the measurements, the average background signal from regions outside the root is measured and subtracted from the pre-branch site signal.

3.4.1 Quantification of the Periodic Time and Amplitude of DR5 Oscillations Required for Pre-Branch Site Formation

"Virtual Stack" tool in Fiji can be used for fast quantification of periodic time of *DR5:Luc* oscillation.

1. Open the 16-bit TIFF image series in Fiji.
2. Adjust the brightness and contrast of the images to enable the visualization of *DR5:Luc* signal along the primary root (> Image > Adjust > Brightness/Contrast), and apply the Look-Up Table (LUT) of choice (e.g., Rainbow RGB) (> Image > Lookup Tables > LUT) to represent the signal intensity along the root (Fig. 2b).
3. Preview the image series, and identify the pre-branch sites that were formed during the primary root elongation (Fig. 2c).
4. Use the "rectangle tool" to select a square region where a pre-branch site is formed, and this region should cover *DR5* oscillation that occurred prior to pre-branch site formation (Fig. 2c) (*see* **Note 11**).
5. Measure the signal intensity changes in this region over all images of the movie. (> Process > Batch > Virtual Stack). The value of signal intensity is copied into an excel sheet (Fig. 2d and e).
6. Select a region of the same size outside the root by moving the same rectangle to a region out of root, and measure the luciferase signal (> Process > Batch > Virtual Stack) to determine the background signal (Fig. 2c and e). Copy the signal values into an excel sheet and subtract these values from the corresponding oscillation signals.
7. Determine the period of the *DR5* oscillations based on the number of frames that space a *DR5:Luc* maximum in the OZ of each seedling root, and multiply with the time per cycle. Use more than 70 time points from at least 15 individual seedlings to make a histogram.
8. The difference between the highest value and lowest value of *DR5:Luc* in the OZ defines the amplitude of *DR5* oscillations (Fig. 2f). The time frames when root display the highest/ maximum or lowest/minimal *DR5:Luc* signal in the OZ can be traced by using "Virtual Stack."

3.4.2 Quantification of DR5:Luc Signal in the OZ Over Time

DR5:Luc oscillates periodically in the OZ, which changes its location in the image frame over time as the primary root elongates. However, "Virtual Stack" is only able to access the signal changes in a fixed region in the image frame, but not capable for quantifying the signal changes in a specific region in an elongating root. Below, we describe the method for precisely quantifying the spatiotemporal changes of *DR5:Luc* signal in OZ over time.

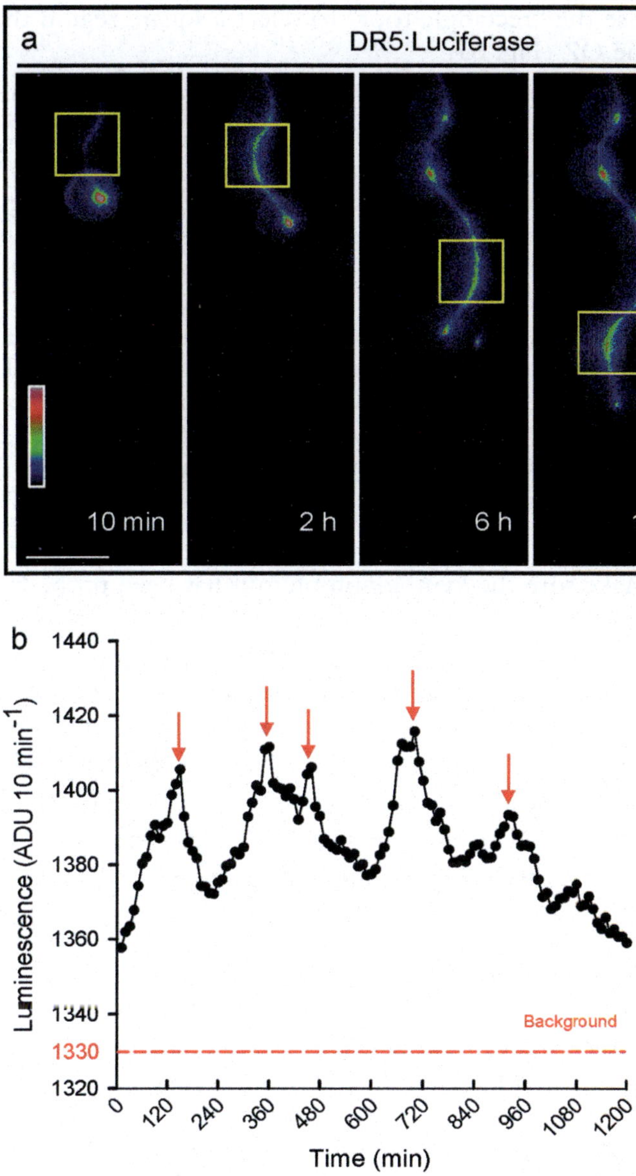

Fig. 3 Quantification of *DR5:Luc* signal in OZ over time. (**a**) Image series of *DR5:Luc* signal in a root tip over 18 h, and OZ is covered by a yellow boxed. (**b**) Quantification of *DR5:Luc* luminescence over time measured in the yellow boxed region shown in (**a**) (reproduced from ref. 7). Scale bar, 0.2 cm

1. Open TIFF image series in Fiji.
2. Adjust the brightness/contrast of the images to enable the visualization of *DR5:Luc* signal along the primary root (> Image > Adjust > Brightness/Contrast), and apply the LUT of choice to represent the signal intensity along the root (> Image > Lookup Tables > appropriate color) (Fig. 3a).

3. Use the "rectangle tool" to select a square region that covers the OZ (Fig. 3a).
4. Measure signal intensity changes over time in this region (> Analyze > Measure), and copy the value in Excel (Fig. 3b).
5. Determine the background by measuring the signals from the region from the same size outside the root by using "Virtual Stack." The signal values are copied into Excel and subtracted from the other signals.

3.4.3 Kymograph

To graphically represent the spatiotemporal *DR5:Luc* signal changes during primary root elongation, a kymograph (known as wave writer, http://www.embl.de/eamnet/html/body_kymograph.html) was generated with Fiji. For this purpose, a real-time movie of more than 20 h was analyzed in Fiji, and the *DR5:Luc* signal from a newly grown root region was presented in a kymograph.

1. Make sure the kymograph plug-in has been properly installed in Fiji.
2. Load the time-lapse movie (TIFF series) into Fiji, and a "Z-projection" of the average intensity over 20 h to have an overview of luciferase signal changes following primary root growth over time (> Image > Stacks > Z Project) (Fig. 4a).
3. Draw a segmented line across a newly formed primary root region of interest in the projection (Fig. 4a).
4. Restore this line in the original TIFF series to generate "MultipleKymograph" with an appropriate line width. Evaluate different line widths to have an optimal luciferase signal in the kymograph. In our setup, wide lines that exceed the root width (>9 px) substantially reduced the average luciferase signal intensity in kymograph (Fig. 4b).
5. Apply the LUT of choice (e.g., blue orange icb) in the kymograph image as no RGB colors are assigned to kymograph images as it is generated from a movie.

3.4.4 Pre-Branch Site Measurement

To monitor the pre-branch site numbers, use 8-day-old seedlings harboring *DR5:Luc* reporter for pre-branch site quantification.

1. Open a single TIFF image (including luciferase image and respective bright field image obtained in Subheading 3.3.5) by Fiji (Fig. 5a and b).
2. Adjust the brightness/contrast of the images, and apply LUT to represent the signal intensity along the root (Fig. 5c).
3. Static luminescence signals that were visible along the primary root outside the OZ were counted as pre-branch sites (Fig. 5c).
4. At least eight seedlings were quantified per each genotype.

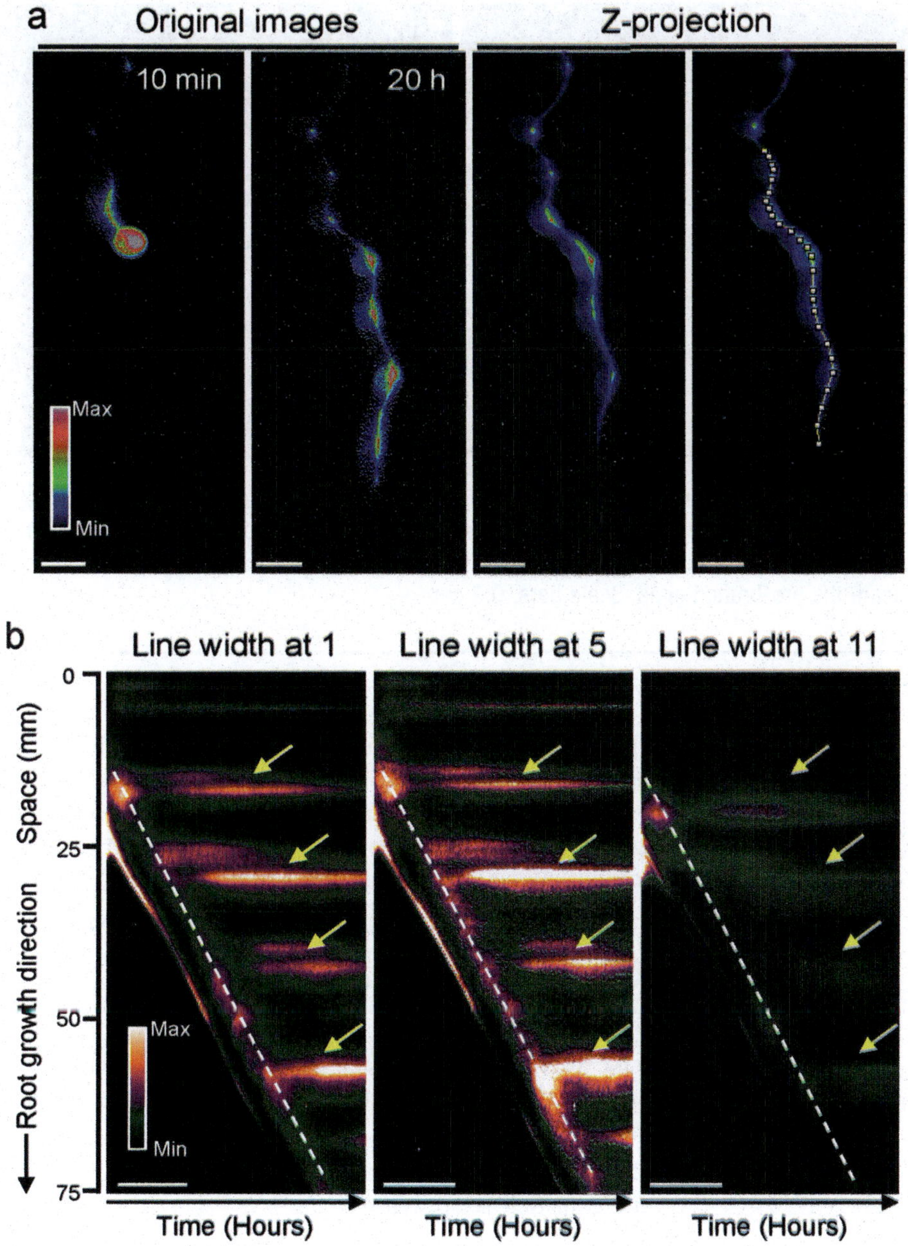

Fig. 4 Kymograph. (**a**) Original images and Z-projection of *DR5:Luc* signal in a root tip over 20 h. Yellow dash line indicates a root region of interest that is used to generate kymograph. (**b**) Kymographs with different line widths. White dash lines indicate the *DR5:Luc* signal in OZ, and yellow arrows indicate the pre-branch sites. Scale bars, 0.1 cm

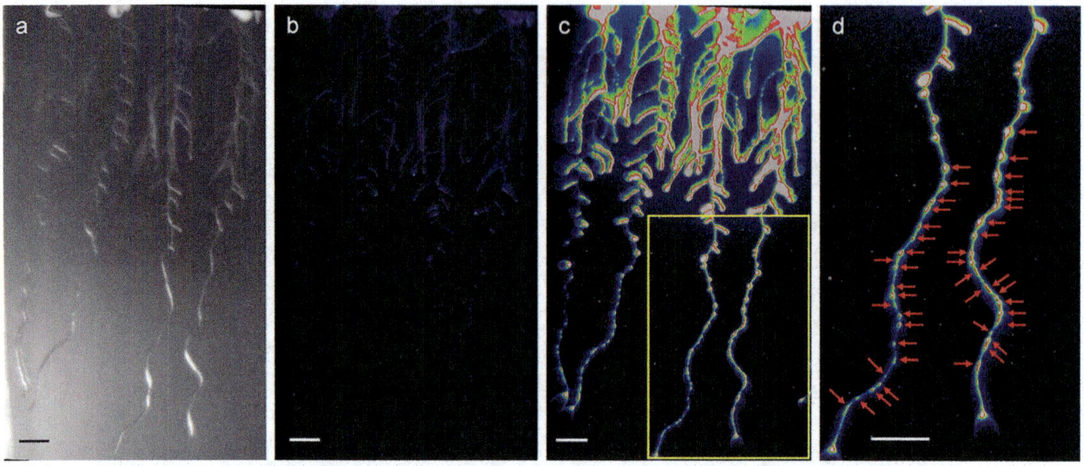

Fig. 5 Quantification of pre-branch site number in 8-day-old *Arabidopsis* seedlings. (**a**) Bright field image of 8-day-old seedlings. (**b**) Luciferase image of the seedlings after 20 min exposure. (**c**) Adjusted the bright/contrast of the image to visualize the pre-branch sites in root. (**d**) Zoom of yellow boxed region in (**c**). Red arrows indicate the pre-branch sites. Scale bars, 0.5 cm

4 Notes

1. A macro lens is important for imaging root tip luciferase signal. In our test, a macro lens from Navitar (50 mm F0.95 TV Lens (put on D0.95, 0.6 m distance)) is not able to obtain luciferase signals, whereas a macro lens from Nikon as mentioned above works for luciferase imaging.
2. Select 3-day-old seedlings for long-term imaging to study expression dynamics; image 8-day-old seedlings at a single time point for rapid quantification of the number of pre-branch sites and their density.
3. Make sure that the surface of the agar is flat (no air bubbles, horizontal) after it is set to facilitate to focus on the root tip for imaging.
4. Any precipitation after defrosting needs to be resuspended by pipetting before spraying.
5. It is important to spray the D-luciferin solution smoothly on the agar at a low speed, which can be accessed by adjusting the sprinkler of the airbrush and the air pressure of the compressor. Make sure that the D-luciferin is equally distributed over the plate. High-speed spraying can break the surface of the agar and affect the distribution of D-luciferin on the agar plate.
6. Ensure that the root tips are aligned in a row on the plate.
7. To keep the root tips in focus during a long-term *DR5:Luc* movie, it is important to keep the plate lid parallel to the macro

lens. Otherwise the root tip might grow out of focus after a few hours.

8. Check whether the luciferase signal in the OZ in root tip is visible in the first image after 10 min, and adjust the settings if needed (Fig. 2b). The exposure time can be changed according to the luciferase signal intensity obtained from the different reporter genes. Make sure the exposure time does not exceed the time period of each cycle.
9. Up to ten seedlings can be put in one 12 × 12 cm plate, to avoid contact with the borders of the plate.
10. Cut off long LRs that overlap with or are in contact with primary roots because the luciferase signal from these LRs might affect the accuracy of the quantification of pre-branch sites. Separate the primary roots that overlap or that are twisted by slightly moving the root tip.
11. Do not use twisted roots for the analysis (Fig. 2g).

Acknowledgments

This work was supported by Research Foundation—Flanders, Bilateral Research Cooperation (project number: G002817 N) with MOST (China) (2016YFE0109900), and China National Natural Science Foundation (No. 31672223).

References

1. Grossmann G, Guo WJ, Ehrhardt DW, Frommer WB, Sit RV, Quake SR, Meier M (2011) The RootChip: an integrated microfluidic chip for plant science. Plant Cell 23:4234–4240
2. Busch W, Moore BT, Martsberger B, Mace DL, Twigg RW, Jung J et al (2012) A microfluidic device and computational platform for high-throughput live imaging of gene expression. Nat Methods 9:1101–1106
3. Kast EJ, Nguyen MD, Lawrence RE, Rabeler C, Kaplinsky NJ (2013) The RootScope: a simple high-throughput screening system for quantitating gene expression dynamics in plant roots. BMC Plant Biol 13:158
4. Rellan-Alvarez R, Lobet G, Lindner H, Pradier PL, Sebastian J, Yee MC et al (2015) GLO-Roots: an imaging platform enabling multidimensional characterization of soil-grown root systems. Elife 4. https://doi.org/10.7554/eLife.07597
5. Zhu X, Feng Y, Liang G, Liu N, Zhu JK (2013) Aequorin-based luminescence imaging reveals stimulus- and tissue-specific Ca^{2+} dynamics in Arabidopsis plants. Mol Plant 6:444–455
6. Paulmurugan R, Umezawa Y, Gambhir SS (2002) Noninvasive imaging of protein-protein interactions in living subjects by using reporter protein complementation and reconstitution strategies. Proc Natl Acad Sci U S A 99:15608–15613
7. Endo M, Shimizu H, Nohales MA, Araki T, Kay SA (2014) Tissue-specific clocks in Arabidopsis show asymmetric coupling. Nature 515:419–422
8. Fujikawa Y, Kato N (2007) Split luciferase complementation assay to study protein-protein interactions in Arabidopsis protoplasts. Plant J 52:185–195
9. Fujimoto SY, Ohta M, Usui A, Shinshi H, Ohme-Takagi M (2000) Arabidopsis ethylene-responsive element binding factors act as transcriptional activators or repressors of GCC box-mediated gene expression. Plant Cell 12:393–404

10. Moreno-Risueno MA, Van Norman JM, Moreno A, Zhang J, Ahnert SE, Benfey PN (2010) Oscillating gene expression determines competence for periodic Arabidopsis root branching. Science 329:1306–1311
11. Van Norman JM, Xuan W, Beeckman T, Benfey PN (2013) To branch or not to branch: the role of pre-patterning in lateral root formation. Development 140:4301–4310
12. Xuan W, Band LR, Kumpf RP, Van Damme D, Parizot B, De Rop G et al (2016) Cyclic programmed cell death stimulates hormone signaling and root development in Arabidopsis. Science 351:384–387
13. Xuan W, Audenaert D, Parizot B, Moller BK, Njo MF, De Rybel B et al (2015) Root cap-derived Auxin pre-patterns the longitudinal Axis of the Arabidopsis root. Curr Biol 25:1381–1388
14. Thompson JF, Hayes LS, Lloyd DB (1991) Modulation of firefly luciferase stability and impact on studies of gene-regulation. Gene 103:171–177
15. Nguyen VT, Morange M, Bensaude O (1989) Protein denaturation during heat-shock and related stress - Escherichia-Coli Beta-Galactosidase and Photinus-Pyralis luciferase inactivation in mouse cells. J Biol Chem 264:10487–10492

Chapter 14

Cortical Cell Length Analysis During Gravitropic Root Growth

Maria Schöller, Jürgen Kleine-Vehn, and Elena Feraru

Abstract

The typical parameter used to evaluate the root growth response to gravity is the degree of root bending in time. This employs the quantification of the root tip angle toward gravity and, hence, does not directly assess the actual differential growth process. Here, we describe the cortical cell length as a parameter to quantify cell elongation during the gravitropic response, using median longitudinal confocal sections. This analysis depicts that root organ bending is a consequence of differential cellular elongation on the upper versus lower side of the gravistimulated root. Moreover, we introduce here a simple mounting setup that is suitable to gravistimulate and subsequently image seedlings on upright microscopes.

Key words Cortical cells, Gravitropic response, Gravistimulation, Cell length quantification, Mounting, Agar slides, Microscopy, CLSM, *Arabidopsis thaliana*

1 Introduction

Root gravitropism is a crucial plant response, allowing plants to anchor themselves into the soil and to explore the availability of water and nutrients. Gravitropism encompasses three crucial events (reviewed in [1]): (1) gravity perception in the columella cells of the root cap [2, 3]; (2) asymmetric distribution, gradient formation, and shootward transport of auxin through the lower side of the gravistimulated root; and (3) differential growth response in the elongation zone [4–11].

Following gravity perception, the direction of auxin transport in columella cells aligns with the new gravity vector and redistributes auxin preferentially to the lower epidermal cell files of the root. The resulting asymmetric auxin distribution across the root tip causes differential cell elongation along the root organ and, eventually, root bending toward gravity [9–11]. Accordingly, the auxin-dependent limitation of cellular elongation in the lower epidermal cell files is crucial for the gravity response [9]. Moreover, the epidermis is a determining tissue for overall root organ growth rates

Daniela Ristova and Elke Barbez (eds.), *Root Development: Methods and Protocols*, Methods in Molecular Biology, vol. 1761,
https://doi.org/10.1007/978-1-4939-7747-5_14,

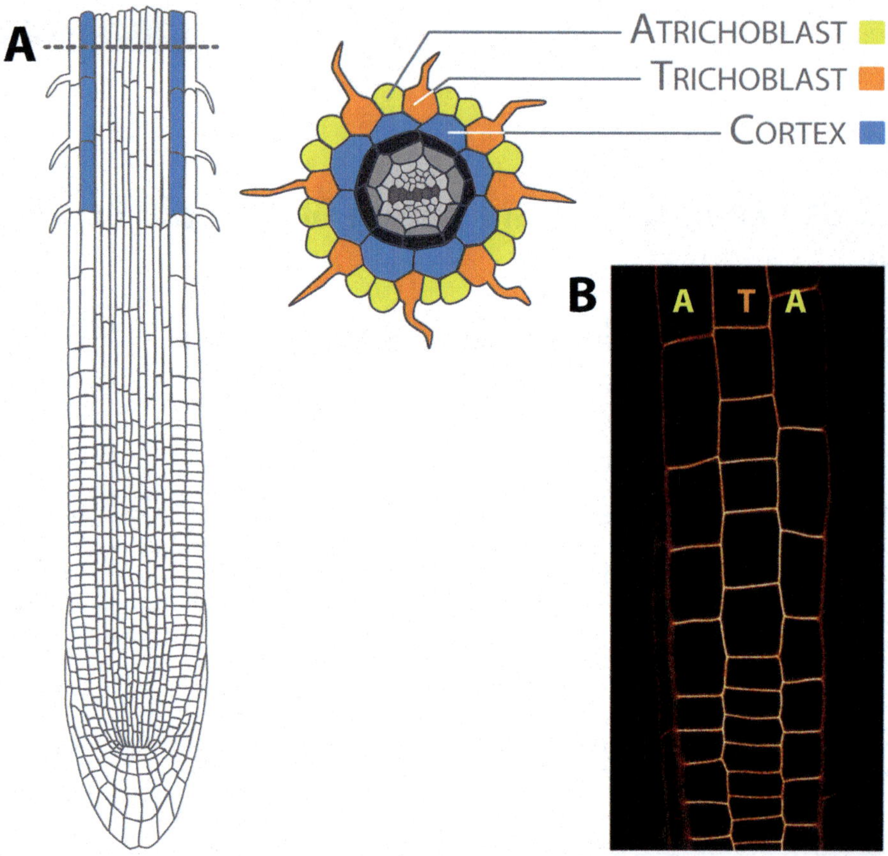

Fig. 1 Root architecture and epidermis patterning. (**a**) Schematic representation of the different tissues in a median longitudinal section of the main root; adapted from [16]. Note the alternation of trichoblast and atrichoblast cells in the cross section of the root. (**b**) Root epidermal section in the early elongation zone showing the plasma membrane localization of NPSN12-YFP. Note the alternation of shorter trichoblast ("T") and longer atrichoblast ("A") files along the longitudinal axis of the root

[12]. It is, thus, tempting to choose the root epidermal cell files for quantifying the cellular growth responses. However, the root epidermis is complex in *Arabidopsis*, with root hair (trichoblast) and non-root hair (atrichoblast) cell files alternating along the longitudinal axis of the root ([13], Fig. 1). Notably, the trichoblast cells are substantially shorter than the atrichoblast cells (Fig. 1b). The cell-type depiction is unfortunately not always obvious in median longitudinal sections, causing heterogeneity in cell size measurements of the epidermis (represented either by both shorter trichoblast and longer atrichoblasts cell files or only by one of the two cell-type files) (Fig. 1). Thus, straightforward cell size quantification in median longitudinal microscopic sections is impeded by the differences in size between epidermal cell types.

Here, we propose the cortical cell length as a suitable parameter for quantifying root bending during the gravitropic response, because this cell population is the most exterior homogenous cell type. We show that cortical cell length is a robust parameter for the quantification of gravitropic growth. In the region of the most prominent bend, we detected shorter cortical cells on the lower side compared to the upper side of the gravistimulated root. This is in full agreement with the assumption that the bending of the root toward gravity is a consequence of differential cell elongation. Additionally, we describe a mounting procedure, called "agar slides", which is suitable for gravistimulation of the seedlings prior investigation using upright microscopes. For inverted microscopes, chamber slides are most suitable [14]. The method requires confocal-based imaging of roots and provides information on the asymmetric cellular elongation after gravistimulation.

2 Materials

2.1 Plant Material and Growth Conditions

1. 4-day-old seedlings of plasma membrane marker *NPSN12-YFP* ([15]; *see* **Note 1**), grown in a plant cabinet under long-day photoperiod.
2. Sterile square Petri dishes (12 × 12 × 1.7 cm) with vents, containing 50 mL plant medium, made of 0.5 g/L MES, 2.3 g/L Murashige and Skoog salt, 10 g/L sucrose, and 10 g/L plant agar (pH 5.9).
3. 70% ethanol for seed sterilization.
4. Breathable tape for sealing the plates.

2.2 Specimen Preparation

1. Petri dishes (12 × 12 × 1.7 cm) containing a thin layer of agar-solidified plant medium (*see* **Note 2**).
2. Empty, square Petri dishes (12 × 12 × 1.7 cm).
3. Scalpel blade for slicing the solid plant medium.
4. Soft touch tweezers for transferring the seedlings.
5. 20 mL spray bottle containing sterile water.
6. Microscope slides.
7. Coverslips (24 × 50 mm #1.5).
8. Breathable tape for sealing.
9. Aluminum foil for gravistimulation in dark conditions.

2.3 Imaging and Analysis

1. Upright confocal laser scanning microscope (CLSM) equipped with 20× glycerol immersion objective.
2. ImageJ 1.41 software (http://rsb.info.nih.gov/ij/) for cell length measurements.
3. Microsoft Excel or similar software.
4. Statistics software.

3 Methods

3.1 Growth

1. Freshly sterilize the seeds using 70% ethanol for 2 min.
2. Allow the seeds to dry well in a laminar flow hood.
3. Sow the seeds on a plant medium-containing Petri dish.
4. Seal the plate with breathable tape.
5. Stratify the plates containing seeds for a minimum of 2 days at 4 °C in the dark.
6. Grow the seedlings for 4 days in a growth chamber or cabinet at 21 °C long-day photoperiod.

3.2 Mounting "Agar Slides"

1. Cut an approximately coverslip-sized block of sterile agar out of a Petri dish with agar-solidified plant medium (Fig. 2a).
2. Place the block of agar on the surface of a microscope slide (Fig. 2a).
3. Carefully transfer five to eight seedlings on the surface of the agar block. The roots should be well in contact with the agar

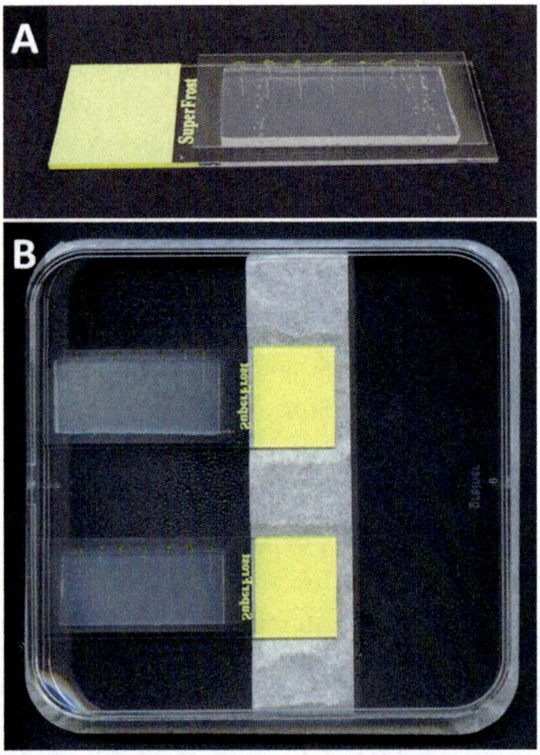

Fig. 2 Preparation and manipulation of "agar slides". (**a**) Seedlings mounted on top of a thin block of agar-solidified plant medium laid between a microscope slide and a coverslip. (**b**) "Agar slides" fixed in an empty Petri dish, containing vertically growing seedlings. The seedlings can be gravistimulated by turning the Petri dish 90°

medium, while the cotyledons should protrude from the agar block (Fig. 2a).

4. Cover the seedlings with a coverslip. Place the coverslip slowly to prevent air bubbles. We called this microscope arrangement "agar slides" (Fig. 2a; *see* **Note 3**).
5. Fix two "agar slides" in an empty square Petri dish using breathable tape (Fig. 2b).
6. Prior closing the Petri dish, spray the lid with sterile water (*see* **Note 4**).
7. Close the Petri dish containing the "agar slides" with breathable tape (Fig. 2b).
8. Increase the number of inspected seedlings by preparing a few of these Petri dishes.

3.3 Gravistimulation

1. Wrap the Petri dishes containing the "agar slides" with aluminum foil.
2. Transfer the Petri dishes back to the previous growth conditions and allow the seedlings to grow in the dark for 6 h (*see* **Note 5**).
3. Gravistimulate the seedlings by rotating the Petri dishes 90°. Image the seedlings after 48 h following gravistimulation (*see* **Note 6**).

3.4 Confocal Imaging and Quantification

1. Acquire images of the median longitudinal section of the region of most prominent bending, focusing on the cortical cells (Fig. 3a; *see* **Note 7**). We used the 20×, glycerol immersion objective.
2. Image at least three cortical cells at the inner and the outer sides of the bend (Fig. 3a, *see* **Note 8**).
3. Measure the length of the three cortical cells per inner and outer files at the region of most prominent bending (*see* **Note 9**).
4. Average and compare the inner and outer values (Fig. 3b; *see* **Note 10**).

4 Notes

1. Any other plasma membrane marker showing a good fluorescence signal in the cortical cells can be used for cell length investigation. We also used propidium iodide (40 μg/mL in H_2O) with very similar results.
2. Pour the Petri dishes with about 35 mL freshly cooked medium (cooled down to about 60–70 °C). Avoid moving the plates during agar solidification as this would lead to a rippled surface, which will interfere later with imaging of the specimen. It is

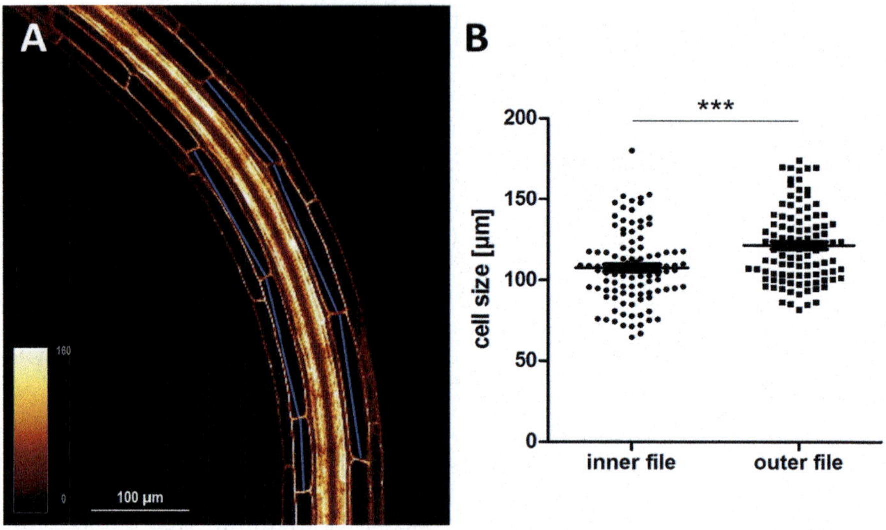

Fig. 3 Quantification of cortical cell length. (**a**) Median longitudinal confocal section showing the bending region of gravistimulated *UBQ10::NPSN12-YFP* root. The blue lines indicate the cells used for quantification. (**b**) Quantification of the cortical cell size at the inner and outer side of the bending region ($n = 34$). *** $p < 0.0001$; unpaired, two-tailed t-test

advisable to pour the plates freshly before use, since a layer of condensed water could make the medium slippery, causing drifts during gravistimulation.

3. Seedlings can be mounted in chambered cover glasses when an inverted microscope is available. We have previously described its usage [14].
4. Spraying water on the lid of the Petri dish prevents the specimen from drying out. This does not disturb the position of the "agar slides".
5. Gravistimulating in the dark prevents a negative phototropic response, which would influence the growth direction of the roots.
6. The 48 h gravistimulation ensures that the cells in the bending region have already reached their maximal length and are fully differentiated. Hence, this method assesses the full differential growth response.
7. When propidium iodide is used (40 µg/mL in H_2O), add the dye to the specimen 30 min prior to imaging. Carefully lift the coverslip and pipet the dye between seedlings and coverslip. Ensure that the dye-containing solution is homogenously administered.
8. If necessary, multiple images in various focal planes (Z-stack imaging option) can be acquired.
9. First identify the region of the root with the most prominent bending and then choose the three most central cells to be measured using ImageJ.

10. For the quantification of differences between the gravitropic response of seedlings with a different genotype or treatment, make a ratio of the average outer and inner cortical cell length per root and compare between treatments.

Acknowledgments

We are grateful to Elisabeth Sarkel for valuable comments on this manuscript, Niko Geldner for the published *NPSN12-YFP* fluorescent marker line, and the BOKU-VIBT Imaging Center for access. This work was supported by the Vienna Science and Technology Fund (WWTF) (Vienna Research Group project to J.K.-V.), Austrian Science Fund (FWF) (Projects: P29754 to J.K.-V. and T-728-B16 to E.F.), and the European Research Council (ERC) (Starting Grant 639478-AuxinER to J.K-V.).

References

1. Sato EM, Hijazi H, Bennet MJ et al (2015) New insights into root gravitropic signalling. J Exp Bot 66(8):2155–2165
2. Blancaflor EB, Fasano JM, Gilroy S (1998) Mapping the functional roles of cap cells in the response of Arabidopsis primary roots to gravity. Plant Physiol 116(1):213–222
3. Leitz G, Kang BH, Schoenwaelder ME et al (2009) Statolith sedimentation kinetics and force transduction to the cortical endoplasmic reticulum in gravity-sensing Arabidopsis columella cells. Plant Cell 21(3):843–860
4. Bennett MJ, Merchant A, Green HG et al (1996) Arabidopsis AUX1 gene: a permease like regulator of root gravitropism. Science 273 (5277):948–950
5. Luschnig C, Gaxiola RA, Grisafi P et al (1998) EIR1, a root-specific protein involved in auxin transport, is required for gravitropism in Arabidopsis thaliana. Genes Dev 12 (14):2175–2187
6. Galweiler L, Guan C, Muller A et al (1998) Regulation of polar auxin transport by AtPIN1 in Arabidopsis vascular tissue. Science 282(5397):2226–2230
7. Mullen JL, Ishikawa H, Evans ML (1998) Analysis of changes in relative elemental growth rate patterns in the elongation zone of Arabidopsis roots upon gravistimulation. Planta 206(4):598–603
8. Rashotte AM, Brady SR, Reed RC et al (2000) Basipetal auxin transport is required for gravitropism in roots of Arabidopsis. Plant Physiol 122(2):481–490
9. Swarup R, Kramer EM, Perry P et al (2005) Root gravitropism requires lateral root cap and epidermal cells for transport and response to a mobile auxin signal. Nat Cell Biol 7 (11):1057–1065
10. Band LR, Wells DM, Larrieu A et al (2012) Root gravitropism is regulated by a transient lateral auxin gradient controlled by a tipping-point mechanism. Proc Natl Acad Sci U S A 109(12):4668–4673
11. Brunoud G, Wells DM, Oliva M et al (2012) A novel sensor to map auxin response and distribution at high spatio-temporal resolution. Nature 482(7383):103–106
12. Savaldi-Goldstein S, Peto C, Chory J (2007) The epidermis both drives and restricts plant shoot growth. Nature 446(7132):199–202
13. Dolan L (1996) Pattern in the root epidermis: an interplay of diffusible signals and cellular geometry. Ann Bot 77(6):547–553
14. Feraru MI, Kleine-Vehn J, Feraru E (2015) Auxin carrier and signaling dynamics during gravitropic root growth. Methods Mol Biol 1309:71–80
15. Geldner N, Denervaud-Tendon V, Hyman DL et al (2009) Rapid, combinatorial analysis of membrane compartments in intact plants with a multicolor marker set. Plant J 59(1):169–178
16. Bouché F (2017) Arabidopsis—root cell types. https://doi.org/10.6084/m9.figshare.4688752.v1. Accessed 28 Apr 2017

Chapter 15

Growth Rate Normalization Method to Assess Gravitropic Root Growth

Maria Schöller, Elizabeth Sarkel, Jürgen Kleine-Vehn, and Elena Feraru

Abstract

Time-lapse imaging of roots is highly suitable for depicting gravitropic growth behaviors. However, roots may show faster or slower bending kinetics when compared to control as a result of differences in overall root growth. Accordingly, conditions that cause differential organ growth require growth rate normalization to compare gravitropic curvature. Here, we describe a simple normalization method for gravitropic root growth evaluation. We exemplify this method by exposing seedlings to distinct environmental conditions or disturbing the cellular auxin contents. This data shows that the method is suitable to discriminate between gravitropic and overall organ growth defects.

Key words *Arabidopsis thaliana*, Root, Gravitropic response, Kinetics, Growth rate, Quantification, Normalization

1 Introduction

Plant response to gravity has fascinated researchers for more than 200 years. Pioneers such as Knight [1] or Charles and Francis Darwin [2] have accurately reported how plants alter their growth in response to gravity. Since then, the mechanism and its underlying molecular players have been intensively studied. Traditionally, gravitropic organ growth has been evaluated using one-time point measurements. More recently, time-lapse imaging has allowed for more sophisticated depiction of gravitropism. Elaborate quantification methods for investigating root growth and curvature employ real-time analysis [3] in combination with image analysis tools [4], such as ARTT [5], RootTrace [6], KineRoot [7], RootLM [8], or Multi-ADAPT [9]. All these quantification methods provide accurate information on the analyzed parameters, allowing detailed comparison of environmental conditions, pharmacological treatments, and/or mutants at many time points throughout the gravitropic response.

Daniela Ristova and Elke Barbez (eds.), *Root Development: Methods and Protocols*, Methods in Molecular Biology, vol. 1761, https://doi.org/10.1007/978-1-4939-7747-5_15,

However, there are some challenges in the analysis of kinetic data. First, the assessment of multiple time points may lead to ambiguous results, with individual time points being significantly different among many others that are not. Moreover, a central question is whether distinct root bending kinetics are actually due to defects in gravitropism or just related to differences in growth rates. Distinct growth kinetics resulting from a mutation, treatment, or environmental condition could erroneously be interpreted as gravitropic defects. Because of this pitfall in data interpretation, we want to draw attention to the necessity of considering growth rates when analyzing root response to gravity. Here, we present a simple method which addresses these apparent issues.

The normalization requires time-lapse imaging of gravistimulated seedlings (Fig. 1a). We employed a custom-made, real-time imaging platform that allows monitoring of seedlings' growth and responses over time (e.g., [10, 11]). However, any setup that records seedling growth over time can be used. The careful assessment of gravitropic growth kinetics is important if the growth rate is distinct between the analyzed samples. To demonstrate the importance of data normalization to growth, we compared the response of seedlings to gravity in dark versus light conditions. These two conditions show distinct growth rates during vertical growth, leading to distinct bending kinetics over time (Fig. 1). When the root tip angles of gravistimulated seedlings are compared, single time point data indicate a highly significant difference between dark and light conditions (Fig. 1b upper), but this could be due to generally slower growth rates in the dark (Fig. 1b lower). The bending kinetics of these conditions look more alike but still show statistically significant differences between dark- and light-grown gravistimulated seedlings (Fig. 2a). Importantly, after normalization of the root tip angles to growth rate, dark- and light-stimulated seedlings show essentially the same gravitropic kinetics (Fig. 2b).

We also applied the normalization to evaluate the gravitropic growth of auxin-treated versus untreated seedlings. The growth of auxin-treated roots is substantially slower than the growth of control roots, which is also reflected in distinct kinetics of gravitropic growth (Fig. 2c). Normalization to growth rate diminishes these differences (Fig. 2d). However, in contrast to the light/dark comparison, the normalized gravitropic growth analysis of auxin-treated seedlings still shows a statistically significant reduction in gravitropic bending (Fig. 2d).

Notably, the asymmetric redistribution of auxin across the root tip reaches its relative maximum (difference between upper and lower sides) at the midpoint of the bending response before returning to symmetric distribution [12]. Thus, root gravitropism could be further subdivided into separated phases. Considering this, we also investigated the suitability of growth normalization on the

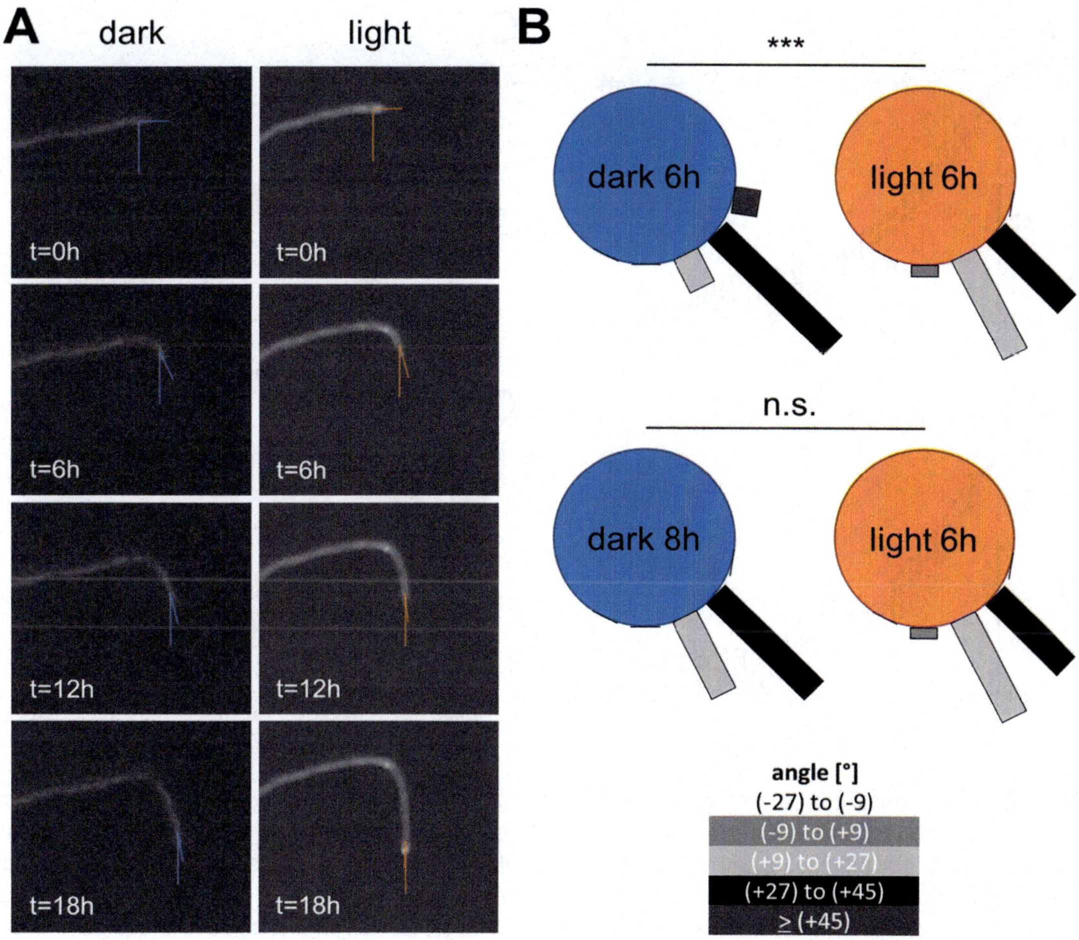

Fig. 1 Data acquisition and conventional visualization of root gravitropic response. (**a**) Representative images of root tips at the indicated time points following gravistimulation under dark or light conditions. (**b**) Diagrams showing the angles of the root tip response to gravity in dark or light conditions at the indicated time points following gravistimulation. Note the significantly different response between dark- and light-grown roots gravistimulated for 6 h. In contrast, upon comparable total growth (6 h in light roughly corresponds to 8 h in dark), there is no significant difference. The length of the bars represents the relative number of roots per category. $n = 18$ (dark), 20 (light). *** $p \leq 0.001$, two-tailed t-test

initial phase of the gravitropic response (half response), defined as the responding root tip reaching an angle of 45° toward the gravity vector (Fig. 3). After normalization to growth, the half response of dark- versus light-gravistimulated root tips shows no difference (Fig. 3a, b). In contrast, the kinetic curves of normalized half response appear distinct for auxin-treated and untreated seedlings (Fig. 3c, d). This analysis emphasizes the possibility to evaluate the kinetics of early gravity response to uncouple initial gravitropism from subsequent responses (e.g., realignment of growth to the gravity vector).

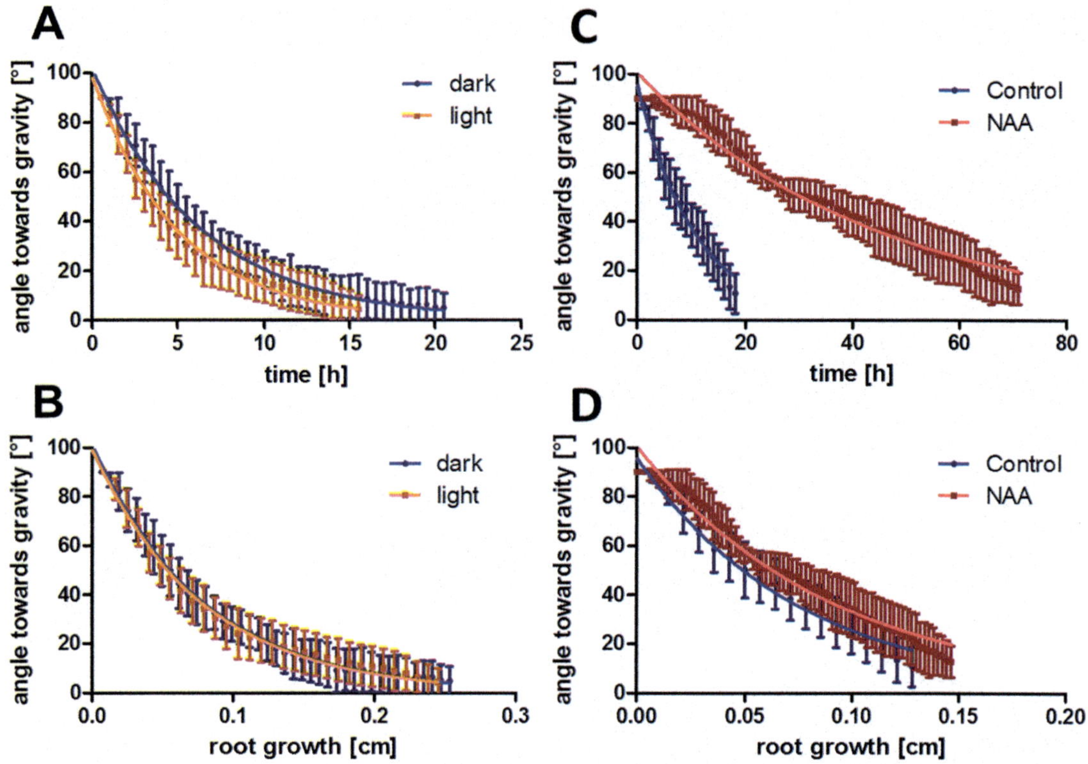

Fig. 2 Effect of normalization to root growth on the kinetics of the full gravitropic response. (**a–d**) Kinetics of full gravitropic response of seedlings grown in dark versus light (**a**, **b**) or treated with 125 nM NAA versus DMSO control (**c**, **d**); before (**a**, **c**) and after (**b**, **d**) normalization. The threshold angles were defined as 5° (**a**, **b**) or 13° (**c**, **d**). $n = 18$ (dark), 20 (light), 6 (NAA), 8 (DMSO); p-values <0.0001 (**a**), $= 0.3652$ (**b**), < 0.0001 (**c**), < 0.0001 (**d**) correspond to the comparison of fitted nonlinear regression with extra sum-of-squares F test

In conclusion, in this paper we show the importance of considering growth rate when investigating gravitropic responses and propose a normalization method that may be a useful tool to characterize environmental conditions, treatments, or mutants, revealing whether they are truly affecting gravitropic responses. We propose combining both the conventional (angle vs. time) and the growth-normalized (angle vs. growth) kinetic analyses for a more comprehensive interpretation of root gravitropic responses.

2 Materials

2.1 Plant Material and Reagents

1. 4–5-day-old seedlings of *Arabidopsis thaliana* Col-0 grown in a plant growth cabinet or chamber under long-day photoperiod (*see* **Note 1**).
2. Sterile square Petri dishes (12 × 12 × 1.7 cm) with vents, containing plant medium made of 0.5 g/L MES, 2.3 g/L Murashige and Skoog salt, 10 g/L sucrose, and 8 g/L plant agar (pH 5.9).

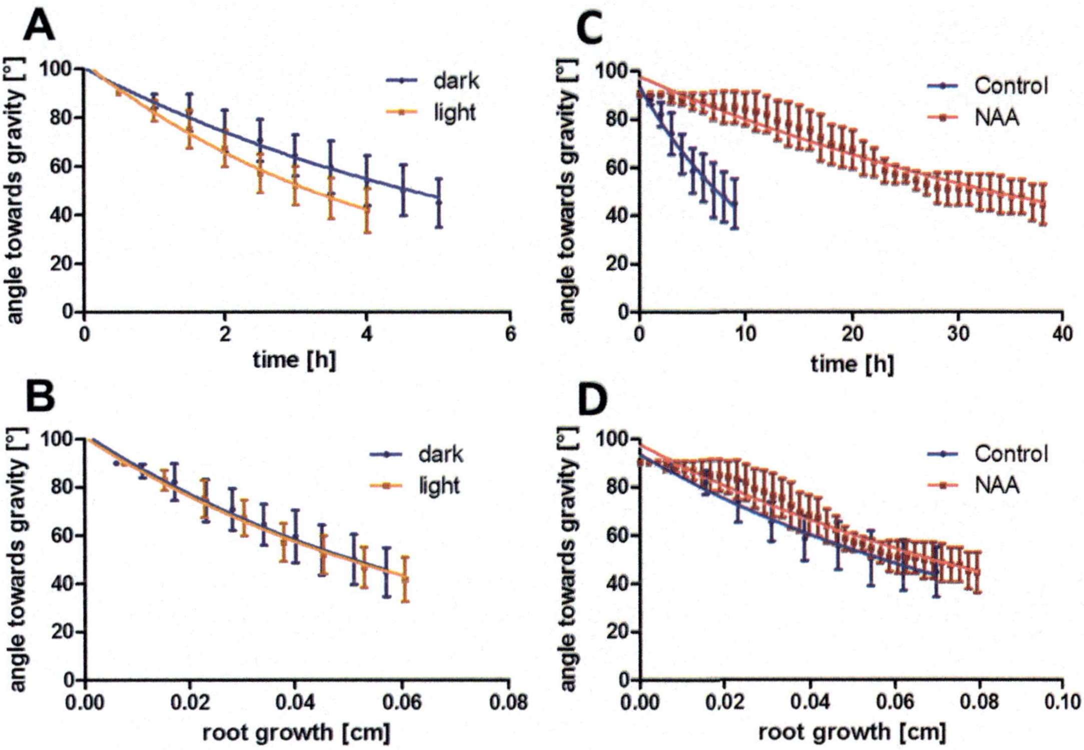

Fig. 3 Effect of the normalization to root growth on the kinetics of the half gravitropic response. (**a–d**) Kinetics of early gravitropic response of seedlings grown in dark versus light (**a**, **b**) or treated with 125 nM NAA versus DMSO control (**c**, **d**); before (**a**, **c**) and after (**b**, **d**) normalization. The threshold angle was defined as 45° for all cases. $n = 6$ (NAA), 8 (DMSO), 18 (dark), 20 (light); p-values <0.0001 (**a**), $= 0.0849$ (**b**), < 0.0001 (**c**), < 0.0449 (**d**) correspond to the comparison of fitted nonlinear regression with extra sum-of-squares F test. Note the similar (**b**) or distinct (**d**) kinetic curves after normalization to root growth, despite ambiguous p-values

3. 70% ethanol for seed sterilization.
4. DMSO (dimethyl sulfoxide) solvent.
5. NAA (1-Naphthaleneacetic acid) 125 mM dissolved in DMSO.
6. Breathable tape for sealing the plates.
7. Aluminum foil for dark conditions.
8. Soft tweezers for manipulation of seedlings.

2.2 Equipment for Real-Time Imaging

1. Custom-made growth chamber for real-time monitoring of in vitro seedling growth equipped with camera and light: infrared for dark conditions and white light for continuous light (*see* **Note 2**).

2.3 Analysis

1. ImageJ 1.41 software (http://rsb.info.nih.gov/ij/) for root angle measurements.
2. Microsoft Excel or similar software.
3. GraphPad Prism or comparable program for statistics.

3 Methods

3.1 Seedling Growth

1. Freshly sterilize the seeds using 70% ethanol for 2 min.
2. Allow the seeds to dry well in a laminar flow hood.
3. Plate the seeds in a line on plant medium-containing Petri dishes.
4. Seal the plates with breathable tape.
5. Stratify the plates containing seeds for a minimum of 2 days at 4 °C in the dark.
6. Grow the seedlings vertically for 4 days in a growth chamber or cabinet at 21 °C long-day photoperiod.

3.2 Root Gravistimulation in Real-Time Imaging Chamber

3.2.1 Gravistimulation in Dark

1. Wrap the Petri dishes containing 4-day-old vertically grown seedlings with aluminum foil and place them back in the same conditions for overnight growth in dark (*see* **Note 3**).
2. The next day, remove the aluminum foil from the overnight dark-accustomed plates with seedlings and place them into the real-time imaging chamber to monitor the gravitropic response.
3. Rotate the plates 90° to gravistimulate the roots (*see* **Note 4**).
4. Turn on the two infrared diodes from the real-time imaging chamber to generate dark conditions.
5. Acquire time series images of gravistimulated seedlings for longer than 24 h following gravistimulation, as described in Subheading 3.3.

3.2.2 Gravistimulation in Light

1. Turn on the white light source at the top front of the imaging platform.
2. Transfer the Petri dishes with vertically grown, 5-day-old seedlings into the real-time imaging chamber to monitor the gravitropic response (*see* **Note 5**).
3. Rotate the plates 90° to alter the gravitropic vector (*see* **Note 4**).
4. Acquire time series images of gravistimulated seedlings for longer than 24 h following gravistimulation, as described in Subheading 3.3.

3.2.3 Auxin Treatment

1. Prepare and pour fresh plant medium supplemented with 125 nM NAA or the equivalent volume of DMSO (solvent control) into Petri dishes.
2. Transfer 4-day-old vertically grown seedlings to NAA- or DMSO-containing Petri dishes.

3. Wrap the Petri dishes in aluminum foil and place them back in the same conditions for overnight growth in dark.
4. Proceed with the gravistimulation as described in Subheading 3.2.1.

3.3 Imaging and Data Analysis

1. Acquire images every 30 min (or more often if a better temporal resolution is needed) upon root gravistimulation in the real-time imaging chamber.
2. Use the angle tool of ImageJ to quantify the angle of the root tip relative to the gravity vector at each time point during the gravitropic response (Fig. 1; *see* **Note 6**).
3. Define a threshold angle for the analysis: full (Fig. 2) or half (Fig. 3) gravitropic response of bending kinetics (*see* **Note** 7).
4. Per treatment, average the tip angle values of all roots for each time point and calculate the standard deviation.
5. Measure the total root growth throughout the analyzed time span and calculate the average growth rate per hour (*see* **Note 8**).
6. For each time point, multiply the time [h] with the average growth rate [cm/h].
7. Use GraphPad Prism (or any other comparable program) to do the data normalization (Figs. 2 and 3). Below we describe the steps in GraphPad Prism 5:

   Create a data table:

   New Table & Graph → XY → Y: check "Enter and plot error values already calculated elsewhere". Press "Create".

   Add your data into "Data Tables": x equals either "time [h]" or "root growth [cm]", A and B correspond to the respective angles of the compared conditions/mutants/treatments, for example, A = dark and B = light (*see* **Note 9**).

   Fit a regression curve to the data and chose parameters for statistical analysis (*see* **Note 10**):

   Analyze → XY analyses → Nonlinear regression (curve fit): click "OK", then chose the following parameters in the tabs:

   - Fit → Exponential (Choose an equation) → One phase decay
   - Compare → Do the best-fit values of selected parameters differ between data sets? (What question are you asking) → Extra sum-of-squares F test (Comparison method) → K (Choose one or more parameters) (*see* **Note 10**).
   - Constrain → Plateau constant equal to (Constrain type): type "0" in the value box (*see* **Note 11**).
   - Click "OK".

8. Find the calculated *p*-value in Results → Nonlin fit → Comparison of fits → *p*-value. This shows the significance of the gravitropic response kinetics normalized to the speed of growth (*see* **Note 12**).

4 Notes

1. Mutants or any other lines subjected to various treatments or environmental conditions could be used.
2. We constructed a real-time imaging chamber as described in [10]. The imaging platform allows the kinetic analysis of seedlings that grow on agar-containing plates installed in the chamber (Fig. 1a, [11]). In addition to the two diodes providing infrared light for dark conditions, we installed a white light bulb in the top front of the box in order to provide continuous white light for gravitropic response and growth rate comparison. By placing the white light diagonally above the imaging platform and not directly above the plates, we aimed to reduce the effect of negative phototropism on root gravitropic response. The time series images were taken using a Canon EOS1200D camera and EOS Utility 2 software for remote control.
3. Allowing the roots to become accustomed to the dark overnight before gravistimulation reduces the variability of the results.
4. When determining whether a particular mutation, treatment, or environmental condition alters the gravity response, it is advisable to include non-gravistimulated controls in order to determine whether the vertical growth rate of altered roots differs from that of control roots.
5. In the case of root gravistimulation in the light, no light pretreatment is needed, since the seedlings are already accustomed to light conditions.
6. Quantify only the roots which have a starting angle of about 90° relative to the gravity vector to ensure analysis of roots with comparable gravistimulation.
7. For describing the full gravitropic response, we propose defining a threshold angle of ±5° toward gravity. In the depicted experiment, we did not record the full response of auxin-treated samples; hence, for this we set a threshold angle of 13°. Alternatively, considering that the crucial events in gravitropism occur immediately after onset of gravistimulation and during the initial bending, we suggest investigating also the half response. For this, we defined a threshold angle of 45°.

8. For example, for the half response, measure the root growth [cm] for each root between the first image after gravistimulation and the first image when the average root tip angle falls below 45°. Then, average the growth of all roots per genotype/condition and divide by the time [h] between those two images.
9. Since the normalized *x*-values for root growth are different for each condition or genotype, the datasets can be inserted below each other in different rows.
10. The "one-phase exponential decay" was the best fit for our data points. This nonlinear regression is based on three variables: (1) Y0 is the starting point of the curve, in our case the starting angle of the root tip toward gravity. (2) The plateau is the value that the curve would approach given an infinite amount of time. In the absence of other stimuli, gravistimulated wild-type seedlings will eventually resume growth toward the gravity vector. Thus, in our setup, we can assume a plateau of 0°. (3) *K* is the rate constant, which describes the speed of a reaction. For the quantification of global gravitropic response kinetics, we are most interested in differences in this variable between different conditions/treatments. However, different questions might require the comparison of other variables (for further depiction, *see* [11]).
11. The results of statistical tests become more precise the more—previously known—variables are defined. In this analysis, only straight-growing seedlings are included (*see* **Note 6**), which would allow constraining the value for the starting angle Y0 to 90. However, we observe a short lag in the beginning of the gravitropic response, which, upon constraining Y0 to 90°, causes a suboptimal fit of the curve to the majority of data points. We achieve the best global fit by leaving Y0 without constraint yet restraining the plateau to 0, assuming that the root tips eventually reorient toward gravity.
12. The average growth serves as a quick and easy way to get an impression of whether there is a real gravitropic defect or merely altered kinetics which can be explained by different growth rates between specimens. For high temporal resolution of growth behavior and how this correlates with curvature over time, the growth rate should be measured for each time point.

Acknowledgments

This work was supported by the Vienna Science and Technology Fund (WWTF) (Vienna Research Group project to J.K.-V.), Austrian Science Fund (FWF) (Projects: P29754 to J.K.-V. and T-728-B16 to E.F.), and the European Research Council (ERC) (Starting Grant 639478-AuxinER to J.K-V.).

References

1. Knight TA (1806) On the direction of the radicle and germen during the vegetation of seeds. Philos Trans R Soc Lond 96:99–108
2. Darwin C, Darwin F (1880) The power of movements in plants. Murray, London
3. Wolverton C (2015) Quantification of root gravitropic response using a constant stimulus feedback system. Methods Mol Biol 1309:23–30
4. Miller ND, Parks BM, Spalding EP (2007) Computer-vision analysis of seedling responses to light and gravity. Plant J 52(2):374–381
5. Russino A, Ascrizzi A, Popova L (2013) A novel tracking tool for the analysis of plant-root tip movements. Bioinspir Biomim 8 (2):025004
6. French A, Ubeda-Tomas S, Holman TJ et al (2009) High-throughput quantification of root growth using a novel image-analysis tool. Plant Physiol 150(4):1784–1795
7. Basu PA, Pal A, Lynch JP et al (2007) A novel image-analysis technique for kinematic study of growth and curvature. Plant Physiol 145 (2):305–316
8. Qi X, Qi J, Wu Y (2007) RootLM: a simple color image analysis program for length measurement of primary roots in Arabidopsis. Plant Root 1:10–16
9. Ishikawa H, Evans M (1997) Novel software for analysis of root gravitropism: comparative response patterns of Arabidopsis wild-type and axr1 seedlings. Plant Cell Environ 20 (7):919–928
10. Zhu Q, Zadnikova P, Smet D et al (2017) Real-time analysis of the apical hook development. Methods Mol Biol 1497:1–8
11. Béziat C, Barbez E, Feraru MI et al (2017) Light triggers PILS-dependent reduction in nuclear auxin signalling for growth transition. Nat Plants 3:17105
12. Band LR, Wells DM, Larrieu A et al (2012) Root gravitropism is regulated by a transient lateral auxin gradient controlled by a tipping-point mechanism. Proc Natl Acad Sci U S A 109(12):4668–4673

Chapter 16

Immunoprecipitation of Membrane Proteins from *Arabidopsis thaliana* Root Tissue

Sascha Waidmann, Lucinda De-Araujo, Jürgen Kleine-Vehn, and Barbara Korbei

Abstract

Here, we present different methods for immunoprecipitating membrane proteins of *Arabidopsis thaliana* root material. We describe two extraction methods for the precipitation either for an integral membrane protein of the endoplasmic reticulum (ER) or a peripheral membrane protein partially localized at the plasma membrane, where we precipitate the protein out of the total membrane as well as total cytosolic fractions.

Key words Immunoprecipitation, Integral membrane proteins, Peripheral membrane proteins, Membrane fraction, Cytosolic fraction, Cross-linked beads, *Arabidopsis thaliana*

1 Introduction

Immunoprecipitation is a method to isolate and concentrate particular proteins out of a cell lysate with the help of specific antibodies coupled to a solid substrate. If possible, the conditions should be chosen in such fashion that the proteins remain in their native conformation, enabling to conserve the complex with their binding partners. Thus, the ideal extraction buffer will leave proteins and interactions intact while at the same time solubilizing adequate amounts from the sample for subsequent analysis. Accordingly, the extraction step is the main challenge for isolating membrane proteins [1]. We present here a method for immunoprecipitating integral or peripheral membrane proteins out of *Arabidopsis thaliana* root material with two different extraction methods for either an integral membrane protein of the ER, the protein PILS2 and PILS3 [2], or a peripheral membrane protein, which is partially found to associate with the plasma membrane, TOL6 [3]. The first method used the zwitterionic detergent CHAPS, which effectively mimics the lipid bilayer. It is especially

Daniela Ristova and Elke Barbez (eds.), *Root Development: Methods and Protocols*, Methods in Molecular Biology, vol. 1761, https://doi.org/10.1007/978-1-4939-7747-5_16,

well suited for protecting the native state of membrane proteins [4] without altering their charge.

The second extraction method builds on previously described studies [5], using the lower relative centrifugal force of a microcentrifuge by manipulation of sample density throughout the procedure, which allows for the isolation of membranes from small amounts of *Arabidopsis thaliana* root material. A single purification step enriches for membrane proteins by removing the more abundant soluble proteins and can therefore also be used to investigate soluble proteins that partially associate with membrane components. This method allows discrimination between the membrane-associated and the cytosolic form of a protein and thus to determine potential cis- and trans-acting factors affecting the localization of the protein of interest (e.g., posttranslational modifications or potential interaction partners).

Both of the described methods are suitable for downstream applications, where the dissociated proteins can be analyzed by electrophoretic methods like SDS-PAGE for Western blot analysis or identified by mass spectrometry [6].

2 Materials

Prepare all solutions using ultrapure water (ddH_2O) and always prepare stock solutions in advance.

2.1 Plant Growth Medium, Seedling Growth, and Plant Root Harvesting

1. *Arabidopsis thaliana* Col-0 (WT), 35S::GFP-PILS2, 35S::GFP-PILS3 [2], and *tol6–1* [3] seeds.
2. Plant growth medium: 2.3 g/L Murashige and Skoog (MS) salt, 0.5 g/L MES, 10 g/L sucrose, 8 g/L plant agar, adjust pH to 5.9, autoclave and store at RT.
3. Sterile square petri dishes (12 × 12 × 1.7 cm) with vents.
4. Breathable paper tape.
5. 70% and 100% ethanol.
6. 0.1% (w/v) agarose in ddH_2O: autoclave and store at RT.
7. Liquid nitrogen.
8. Mortar and pestle.
9. Safe lock 2 mL reaction tubes.
10. Razor blade or scalpel.

2.2 Protein Extraction from Arabidopsis thaliana Roots

2.2.1 Extraction for Integral Membrane Protein from the ER

1. Protein extraction buffer: 25 mM 2-amino-2-(hydroxymethyl)-1,3-propanediol hydrochloride (Tris–HCl) pH 7.8, 10 mM $MgCl_2$, 15 mM ethylene glycol-bis(2-aminoethylether)tetraacetic acid (EGTA), 75 mM NaCl, 1 mM dithiothreitol (DTT), 1 mM NaF, 0.5 mM $NaVO_3$, 15 mM β-glycerophosphate, 0.5 mM phenylmethanesulfonyl fluoride (PMSF), 5 μg/mL leupeptin, 5 μg/mL aprotinin, 0.1% Tween 20 (*see* **Note 1**). Store at −20 °C.
2. 10% CHAPS solution (*see* **Note 2**).
3. Orbital shaker.
4. Mid bench centrifuge.
5. Bradford solution and BSA standard solution (1 μg/μL).

2.2.2 Extraction for Peripheral Membrane Proteins from the Plasma Membrane

1. 1× extraction buffer: 30 mM Tris–HCl (pH 7.3), 10 mM Tris–HCl (pH 8.8), 10 mM KH_2PO_4 buffer (pH 7.4), 6 mM EGTA, 6 mM ethylenediaminetetraacetic acid (EDTA), 2 M D-Sorbitol (*see* **Note 3**), 0.2% (w/v) Casein, 5 mM β-Glycerophosphate, and 2 mM NaF, 2.4 mM benzamidine, 2 μg/mL aprotinin/leupeptin, 4 μg/mL E64, 2 μg/mL pepstatin, 2 mM PMSF, 2 nM okadaic acid, 1 mM DTE, 2 mM Na_2MoO_4, 9.4 mM Na-borate, and 9.4 mM ascorbic acid to have the ready to use buffer (*see* **Note 4**). For a more detailed description please refer to [5].
2. RIPA buffer: 50 mM Tris–HCl (pH 7.4), 150 mM NaCl, 1% (v/v) NP-40, 1% (w/v) sodium deoxycholate, 0.1% (w/v) SDS (0.1%), 1× Complete Mini Protease Inhibitor Cocktail, 1.2 mM PMSF in ddH_2O.
3. 10% (w/v) polyvinylpyrrolidon (PVPP): resuspend insoluble PVPP in ddH_2O. Keep at 4 °C (*see* **Note 5**).

2.3 Cross-Linking of Antibodies to Beads and Immunoprecipitation

1. Protein A or G beads.
2. α-GFP antibody (ab290) for detection of PILS reporter proteins with a GFP-tag or affinity-purified rabbit polyclonal α-TOL6 antibody for detection of endogenous TOLs.
3. 10× TBS: 1.5 M NaCl, 100 mM Tris–HCl, pH 8.0. Autoclave and store at RT.
4. TBS-T: 1× TBS + 0.05% Triton X-100.
5. 0.2 M Na-borate buffer: dissolve 0.2 M boric acid and adjust pH to 9.2 with NaOH.
6. 20 mM dimethyl pimelimidate dihydrochloride (DMP), always prepare fresh (*see* **Note 6**). Take DMP out of −20 °C freezer 30 min before use to equilibrate to RT.
7. 0.2 M Tris pH 8.0.
8. 0.1 M glycine pH 2.0.

9. 10% NaN_3.
10. Low-retention tube.
11. Rotator.
12. IP-buffer: 20 mM Tris pH 7.5, 150 mM NaCl, 10% glycerol, 2 mM EDTA.
13. Detergents: sodium deoxycholate, Triton X-100.
14. 0.1 M glycine pH 2.0.
15. 1.5 M Tris pH 9.2.
16. 1× sample buffer: 0.5% (w/v) CHAPS, 3% (w/v) SDS, 60 mM DTE, 125 mM Tris–HCl (pH 6.8), 30% glycerol, add bromphenol blue with a toothpick. Store the 1× sample buffer at −20 °C.

3 Methods

3.1 Plant Material Growth and Harvesting

1. *Arabidopsis* seedlings are grown in a plant growth chamber at 22 °C in a 16-h-light/8-h-dark regime at 150 μmol/m^2/s light intensity and 60% relative humidity for 5 days.
2. In vitro growth is carried out in standard, sterile square petri dishes (120 × 120 × 17 mm) with vents. Pour enough media into plates to cover approximately half of the depth of the plate. Allow the plates to cool at room temperature for about an hour to allow the agar to solidify. If the plates are not to be used immediately, wrap them in plastic and store at 4 °C.
3. Sterilize the seeds by soaking them first in 70% and then in 100% ethanol.
4. Remove ethanol and allow the seeds to dry in a laminar flow hood.
5. Mix the seeds with the sterile 0.1% agarose. Volume of agarose needs to be adjusted to the amount of seeds (roughly four times). Sow the seeds by pipetting them with cut pipette tips in a horizontal line on the plate as shown in Fig. 1a. Up to three lines per plate can be sown, which would be the equivalent of an "agar-seed" volume of approximately 300 μL. Allow the top agar to dry slightly before placing lid onto plate (*see* **Note 7**).
6. Seal the plates with air-permeable breathable paper tape to prevent desiccation while allowing slight aeration.
7. Stratify the seeds for 1–2 days by keeping the plates at 4 °C in the dark.
8. Grow the seedlings in a growth chamber for 5 days. The plates should be placed vertically into the plant incubator to avoid roots growing into the agar (*see* **Note 8**).

Fig. 1 (**a**) Example for plating of *Arabidopsis thaliana* seeds on solid agar plates and for (**b**) for root harvesting 5 dag (days after germination) with a razor

9. For the harvesting of plant root material, the roots should be excised at the root hypocotyl junction (Fig. 1b) in the petri dish with a razor blade or scalpel and the tissue collected with using flat tweezers (*see* **Note 9**). Remove excess of liquid from root tissue by dabbing gently with absorbent tissue but avoiding too much pressure and freeze immediately in liquid nitrogen.
10. Thoroughly grind the plant material in liquid nitrogen with mortar and pestle. Transfer the ground material to a precooled 2 mL safe lock reaction, and place it into liquid nitrogen.
11. For each extraction, weigh between 100 and 200 mg of plant powder on a scale in a precooled 2 mL safe lock reaction tube (*see* **Note 10**).
12. Continue with extraction or store plant powder at −80 °C.

3.2 Total Membrane Protein Extraction from Arabidopsis thaliana *Roots*

All steps have to be carried out on ice or in a precooled (4 °C) centrifuge to avoid protein degradation.

3.2.1 Method for Integral Membrane Protein of the ER

1. Take sample from liquid nitrogen and place on ice.
2. Add the CHAPS solution to the protein extraction buffer to a final concentration of 1%.
3. Add the same amount of protein extraction buffer with CHAPS as plant powder to the reaction tube on ice and mix thoroughly.
4. Place the reaction tube for 20 min on an orbital shaker at 1600 rpm and 4 °C.
5. Spin down for 30 min with 16,000 × *g* at 4 °C.

6. Transfer supernatant to a fresh 1.5 mL reaction tube and keep on ice.
7. Measure and calculate protein concentration, using a Bradford solution against a BSA standard row.

3.2.2 Method for Peripheral Membrane Proteins from the Plasma Membrane

The procedure for total membrane extraction is based on the method originally developed by [5, 7].

1. Add 1× extraction buffer (1.7 μL per 1 mg of root material) to the homogenized plant material. Vortex and keep tubes on ice.
2. For pre-clearing, transfer the root cell homogenate to the tube with the equilibrated PVPP; vortex and incubate on ice for 5 min to ensure efficient extraction.
3. Equilibrated 10% PVPP suspension with 1× extraction buffer: transfer 0.5 μL of the PVPP suspension per mg root material to a fresh 1.5 mL reaction tube (*see* **Note 11**), and add 500 μL 1× extraction buffer. Vortex briefly, centrifuge for a few seconds, and leave on ice for 10 min. Decant and discard the excess 1× extraction buffer. Keep the equilibrated PVPP on ice.
4. Centrifuge the sample at 4 °C (2 min, 500 × *g*).
5. Transfer the supernatant to a fresh 1.5 mL reaction tube.
6. Re-extract the PVPP/root material pellet with the same amount of 1× extraction buffer for a further 5 min, and centrifuge again at 4 °C (2 min, 500 × *g*). Combine the supernatants and vortex. Clear the supernatant from minor traces of PVPP by centrifugation at 4 °C (2 min, 500 × *g*). Transfer the approximately 400 μL of the supernatant into two separate fresh 1.5 mL reaction tubes (*see* **Note 12**).
7. Centrifuge for 2 h at 4 °C 16,000 × *g* in a mid bench centrifuge to pellet the membrane fraction. After centrifugation transfer the supernatant (cytosolic fraction) to a fresh 2 mL reaction tube, and add RIPA buffer to a final volume of 1 mL.
8. Resuspend the membrane fraction pellet (recombine the pellets from two tubes) in 1 mL RIPA buffer (*see* **Note 13**).
9. Take a 25 μL from each sample and mix with 2× LB; boil for 10 min at 98 °C to analyze the membrane or cytosolic fraction before IP.
10. Shake both samples at 4 °C on an orbital shaker at 1600 rpm for 1 h.
11. Centrifuge for 1 min at 4 °C (16,000 × *g*).
12. Transfer the supernatants of both the cytosolic and the membrane fraction into separate fresh 1.5 mL reaction tubes. Place on ice for immunoprecipitation.

3.3 Cross-Linking of Antibodies to Beads and Immunoprecipitation

3.3.1 Optional Step: Cross-Linking of Antibodies to Beads (See Note 14)

Expect for steps 4–7, all steps should be performed on ice or at 4 °C.

1. Wash 50 μL beads twice briefly with 1 mL TBS-T in a 1.5 mL reaction tube: resuspend beads by flicking. Pellet beads by spinning for 1 min at 2000 rpm at 4 °C.
2. Add 1 mL TBS-T to the washed beads, then add antibody to desired concentration (*see* **Note 15**). We used 0.5 μL of α-GFP antibody or 15 μL of the purified rabbit serum.
3. Incubate overnight on a rotator at 4 °C at 12 rpm.
4. Wash beads three times briefly with 1 mL TBS-T.
5. Wash beads three times briefly with 1 mL 0.2 M Na-borate, pH 9.2.
6. Add 1 mL of 20 mM DMP solution to the beads and mix at RT for 30 min on a rotator.
7. Stop cross-linking reaction by washing the beads twice for 10 min with 1 mL 0.2 M Tris–HCl (pH 8.0).
8. Wash beads twice briefly with 1 mL TBS-T.
9. Wash beads twice briefly with 0.1 M glycine pH 2.0.
10. Wash beads three times briefly with 1 mL TBS-T to re-equilibrate.
11. Antibody-coupled beads can be stored at 4 °C in TBS-T with 0.05% NaN_3.

3.3.2 Immuno-precipitation

1. For pre-clearing transfer protein, extract to a 1.5 mL reaction tube and add 50 μL of protein A or G beads (*see* **Note 16**). Incubate at 4 °C for 1 h. Pellet beads by spinning for 1 min at 2000 rpm at 4 °C.
2. Transfer pre-cleared extract to a new 1.5 mL reaction tube and add antibody-coupled beads.
3. Incubate with incubate overnight on a rotator at 4 °C at 12 rpm (*see* **Note 17**).
4. Pellet the beads and remove supernatant (keep for analysis if desired).

3.3.3 Washing (See Note 18)

1. Wash five times with 1 mL ice-cold IP-buffer containing 0.1% sodium deoxycholate and 0.5% Triton X-100. Resuspend beads by flicking, and incubate with rotation at 4 °C for 10 min. Pellet beads by spinning for 1 min at 2000 rpm at 4 °C. Change tube after the first washing.
2. Wash ten times with 1 mL ice-cold IP-buffer with no additions of detergents. Resuspend beads by flicking, and incubate with rotation at 4 °C for 10 min. Pellet beads by spinning for 1 min at 2000 rpm at 4 °C (*see* **Note 19**).
3. To remove all pH-buffering agents, wash beads twice briefly with 1 mL 150 mM ice-cold NaCl.

3.3.4 Elution and Bead Recovery (See Note 20)

1. Under the flow: add 100 μL of 0.1 M glycine (pH 2.0) to the beads, mix by flicking for 1 min. Pellet beads and transfer the supernatant to a fresh low-retention tube containing 5 μL of 1.5 M Tris (pH 9.2). Repeat this step an additional time.
2. Use eluate directly or freeze for further analysis in liquid nitrogen and store at −80 °C.
3. For the recovery of the beads, add 1 mL of 0.1 M glycine (pH 2.0) to the beads, mix by flicking for 1 min. Pellet beads and remove supernatant. Repeat once. Neutralize the beads by washing three times briefly with 1 mL TBS-T, and store beads at 4 °C in TBS-T with 0.05% NaN3 for reuse.
4. Alternatively, if you do not wish to recover the beads, at **step 19**, add to 50 μL of 1× sample buffer (freshly made), and incubate for 10 min at 98 °C (*see* **Note 21**).
5. Centrifuge and transfer the supernatant (eluted proteins from the beads) into a fresh tube.

3.3.5 Western Blot and/or Silver Staining

1. To verify the success of the IP, a Western blot or silver staining with 10–20 μL of the elution can be performed (Fig. 2 for integral membrane protein and Fig. 3 for the peripheral membrane protein).

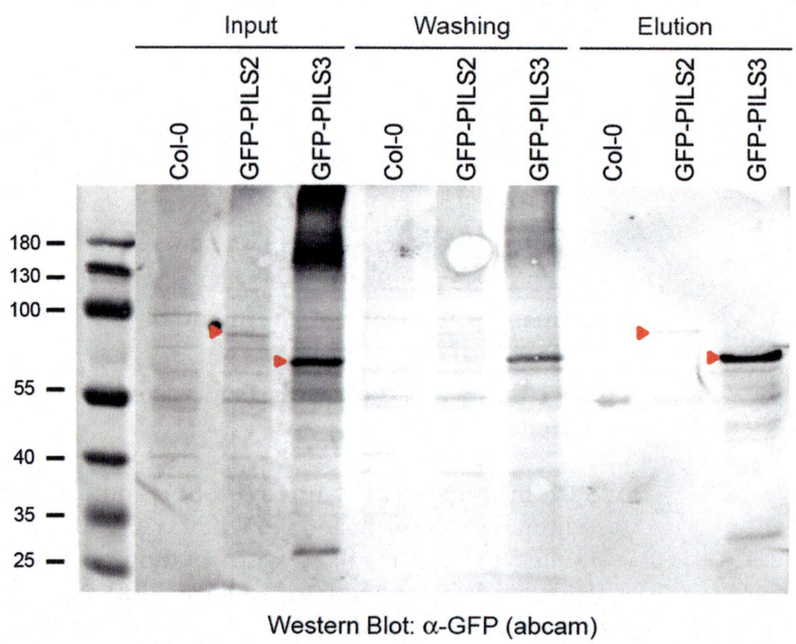

Fig. 2 Western blot of PILS2 and PILS3 immunoprecipitation out of the integral membrane fraction. IPs were performed using GFP-PILS2, GFP-PILS2, and Col-0 *Arabidopsis thaliana* root extracts with GFP antibodies coupled to beads. The left lanes show the input, where 1/100 of the total input was loaded. The middle lanes show the washings, and the right lanes represent 1/5 of the total eluted protein from the beads. Red arrowheads show the proteins in the input as well as after immunoprecipitation

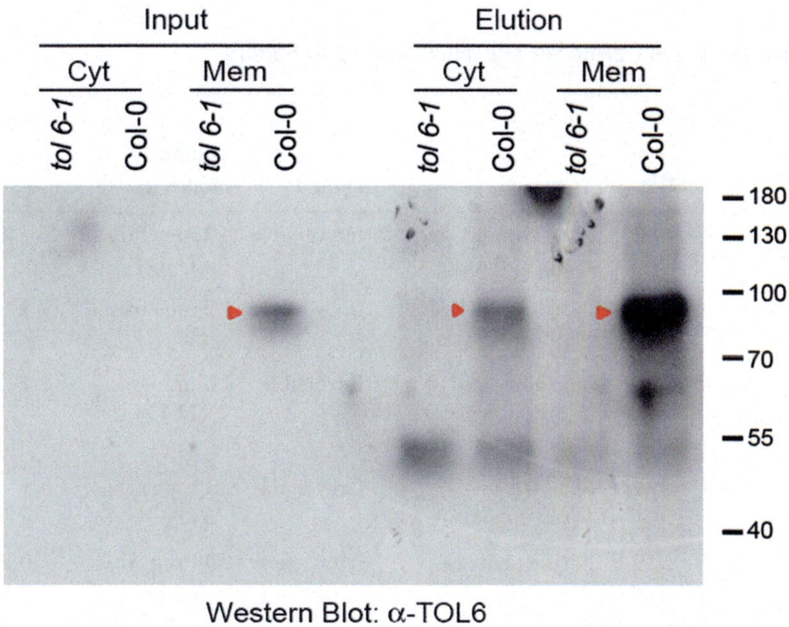

Fig. 3 Western blots of TOL6 immunoprecipitation out of the membrane (Mem) and cytosolic (Cyt) fractions. IPs were performed from *tol6–1* and Col-0 *Arabidopsis thaliana* root extracts with beads coupled to anti-TOL6 rabbit serum. The left lanes show the input amount, where 1/100 of the total input was loaded, while right lanes represent 1/5 of the total eluted protein from the beads. Endogenous TOL6 is clearly visible in the Col0 membrane lysate (red arrowhead) as well as after immunoprecipitation in both Col0 membrane and cytosolic fractions (red arrowhead), which are missing in *tol6–1* control extracts

4 Notes

1. Concentration and composition of protease and phosphatase inhibitors can vary dependent on the specific needs. An overview of the most commonly used inhibitors is displayed in Table 1.
2. Concentration and composition of detergents can vary dependent on the specific protein of interest. A good overview can be found here:

   https://www.embl.de/pepcore/pepcore_services/protein_purification/extraction_clarification/solubilisation_membrane_proteins/.
3. The sorbitol is added to create a high-density extraction buffer thus enabling minimal volume of extraction buffer without ultracentrifugation [7].
4. At this step, all appropriate inhibitors that are relevant for the assay can be added, which includes protease and phosphatase (*see* Table 1) but also other inhibitors like NEM (N-ethylmaleimide) [8], which inhibits deubiquitinating

Table 1
Overview of commonly used protease and phosphatase inhibitors

Inhibitor	MW	Target class	Type	Solubility (solvent)	Typical working (1 ×) conc.
Sodium fluoride	42	Ser/Thr and acidic phosphatases	Irreversible	40 mg/mL (H_2O)	1–20 mM
Sodium orthovanadate	183.9	Tyr and alkaline phosphatases	Irreversible	20 mg/mL (H_2O)	1–100 mM
Beta-glycerophosphate (disodium salt)	216	Ser/Thr phosphatases	Reversible	10 mg/mL (H_2O)	1–100 mM
Sodium pyrophosphate	221.9	Ser/Thr phosphatases	Irreversible	65 mg/mL (H_2O)	1–100 mM
AEBSF•HCl	239.5	Serine proteases	Irreversible	200 mg/mL (H_2O)	0.2–1.0 mM
Aprotinin	6511.5	Serine proteases	Reversible	10 mg/mL (H_2O)	100–200 nM
Bestatin	308.4	Amino-peptidases	Reversible	5 mg/mL (MeOH)	1–10 μM
E-64	357.4	Cysteine proteases	Irreversible	20 mg/mL (1:1 EtOH:H_2O)	1–20 μM
EDTA	372.2	Metalloproteases (chelates cations)	Reversible	10 g/100 mL (H_2O)	2–10 mM
Leupeptin	475.6	Serine and cysteine proteases	Reversible	1 mg/mL (H_2O)	10–100 μM
Pepstatin A	685.9	Aspartic acid proteases	Reversible	1 mg/mL (MeOH)	1–20 μM
PMSF	174.2	Serine proteases	Reversible	18 mg/mL (MeOH)	0.1–1.0 mM

enzymes or PUGNAc, an inhibitor of a variety of N-acetylhexosaminidases [9].

5. PVPP is insoluble in water, and thus the powder doesn't dissolve completely. Therefore, vortex the 10% (w/v) PVPP stock vigorously before use and take out the required amount.
6. The pH for DMP should be between pH 8 and 9. Cross-linking efficiency is greatly reduced outside this pH range.
7. Avoid overdrying the plates as this can affect the plant growth; in an optimal scenario all roots should exhibit homogeneous growth.

8. Seedlings can be younger or older, depending on the biological processes and reporter genes that are being investigated.
9. During harvesting, work quickly and be sure to always harvest at same time point of the day for optimal comparison of results of biological repeats.
10. Depending on the expression level of the target protein and further processing after IP, the amount of plant material can be scaled up.
11. The 200 μL tips are cut off approximately 1 cm at the end with scissors. This allows for easier pipetting of the insoluble PVPP.
12. Do not centrifuge more than 200 μL, as this may cause incomplete [7]. Recombine the separated membrane pellet fractions after centrifugation during the resuspension step with RIPA buffer.
13. Efficient resuspension of the pellet fraction is crucial for the further success of the immunoprecipitation.
14. Cross-linking of antibody is necessary to avoid the elution of the antibody, which would prevent sample carry-over for further analysis. Moreover, antibody cross-linked to beads can be reused several times to upscale your IP.
15. The concentration of antibody needs to be tested individually. However, we recommend to initially test 1 mg of antibody per mL of beads.
16. The amount of protein used for the IP depends on the expression level of the target protein and efficiency of antibody binding. We recommend starting with 2–5 mg of protein extract per IP.
17. For some proteins it might be necessary to shorten IP time to reduce unspecific interactions with the bait protein.
18. Washing parameters may vary according to antibody, bait, and further processing after IP. It is optionally possible to add three additional washes with IP-buffer, containing higher salt and detergent concentrations. This step increases the stringency for binding.
19. These steps can be omitted for Co-IP experiments.
20. There are several possibilities to elute the proteins from the beads. Besides elution with glycine, it is possible to elute with SDS and urea. Elution with glycine will enable reuse of the beads, while with an SDS buffer, you will obtain a higher yield [10].
21. For protein complex identification, it is possible to directly digest the proteins with proteases (e.g., trypsin, chymotrypsin) on the beads.

Acknowledgments

Establishment for this protocol was supported by a grant from the Austrian science fund FWF (P 29754 to SW, JK-V; V382 to BK), the Austrian Academy of Science to (LD-A (DOC-ÖAW 24195), the Hochschuljubiläumsstiftung der Stadt Wien (H-235701/2014 to BK), the Vienna Science and Technology Fund (WWTF) (to JK-V), and the European Research Council (ERC) (Starting Grant 639478-AuxinER to JK-V).

References

1. Avila JR, Lee JS, Torii KU (2015) Co-Immunoprecipitation of membrane-bound receptors. Arabidopsis Book 13:e0180
2. Barbez E, Kubes M, Rolcik J, Beziat C, Pencik A, Wang B, Rosquete MR, Zhu J, Dobrev PI, Lee Y, Zazimalova E, Petrasek J, Geisler M, Friml J, Kleine-Vehn J (2012) A novel putative auxin carrier family regulates intracellular auxin homeostasis in plants. Nature 485:119–122
3. Korbei B, Moulinier-Anzola J, De-Araujo L, Lucyshyn D, Retzer K, Khan MA, Luschnig C (2013) Arabidopsis TOL proteins act as gatekeepers for vacuolar sorting of PIN2 plasma membrane protein. Curr Biol 23:2500–2505
4. Labeta MO, Fernandez N, Festenstein H (1988) Solubilisation effect of Nonidet P-40, triton X-100 and CHAPS in the detection of MHC-like glycoproteins. J Immunol Methods 112:133–138
5. Leitner J, Luschnig C (2014) Ubiquitylation-mediated control of polar auxin transport: analysis of Arabidopsis PIN2 auxin transport protein. Methods Mol Biol 1209:233–249
6. Wendrich JR, Boeren S, Moller BK, Weijers D, De Rybel B (2017) In vivo identification of plant protein complexes using IP-MS/MS. Methods Mol Biol 1497:147–158
7. Abas L, Luschnig C (2010) Maximum yields of microsomal-type membranes from small amounts of plant material without requiring ultracentrifugation. Anal Biochem 401:217–227
8. Berleth ES, Pickart CM (1996) Mechanism of ubiquitin conjugating enzyme E2-230K: catalysis involving a thiol relay? Biochemistry 35:1664–1671
9. Haltiwanger RS, Grove K, Philipsberg GA (1998) Modulation of O-linked N-acetylglucosamine levels on nuclear and cytoplasmic proteins in vivo using the peptide O-GlcNAc-beta-N-acetylglucosaminidase inhibitor O-(2-acetamido-2-deoxy-D-glucopyranosylidene)amino-N-phenylcarbamate. J Biol Chem 273:3611–3617
10. Sousa MM, Steen KW, Hagen L, Slupphaug G (2011) Antibody cross-linking and target elution protocols used for immunoprecipitation significantly modulate signal-to noise ratio in downstream 2D-PAGE analysis. Proteome Sci 9:45

Part III

Genome-Wide Methods and Systems Biology

Chapter 17

Monitoring Transcriptomic Changes in Soil-Grown Roots and Shoots of *Arabidopsis thaliana* Subjected to a Progressive Drought Stress

Khurram Bashir, Sultana Rasheed, Akihiro Matsui, Kei Iida, Maho Tanaka, and Motoaki Seki

Abstract

Numerous experiments have been performed in *Arabidopsis* to monitor changes in gene expression that occur in response to a variety of abiotic and biotic stresses, different growth conditions, and at various developmental stages. In addition, gene expression patterns have also been characterized among wild-type and mutant genotypes. Despite these numerous reports, transcriptional changes occurring in roots of soil-grown plants subjected to a progressive drought stress have remained undocumented. To fill this gap, we established a system that allows one to establish water-deficit conditions and to collect root and shoot samples with minimal damage to the root system. *Arabidopsis* plants are grown in a ceramic-based granular soil and subjected to progressive drought stress by withholding water. Root and shoot samples were collected separately, RNA was purified, and a microarray analysis of drought-stressed roots and shoots was performed at 0, 1, 3, 5, 7, and 9 days after the onset of drought stress treatment. Here, we describe the detailed protocol used to analyze the transcriptomic changes occurring in roots and shoots of soil-grown *Arabidopsis* subjected to a progressive drought stress.

Key words Abiotic stresses, *Arabidopsis thaliana*, Ceramic-based granular growth medium, Drought stress, Microarray, Roots

1 Introduction

Plant growth is tremendously affected by different environmental factors. Drought stress represents a critical problem for the majority of crop plants grown each year and results in compromised yields and product quality [1–3]. Global changes in gene expression occur when plants are exposed to a water deficit, allowing plants to adapt to the new environmental condition and minimize injury [4]. Thus, understanding the molecular mechanisms underlying the early perception of stress signals is essential for creating strategies to develop crop plants that are tolerant to drought stress. Microarrays are commonly used to study transcriptomic changes that occur in

Daniela Ristova and Elke Barbez (eds.), *Root Development: Methods and Protocols*, Methods in Molecular Biology, vol. 1761, https://doi.org/10.1007/978-1-4939-7747-5_17,

plants subjected to drought stress [5]. In comparison to shoots, roots are more capable of continuing growth under water-limited conditions. This allows plants to increase root surface area, resulting in a significant increase in water uptake [6]. Roots, as an underground tissue, sense a water deficit earlier than aerial portions of the plant. Therefore, it is extremely important to understand the changes in gene expression that occur in roots during the early stages of a progressive drought stress. The majority of the studies in *Arabidopsis*, however, have focused on transcriptomic changes in shoot tissue [5, 7]. Those that have studied roots have primarily utilized air-dried roots to monitor early changes in gene expression by pulling the plants and subtending root systems from a tissue-culture medium and exposing them to air [8]. In contrast, studies examining the roots of soil-grown plants subjected to a progressive drought stress remain scarce. The major limiting factor preventing these latter studies has been the inability to isolate roots from soil-grown plants subjected to a water deficit that are suitable for RNA extraction and microarray analysis [5].

In order to fill this gap, a plant growth system was established using a ceramic-based, granular growth medium (soil) to mimic a soil that permitted the separate collection of both roots and shoots from plants subjected to a progressive drought stress. Roots suitable for RNA extraction and microarray analysis can easily be recovered from the ceramic-based soil. This system allowed us to monitor changes in gene expression in roots and shoots of soil-grown plants in response to a progressive drought. Root and shoot samples were collected separately at 0, 1, 3, 5, 7, and 9 days of a drought stress regime [5]. This provided the ability to dissect the molecular response of shoots and roots to a progressive drought stress and to analyze the differentially regulated genes and small open reading frames (sORFs) in a tissue-specific manner [5, 9]. Here we describe the plant growth system and microarray analysis procedure that was used. Since the described protocol utilizes Agilent microarray slides, this aspect of our protocol overlaps the previously described procedures [10].

2 Materials

2.1 Seed Sterilization

1. *Arabidopsis thaliana* seeds (Col-0 ecotype).
2. 70% ethanol.
3. Sterilization solution (10 ml, 500 μl Antiformin (NaClO), 5 μl Tween 20).
4. Sterilized water.

2.2 Soil Preparation and Transplanting

1. 2.5 L ceramic-based granular soil (Sakatanotane, Japan).
2. 2.5 mm strainer.
3. Hyponex.

2.3 RNA Isolation

1. RNA isolation kit.
2. Liquid nitrogen.
3. Latex gloves.
4. Multi-beads shocker.
5. 2 ml Eppendorf tubes.
6. Vortex mixer.
7. Centrifuge.
8. Nanodrop ND-1000 UV-VIS spectrophotometer.
9. Ultrapure water.
10. 70% ethanol.
11. Isopropanol.
12. RNA quality check (Bioanalyzer).

2.4 Microarray Analysis

1. Low Input Quick Amp Labeling Kit, one-color, Cyanine 3-CTP.
2. *Arabidopsis* custom microarray.
3. G2505B microarray scanner.
4. R 3.3.2 software program (R Core Team).

3 Methods

3.1 Sterilization of Plant Material

1. Place 500 μl 70–75% ethanol in an Eppendorf tube containing *Arabidopsis* seeds, then spin and remove the ethanol.
2. Add 1 ml sterilization solution, let the seeds settle to the bottom and then discard the solution.
3. Add a fresh 1 ml sterilization solution to the tube, and incubate the tube for 5 min at room temperature with continuous vortexing.
4. Spin and remove the solution.
5. Finally, wash with sterile, distilled water (SDW) at least five times to remove all traces of the sterilization solution from the tube.

3.2 Plant Growth Conditions

Direct sowing of seeds in the ceramic-based, granular soil medium is not recommended as it results in poor germination. Instead, first sow the seeds on Murashige and Skoog (MS) sterile plates, and

Fig. 1 Ceramic-based, granular soil medium used to grow plants subjected to a progressive drought stress. (**a**) Ceramic-based, granular soil. (**b**, **c**) *Arabidopsis* plants transferred from MS medium to the ceramic-based, granular soil

then move the germinated seedlings to the ceramic soil medium after about 9–10 days.

1. After sterilization, incubate seeds of *Arabidopsis thaliana* (Col-0 ecotype) in the dark for at least 2 days, and then germinate the seeds on MS medium at 22 °C under 16-h light/8-h dark (40–80 μmol photons/m^2/s), and grow the seedlings for 9–10 days.
2. Filter the ceramic-based soil medium to remove small particles. Then place 50 g of the ceramic soil in each pot (*see* **Note 1**) (Fig. 1a).
3. Place the soil-filled pots in trays half-filled with water for 24 h to ensure that the soil is completely wet. Use 2 l water containing 1 ml hyponex as a nutrient source per tray. Each tray should contain 24 small pots. Subsequently, transfer the plantlets from the MS medium to the ceramic soil and grow the plants at 22 °C (16-h light/8-h dark cycle, 60 μmol/m^2/s photon flux density).
4. Avoid exposing the trays directly to blowing air inside the environmental chamber, which will cause the plants to dry out before they become established.
5. Cover the trays with plastic wrap for 2 days in order to maintain a high humidity during which time the plants can acclimate to their new growing conditions.
6. Remove the plastic wrap after 2 days and allow the plants to continue to grow for another 5–6 days (*see* **Note 2**) (Fig. 1b, c).
7. Start the drought treatment by removing excess water from the trays and ceasing any subsequent watering. This marks the onset of the drought treatment, which should continue for at least 9 days (Fig. 2).
8. Remove plants from the ceramic soil (*see* **Note 3**) (Fig. 2g) and harvest the roots and shoots separately. Collections of plant

Fig. 2 *Arabidopsis* plants grown in a ceramic soil and subjected to a progressive drought stress. (**a**) Plants growing under well-watered conditions. (**b**) Plants at 3 days after the onset of the progressive drought stress treatment. (**c**) Five days after the onset of the drought stress treatment. (**d**) Seven days after the onset of the drought stress treatment. (**e**) Nine days after the onset of the drought stress treatment. (**f**) Plants rewatered after 9 days of the progressive drought stress treatment. (**g**) Plants that recovered from the drought stress treatment after rewatering

material can be made at 0, 1, 3, 5, 7, and 9 days after the onset of drought treatment (*see* **Note 4**). Typically, shoots and roots from 12 plants are harvested from three pots to comprise a single biological replicate (*see* **Note 5**).

3.3 RNA Extraction

1. Grind samples to a fine powder using a Multi-beads shocker.
2. Extract RNA from all of the biological replicates using RNA Isolation Kit according to the manufacturer's instructions (*see* **Note 6**).
3. Quantify the RNA and dilute all samples to 100 ng RNA/μl.
4. Analyze and confirm the quality of the extracted RNA on bioanalyzer.

3.4 Microarray Analysis

A customized array, containing AGI-annotated genes, small open reading frames (sORFs), and noncoding RNAs [5, 9, 11], was utilized in the microarray analysis. The protocol used for the microarray analysis was similar to that described by [10] with the exception that one-color microarray was used. Briefly, fluorescent-labeled (Cy3) cRNAs were prepared from 200 ng total RNA from each sample using a labeling kit and subsequently hybridized to an *Arabidopsis* custom microarray. Three biological replicates were processed for each treatment at each sampling point.

Arrays were scanned with a microarray scanner [10] and the R 2.12.1 software program (R Core Team). RMA normalization was performed for the obtained signals of the microarray probes using the limma package [12]. A student's t-test (p-value) was used as a parametric test, and the Benjamini and Hochberg false discovery rate (FDR; q-value) procedure was used to control the certainty level [13]. Genes with at least a twofold change in expression and having a q-value < 0.1 were considered to be differentially expressed.

3.5 MapMan and Gene Ontology (GO) Analysis

1. The average $\log_2$ value of all biological replicates was calculated for all samples (roots and shoots) for individual annotations at each time point.
2. Gene ontology analyses were carried out using the PANTHER (protein annotation through evolutionary relationship) classification system database maintained at http://pantherdb.org/ [14]. The GO analyses were performed for molecular function, protein classification, and biological pathway analysis.
3. To further validate the results, the normalized $\log_2$ values were then used to compare the transcriptomic changes using MapMan 3.6.0RC1 [15].
4. PageMan analysis was also performed using MapMan 3.6.0RC1 which included a Wilcoxon test with BH correction [15].

4 Notes

1. Removal of small-sized particles is extremely important as small particles of ceramic granular soil adhere to the root tissue. When removing these particles from root samples, care should be taken to avoid injuring the roots.
2. Plants germinated on MS medium first and then transferred to soil grow bigger than plants sown directly on the ceramic soil. The former plants also flower earlier than plants sown directly on the ceramic soil.
3. Roots that were suitable for RNA extraction and microarray analysis were easily recovered from the ceramic-based, granular soil without causing any root damage.
4. Ceramic soil particles adhering to roots can be gently removed by hand or tweezers, without injuring the roots. All samples were collected at noon. Samples were immediately placed in liquid nitrogen after harvesting and stored at -80 °C until RNA extraction.
5. During the latter stages of the progressive drought stress treatment, roots become tougher due to cortex thickening

(hardening), and care should be taken to ensure complete grinding of these roots during sample preparation in order to avoid the degradation of RNA. This could be achieved by repeating the grinding step. To avoid degradation of RNA, samples must be stored in liquid nitrogen for 3–5 min between grinding.

6. It is recommended to use kit that has the following properties: (1) The kit utilizes a filter cartridge that binds RNA and efficiently extracts high-quality RNA from roots of soil-grown plants, even in the presence of extremely small soil particles which are not easily visible and thus difficult to remove during sampling. (2) The kit extracts total RNA, including very small-sized RNA, and thus the extracted RNA could also be used to monitor the expression of small open reading frames (sORFs).

Acknowledgments

This work was supported by a grant from the RIKEN Center for Sustainable Resource Science (to M.S.), the Japan Science and Technology Agency (JST) [Core Research for Evolutionary Science and Technology (CREST)] (to M.S.), and Grants-in-Aid for Scientific Research in Innovative Areas (no. 16H01476) by the Ministry of Education, Culture, Sports, and Technology of Japan (to M.S.). K.B. is supported through the RIKEN FPR program and S.R. is supported by the RIKEN JRA program.

References

1. Seki M, Umezawa T, Urano K, Shinozaki K (2007) Regulatory metabolic networks in drought stress responses. Curr Opin Plant Biol 10(3):296–302. https://doi.org/10.1016/j.pbi.2007.04.014
2. Boyer JS (1982) Plant productivity and environment. Science 218(4571):443–448
3. Datta K, Baisakh N, Ganguly M, Krishnan S, Yamaguchi Shinozaki K, Datta SK (2012) Overexpression of arabidopsis and rice stress genes' inducible transcription factor confers drought and salinity tolerance to rice. Plant Biotechnol J 10(5):579–586. https://doi.org/10.1111/j.1467-7652.2012.00688.x
4. Seki M, Narusaka M, Ishida J, Nanjo T, Fujita M, Oono Y, Kamiya A, Nakajima M, Enju A, Sakurai T (2002) Monitoring the expression profiles of 7000 *Arabidopsis* genes under drought, cold and high-salinity stresses using a full-length cDNA microarray. Plant J 31(3):279–292. https://doi.org/10.1046/j.1365-313X.2002.01359.x
5. Rasheed S, Bashir K, Matsui A, Tanaka M, Seki M (2016) Transcriptomic analysis of soil-grown *Arabidopsis thaliana* roots and shoots in response to a drought stress. Front Plant Sci 7:180. https://doi.org/10.3389/fpls.2016.00180
6. Sharp RE, Poroyko V, Hejlek LG, Spollen WG, Springer GK, Bohnert HJ, Nguyen HT (2004) Root growth maintenance during water deficits: physiology to functional genomics. J Exp Bot 55(407):2343–2351. https://doi.org/10.1093/jxb/erh276
7. Harb A, Krishnan A, Ambavaram MM, Pereira A (2010) Molecular and physiological analysis of drought stress in Arabidopsis reveals early responses leading to acclimation in plant growth. Plant Physiol 154(3):1254–1271. https://doi.org/10.1104/pp.110.161752
8. Kilian J, Whitehead D, Horak J, Wanke D, Weinl S, Batistic O, D'Angelo C, Bornberg-Bauer E, Kudla J, Harter K (2007) The AtGenExpress global stress expression data set:

protocols, evaluation and model data analysis of UV-B light, drought and cold stress responses. Plant J 50(2):347–363. https://doi.org/10.1111/j.1365-313X.2007.03052.x

9. Rasheed S, Bashir K, Nakaminami K, Hanada K, Matsui A, Seki M (2016) Drought stress differentially regulates the expression of small open reading frames (sORFs) in *Arabidopsis* roots and shoots. Plant Signal Behav 11 (8):e1215792. https://doi.org/10.1080/15592324.2016.1215792
10. Maruyama K, Yamaguchi-Shinozaki K, Shinozaki K (2014) Gene expression profiling using DNA microarrays. Methods Mol Biol 1062:381–391
11. Nguyen AH, Matsui A, Tanaka M, Mizunashi K, Nakaminami K, Hayashi M, Iida K, Toyoda T, Nguyen DV, Seki M (2015) Loss of Arabidopsis 5′–3′ exoribonuclease AtXRN4 function enhances heat stress tolerance of plants subjected to severe heat stress. Plant Cell Physiol 56(9):1762–1772. https://doi.org/10.1093/pcp/pcv096
12. Ritchie ME, Phipson B, Wu D, Hu Y, Law CW, Shi W, Smyth GK (2015) limma powers differential expression analyses for RNA-sequencing and microarray studies. Nucleic Acids Res 43 (7):e47. https://doi.org/10.1093/nar/gkv007
13. Benjamini Y, Hochberg Y (1995) Controlling the false discovery rate: a practical and powerful approach to multiple testing. J R Stat Soc Series B Methodol 57:289–300. https://doi.org/10.1105/tpc.111.087866
14. Mi H, Muruganujan A, Thomas PD (2013) PANTHER in 2013: modeling the evolution of gene function, and other gene attributes, in the context of phylogenetic trees. Nucleic Acids Res 41(D1):D377–D386. https://doi.org/10.1093/nar/gks1118
15. Thimm O, Bläsing O, Gibon Y, Nagel A, Meyer S, Krüger P, Selbig J, Müller LA, Rhee SY, Stitt M (2004) Mapman: a user-driven tool to display genomics data sets onto diagrams of metabolic pathways and other biological processes. Plant J 37(6):914–939. https://doi.org/10.1111/j.1365-313X.2004.02016.x

Chapter 18

Chromatin Immunoprecipitation Sequencing (ChIP-Seq) for Transcription Factors and Chromatin Factors in *Arabidopsis thaliana* Roots: From Material Collection to Data Analysis

Sandra Cortijo, Varodom Charoensawan, François Roudier, and Philip A. Wigge

Abstract

Chromatin immunoprecipitation combined with next-generation sequencing (ChIP-seq) is a powerful technique to investigate in vivo transcription factor (TF) binding to DNA, as well as chromatin marks. Here we provide a detailed protocol for all the key steps to perform ChIP-seq in *Arabidopsis thaliana* roots, also working on other *A. thaliana* tissues and in most non-ligneous plants. We detail all steps from material collection, fixation, chromatin preparation, immunoprecipitation, library preparation, and finally computational analysis based on a combination of publicly available tools.

Key words Chromatin immunoprecipitation (ChIP), Chromatin, Transcription factor (TF), Next-generation sequencing, *Arabidopsis*, Root, Bioinformatics

1 Introduction

The dynamics of interactions between proteins and DNA is central to understand the regulation of transcription, replication, and other DNA-dependent processes in development and environmental responses in plants [1–3]. Chromatin immunoprecipitation followed by next-generation sequencing (ChIP-seq) is a useful technique to assess in vivo TF binding and histone modifications in a genome-wide manner [4, 5]. The first step of ChIP-seq is to fix the plant material right at the moment of collection, using chemicals such as formaldehyde [6]. Then chromatin is extracted and fragmented either by sonication or by digestion with the micrococcal nuclease (MNase) enzyme. Sonication is the most commonly used

Sandra Cortijo and Varodom Charoensawan contributed equally to this work.

Daniela Ristova and Elke Barbez (eds.), *Root Development: Methods and Protocols*, Methods in Molecular Biology, vol. 1761, https://doi.org/10.1007/978-1-4939-7747-5_18, © Springer Science+Business Media, LLC 2018

method of fragmentation and results in fragment of ~300 bp on average, ranging from 100 bp to 1000 bp. The average length of sonicated DNA depends on the duration and strength of sonication [7]. MNase digestion is recommended when analyzing nucleosome stability or when performing ChIP-seq for chromatin marks with a mono-nucleosome resolution [8]. Once the chromatin is fragmented, protein/DNA complexes are immunoprecipitated by using antibodies coupled with paramagnetic beads. Washing steps are included to remove non-specific binding of proteins or DNA to the beads before elution of protein/DNA complexes. After reverse cross-linking and protein digestion, the DNA is extracted using solid-phase reversible immobilization (SPRI) beads. SPRI beads are paramagnetic beads that reversibly bind to DNA in the presence of polyethylene glycol (PEG). DNA size selection is performed using SPRI beads, since increasing the ratio of beads to DNA is reducing the minimum size of DNA fragments captured by the beads. Sequencing libraries are prepared from the immunoprecipitated (IP) DNA as well as for the INPUT (control). Sequenced data of the IP and INPUT are then analyzed and compared using bioinformatics. Here we propose two bioinformatics workflows, to detect peaks with enriched TF binding on one hand and nucleosome positioning on the other hand. These workflows are composed of a combination of publicly available tools, and the link to the manuals or websites of these tools is provided, as well as examples of pseudo scripts including parameters that are typical of ChIP-seq data in *A. thaliana*.

2 Materials

2.1 Materials, Consumables, and Chemicals

1. Formaldehyde.
2. 2 M glycine.
3. Liquid nitrogen.
4. Miracloth.
5. Extraction buffer 1: 0.4 M sucrose, 10 mM Tris–HCl pH 8, 10 mM $MgCl_2$, 5 mM ß-mercaptoethanol, 0.1 mM phenylmethylsulfonyl fluoride (PMSF), protease inhibitor.
6. Extraction buffer 2: 0.24 M sucrose, 10 mM Tris–HCl pH 8, 10 mM $MgCl_2$, 1% Triton X-100, 5 mM ß-mercaptoethanol, 0.1 mM phenylmethylsulfonyl fluoride (PMSF), protease inhibitor.
7. Extraction buffer 3: 1.7 M sucrose, 10 mM Tris–HCl pH 8, 10 mM $MgCl_2$, 0.15% Triton X-100, 5 mM ß-mercaptoethanol, 0.1 mM phenylmethylsulfonyl fluoride (PMSF), protease inhibitor.

8. Nuclei lysis buffer (make fresh): 50 mM Tris–HCl pH 8, 10 mM EDTA, 1% SDS, protease inhibitor.
9. 5 M NaCl.
10. Proteinase K.
11. Parafilm.
12. SPRI beads.
13. 70% ethanol (vol/vol).
14. RNase.
15. MNase enzyme.
16. 0.5 M EDTA.
17. MNase digestion buffer: 20 mM Tris–HCl pH 8, 50 mM NaCl, 1 mM DTT, 0.5% NP-40, 1 mM $CaCl_2$, 0.5 mM phenylmethylsulfonyl fluoride (PMSF), one mini-tablet protease inhibitor for 10 ml of buffer.
18. ChIP dilution buffer: 20 mM Tris–HCl pH 8, 180 mM NaCl, 1 mM EDTA, 1.3% Triton X-100.
19. Magnetic rack for Eppendorf tubes and for 96-well plates.
20. Dynabeads (protein A and protein G).
21. ChIP dilution buffer (make fresh): 16.7 mM Tris–HCl pH 8, 167 mM NaCl, 1.2 mM EDTA, 1.1% Triton X-100.
22. Low-salt wash buffer: 150 mM NaCl, 0.1% SDS, 1% Triton X-100, 2 mM EDTA, 20 mM Tris–HCl pH 8.
23. High-salt wash buffer: 500 mM NaCl, 0.1% SDS, 1% Triton X-100, 2 mM EDTA, 20 mM Tris–HCl pH 8.
24. LiCl wash buffer: 0.25 M LiCl, 1% NP-40 (IGEPAL), 1% sodium deoxycholate, 1 mM EDTA, 10 mM Tris–HCl pH 8.
25. TE buffer: 10 mM Tris–HCl pH 8, 1 m M EDTA.
26. Elution buffer (make fresh): 1% SDS, 0.1 M $NaHCO_3$.
27. 1.5-ml and 2-ml tubes with low absorption of nucleic acids on the tube well.
28. ChIP library preparation kit.
29. Library quantification kit.

2.2 Equipment

1. Sonicator.
2. Rotating wheel.
3. Fluorometric quantification system for low concentrations of DNA.
4. Thermomixer.
5. Automated sample processing system to check the quality of libraries. Here, we used a 4200 TapeStation or 2100 Bioanalyzer instrument.

3 Methods

3.1 Material Collection and Cross-Linking

1. *A. thaliana* seedlings are grown for 7 days on a mesh on solid 1× Murashige and Skoog (MS) media. The roots are detached from the aerial parts using a scalpel (*see* **Notes 1** and **2**).
2. Collect 1–2 g of root tissues into a 50-ml Falcon tube and add 37 ml of 1% formaldehyde for cross-linking (dilution in water freshly prepared, *see* **Notes 3** and **4**).
3. Cross-link under vacuum for 15 min (release the vacuum once after 7 min and reapply). Stop the reaction by adding glycine to a final concentration of 0.125 M. Put under vacuum for an additional 5 min to neutralize the formaldehyde.
4. Remove the seedlings from the vacuum and rinse two times with water. Dry samples as much as possible on paper towel, transfer in an Eppendorf or Falcon tube, and freeze in liquid nitrogen.

3.2 Cell Lysis/Nuclei Isolation

Perform all steps at 2–8 °C with precooled buffers and centrifuges. Add BME, PMSF, and protease inhibitors to solutions just before use.

1. Grind seedlings with mortar and pestle in liquid nitrogen until they become fine powder. Add three times liquid nitrogen during the procedure to help in breaking the cells.
2. Transfer the powder in a Falcon tube and add 30 ml of extraction buffer 1 to the powder in a 50-ml Falcon tube (see **Note 5**) and leave on ice for 5 min. Use similar amounts of powder for all extractions.
3. Filter the solution through the Miracloth into a fresh 50-ml Falcon tube. Repeat once.
4. Spin the filtered solution for 20 min at 1800 × *g* at 4 °C.
5. Gently remove the supernatant by inverting the Falcon and resuspend the pellet in 1 ml of extraction buffer 2. Transfer the solution to a new 1.5-ml Eppendorf tube.
6. Centrifuge at 13,500 × *g* for 10 min at 4 °C.
7. Remove the supernatant and resuspend the pellet in 300 μl of extraction buffer 3.
8. In a clean Eppendorf, add 300 μl of extraction buffer 3. Take the 300 μl solution (resuspended pellet) from **step 7** and carefully transfer it on top of the clean 300 μl of extraction buffer 3 (*see* **Note 6**).
9. Spin for 1 h at 21,000 × *g* at 4 °C. If the pellet is still green, repeat **steps 7–9**. From this step, you can fragment the chromatin either by sonication (Subheading 3.2, **steps 10–24**) or MNase digestion (Subheading 3.2, **steps 25–31**). Sonication is

the most commonly used method of fragmentation and results in fragment of ~300 bp on average, ranging from 100 bp to 1000 bp. MNase digestion is recommended when analyzing nucleosome stability or when performing ChIP-seq for chromatin marks with a mono-nucleosome resolution.

Sonication

10. Remove the supernatant and resuspend the pellet in 300 μl of nuclei lysis buffer.
11. Flash freeze two times in liquid nitrogen to break the nuclear membrane.
12. Keep 5 μl aliquot of non-sonicated chromatin to test sonication efficiency (Subheading 3.2, **step 15**).
13. Sonicate the chromatin by doing two times seven cycles of sonication (alternating 30 s *on*, 30 s *off*) at a high sonication intensity (*see* **Note** 7). Cool down the samples on ice for 10 min between the two sequences of seven cycles.
14. Take 10 μl of sonicated chromatin to test sonication efficiency.
15. Perform reverse cross-linking and DNA extraction on the non-sonicated and sonicated aliquots.
16. Complete non-sonicated and sonicated aliquots to 57.6 μl with nuclei lysis buffer.
17. Add 2.4 μl of 5 M NaCl and incubate overnight at 65 °C to reverse cross-link.
18. Add 1 μl proteinase K, for 30 min at 50 °C.
19. Remove SPRI beads from 4 °C and leave at room temperature for 30 min.
20. Extract DNA using SPRI beads: Add 120 μl of SPRI beads to 60 μl of sample (2× ratio of beads/DNA) and mix well by pipetting up and down at least ten times. Incubate for 2 min at room temperature and place on a magnetic rack for 4 min to capture the beads. Discard the supernatant (*see* **Note 8**).
21. Keep the tubes on the magnetic rack and add 500 μl of freshly prepared 70% ethanol (vol/vol). Incubate for 30 s without disturbing the beads and discard the supernatant. Repeat this step once more and remove all ethanol. Let the beads dry for 10–15 min at room temperature (*see* **Note 9**).
22. Remove the tubes from the magnetic rack. To elute DNA, resuspend the beads in 12 μl of 10 mM Tris–HCl (pH 8) by pipetting up and down at least ten times. Incubate for 2 min and place on the magnetic rack for 4 min to capture the beads. Transfer 10 μl of the supernatant to a new tube.
23. Incubate for 10 min at 37 °C with 1 μl of RNase.

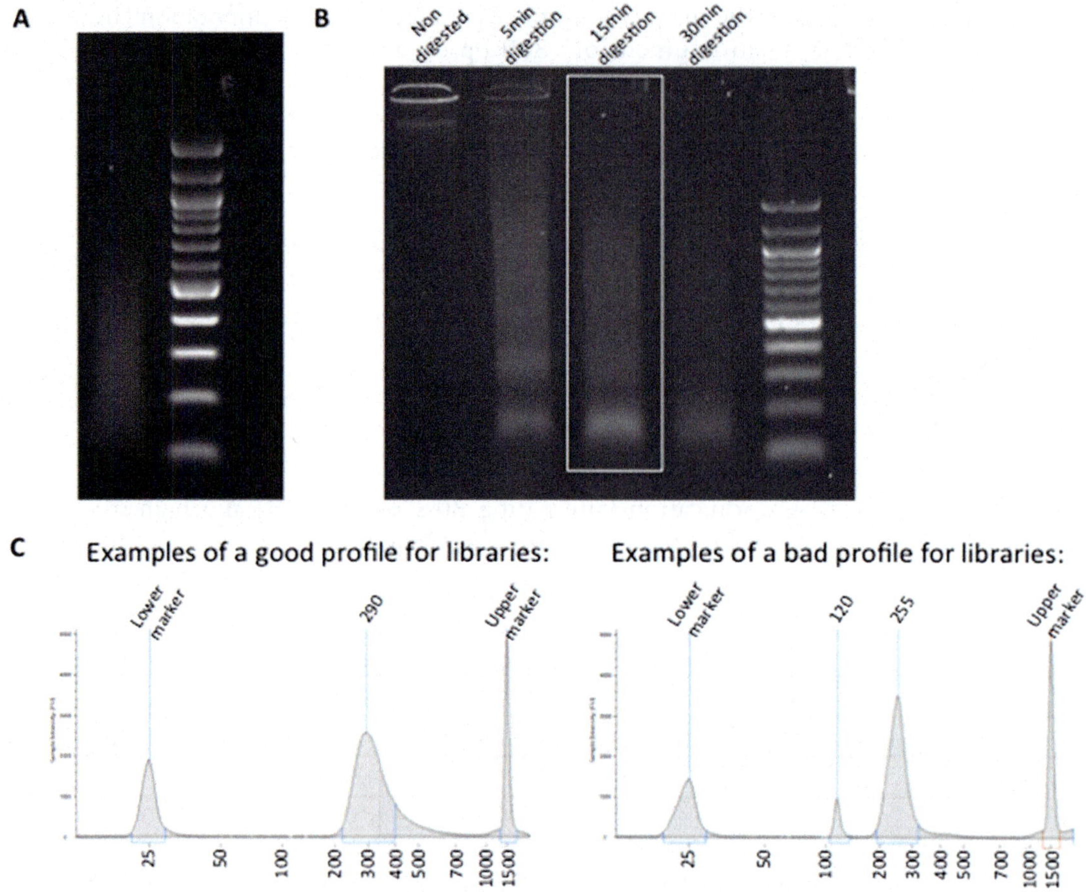

Fig. 1 (**a**) Example of a DNA gel of sonicated chromatin. The DNA smear is going from 100 bp to 1000 bp, with the maximum intensity around 300 bp. (**b**) Example of a DNA gel of MNase digestion test. The best profile for MNase digestion is the middle one, corresponding to 15 min of digestion. Mono-nucleosomes, which are at ~150 bp, are the most abundant. However, di-nucleosomes are still observed, indicating that the chromatin is not over-digested. (**c**) Examples of good (left) and bad (right) profiles of libraries on a 4200 TapeStation. The peak at 120 in the example of bad profile corresponds to adaptor contaminations, one of the most common problems occurring during library preparation

24. Run on a 1.6% gel at 60 V for 90 min to test the sonication efficiency. *See* Fig. 1a for an example of sonication profile.

MNase digestion

25. After the 1-h centrifugation at 21,000 × *g* at 4 °C, remove the supernatant and resuspend the nuclei pellet in 1 ml of MNase digestion buffer. Perform two flash freeze in liquid nitrogen to break the nuclear membrane.
26. Take an aliquot of 40 μl and complete to 57.6 μl with MNase digestion buffer and extract DNA using SPRI beads as explained in Subheading 3.2, **step 15** to estimate the concentration of chromatin by qPCR.

27. Quantify DNA with a fluorometric quantification system for low concentrations of DNA following manufacturer's instructions.
28. Adjust all the samples to the same concentration to a final volume of 500 μl (only half of the chromatin is used, keep the rest at −80 °C in case the digestion is not working).
29. Take 15 μl of non-digested chromatin to test the efficiency of the digestion and complete to 60 μl with MNase digestion buffer. Take 60 μl of non-digested chromatin as a control (first INPUT).
30. Perform MNase digestion on the rest of the samples (425 μl). Warm the tubes to 37 °C for 1–2 min and digest by addition of 0.4 U/ml of MNase (Sigma, N3755) for 15 min at 37 °C (*see* **Note 10**). Stop the reaction by adding EDTA to a final concentration of 5 mM. Take an aliquot of 40 μl of digested chromatin and complete with MNase digestion buffer to 60 μl.
31. Extract DNA for the non-digested and digested aliquots to test the digestion (*see* Subheading 3.2, **step 15** for DNA extraction) and run on a 1.6% gel at 60 V for 1H30. *See* Fig. 1b for an example of MNase digestion profile.

3.3 ChIP

1. Prepare the ChIP dilution buffer. Centrifuge the chromatin at 16,000°× *g* for 5 min at 4 °C to get rid of debris. Carefully transfer the supernatant in a 2-ml tube.
2. If you have performed sonication (Subheading 3.2, **steps 10–24**), use 100 μl of this chromatin for the IP and 50 μl as an INPUT. Keep the rest at −80 °C. Add 900 μl of ChIP dilution buffer to the IP fraction. If you have done MNase digestion (Subheading 3.2, **steps 25–31**), use 250 μl of this chromatin for the IP and use the rest as a digested INPUT. Add 750 μl of ChIP dilution buffer to the IP fraction.
3. Prepare antibody-beads complexes: Use 50 μl of Dynabeads per IP (50% protein A, 50% protein G) and 5 μg of antibody. Prepare a tube with the total amount of beads required for the experiment. Wash the beads three times with 1 ml of ChIP dilution buffer. At each wash, incubate the tube on the magnetic rack for 1 min to capture the beads (Magtration) and remove the supernatant. Remove the tube from the magnetic rack and resuspend the beads with 1 ml of ChIP dilution buffer. After the three washes, resuspend the beads in ChIP dilution buffer and add the appropriate amount of antibody (5 μg in 250 μl of ChIP dilution buffers per IP).
4. Incubate for 2 h at 4 °C with gentle agitation on a rotating wheel. Wash the beads one time with 1 ml of ChIP dilution buffer and resuspend in the initial volume of beads.

5. Combine beads (50 μl) and the diluted chromatin prepared in Subheading 3.3, **step 2** (1 ml) in a 1.5-ml tube. Incubate overnight at 4 °C with gentle agitation on a rotating wheel.
6. Collect immune complexes by Magtration and wash by incubating at 4 °C on a rotating wheel with 1 ml of the following wash buffers (*see* **Note 11**):
   (a) Low-salt wash buffer: two washes of 5 min each
   (b) High-salt wash buffer: two washes of 5 min each
   (c) LiCl wash buffer: one wash of 5 min
   (d) TE: two washes of 5 min each
5. After the second wash with TE, resuspend the beads in 100 μl of elution buffer and transfer into a new tube (*see* **Note 12**).
6. To elute the chromatin from the beads, incubate at 65 °C for 15 min under constant agitation using a thermomixer (1200 rpm). Carefully transfer the supernatant fraction (eluate) to another tube by Magtration. Repeat the elution once more and combine the two eluates.
7. Complete the INPUTs to 200 μl with elution buffer. Add 8 μl of 5 M NaCl to the eluate and INPUTs, close well with parafilm, and reverse cross-link at 65 °C overnight.
8. Add 4 μl of 0.5 M EDTA, 8 μl 1 M Tris–HCl, pH 6.5, and 1 μl of 14 mg/ml proteinase, close well the tubes with parafilm, and incubate for 1 h at 45 °C.
9. Extract DNA from the protein/DNA complexes with SPRI beads.
10. Place the SPRI beads at room temperature for 30 min. Add 442 μl of SPRI beads to 221 μl of sample (2× ratio) and mix well by pipetting up and down at least ten times. Incubate for 2 min at room temperature and place on a magnetic rack for 4 min to capture the beads. Discard the supernatant.
11. Keep the tubes on the magnetic rack and add 700 μl of freshly prepared 70% ethanol (vol/vol). Incubate for 30 s without disturbing the beads and discard the supernatant. Repeat this step once more and then remove the ethanol completely. Let the beads dry for 10 min at room temperature.
12. Remove the tubes from the magnetic rack. To elute DNA, resuspend the beads in 52 μl of 10 mM Tris–HCl (pH 8) by pipetting up and down at least ten times. Incubate for 2 min at room temperature and place on the magnetic rack for 4 min to capture the beads. Transfer 50 μl to a new tube.
13. Incubate for 10 min at 37 °C with 1 μl of RNase.
14. Use 1 μl of IP and 1 μl of 1/10 INPUT to test the IP efficiency by qPCR. In the absence of positive control (known position of binding), proceed directly to sequencing. If MNase digestion is

performed, test MNase digestion efficiency and nucleosome stability by comparing qPCR results for the non-digested and digested INPUTs.

3.4 Library Preparation, Size Selection, and Quantification

Library preparation is done using the TruSeq ChIP sample prep kit with minor modifications (*see* **Note 13**).

1. Quantify the INPUT with a fluorometric quantification system for low concentrations of DNA and use 10–20 ng for library preparation. Use the entire IP fraction for library preparation.
2. Follow the TruSeq ChIP sample prep kit instructions from the end repair step to adaptor ligation.
3. Skip the size selection step of the TruSeq ChIP sample prep manual and follow instructions for the PCR amplification. The step is instead performed after amplification to avoid too important loss of DNA.
4. Size selection is performed after library preparation using SPRI beads to remove the adaptors (normally less than 100 bp in length) as well as oversized fragments. The following example is to remove fragments that are more than 500 bp for MNase-digested samples. It is based on the fact that at higher beads/DNA ratio, the minimum size of DNA fragments captured by the beads is reduced. To remove fragment of another size, read the following documentation on how to perform size selection with SPRI beads and recommended beads/DNA ratios: https://www.beckmancoulter.com/wsrportal/bibliography?docname=SPRIselect.pdf.
5. Add resuspension buffer to 100 μl on libraries.
6. Place the SPRI beads at room temperature for 30 min. Add 65 μl of SPRI beads (0.65× ratio of beads/DNA to remove fragments >500 bp).
7. Incubate at room temperature for 5 min.
8. Incubate on the magnetic stand for 5 min. Transfer 155 μl of supernatant in a new plate (contains fragments <500 bp) and add 50 μl of beads (1.15× ratio of beads/DNA to keep fragments >100 bp). Incubate at room temperature for 5 min.
9. Incubate on the magnetic stand for 5 min and discard the supernatant.
10. With the PCR plate on the magnetic stand, add 200 μl freshly prepared 70% ethanol without disturbing the beads. Incubate the PCR plate at room temperature for 30 s, and then remove supernatant. Repeat once for a total of two ethanol washes, and remove all ethanol.
11. With the PCR plate on the magnetic stand, let the samples air-dry at room temperature for 15 min.

12. Resuspend the dried pellet with 17.5 μl resuspension buffer by pipetting the entire volume up and down ten times.
13. Incubate the PCR plate at room temperature for 2 min. Place the PCR plate on the magnetic stand at room temperature for 5 min.
14. Transfer 15 μl of the clear supernatant into a new 96-well 0.3-ml PCR plate.
15. Check the quality of libraries using an automated sample processing system, like the 4200 TapeStation or 2100 Bioanalyzer instrument, following manufacturer's instructions. Examples of good- and bad-quality libraries are represented in Fig. 1c.
16. Quantify the libraries using a library quantification kit following manufacturer's instructions.
17. Dilute libraries to 10 nM and pool 5 μl of the libraries that are in the same lane.
18. Quantify the pool by fluorometric quantification system for low concentrations of DNA and convert from ng/μl to nM by using this website: http://molbiol.edu.ru/eng/scripts/01_07.html. Dilute the pool to the concentration required by the sequencer or sequencing platform (*see* **Note 14**).

3.5 Bioinformatics Analysis of ChIP-Seq (See Note 15)

See Fig. 2 *for the workflow of the bioinformatics analyses. Some tools might require installation of java and python in order to run the scripts described here.*

3.5.1 Quality Check and Alignment on the Genome

1. This first step is to check the quality of the sequenced reads with FastQC and to check for sequencing adaptors and low-quality bases that may affect mapping (http://www.bioinformatics.babraham.ac.uk/projects/fastqc/). Low-quality reads might be due to over-clustering or other problems during sequencing. To check the quality of sequencing reads with FastQC, type:

```
fastqc mydata.fastq
```

2. The genome of *Arabidopsis thaliana* (e.g., TAIR10) or other genome of interest in the FASTA format needs to be "indexed" before it can be used as reference for mapping. Index the latest version of the *A. thaliana* genome to use it later for alignment with bowtie2 [9] (http://computing.bio.cam.ac.uk/local/doc/bowtie2.html):

```
bowtie2-build genome.fa genome
```

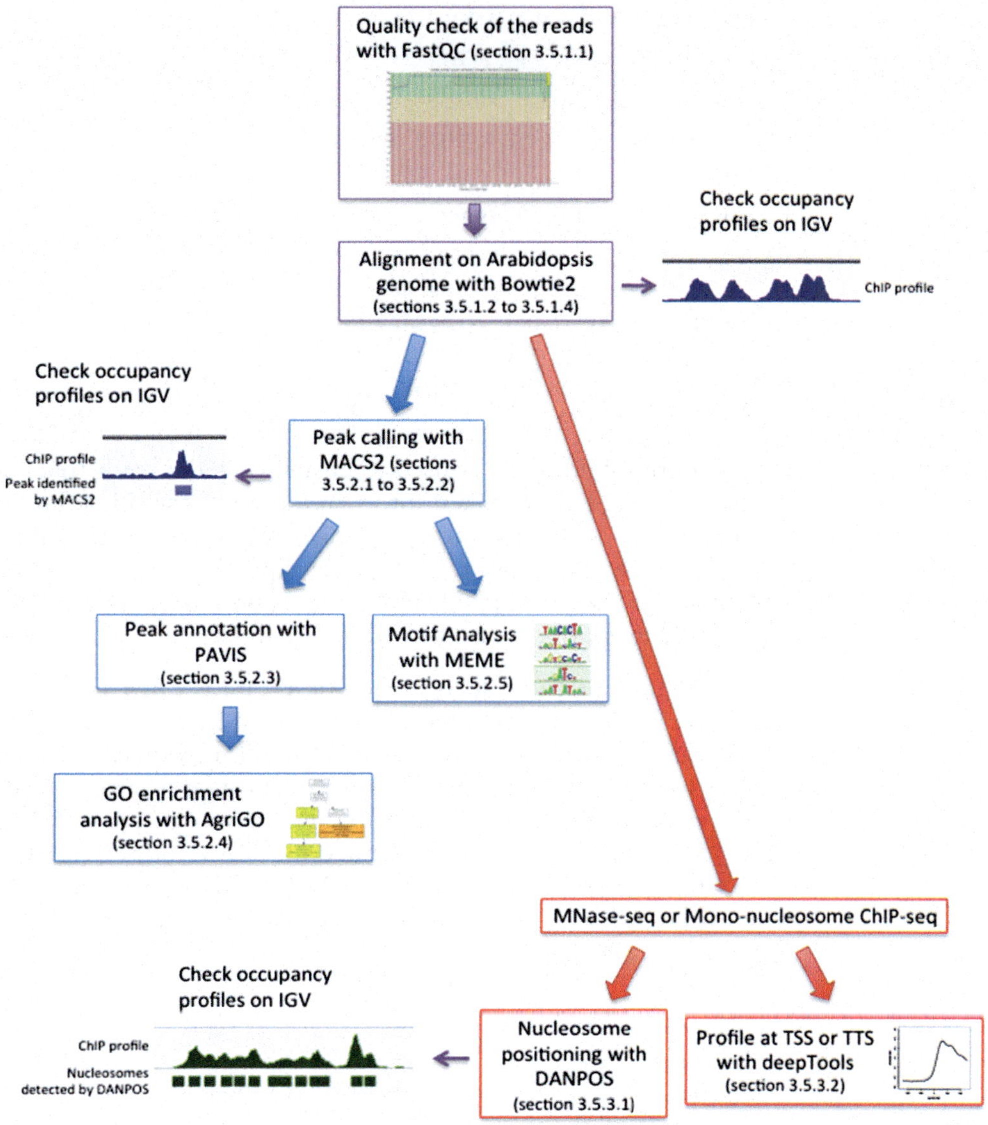

Fig. 2 Workflow for bioinformatics analyses of ChIP-seq data with examples of results. The steps that are common for TF and nucleosome ChIP-seq are in purple, while steps that are recommended for TF or chromatin marks with sharp peaks are in blue and the ones for MNase-seq or mono-nucleosome ChIP-seq are in red. Motif examples are from Franco-Zorrilla and colleagues [21] and are in the *Arabidopsis* MEME motif database

3. Perform alignment of the sequencing result on the *A. thaliana* genome using bowtie2 (*see* **Note 16**). A SAM file, which is a TAB-delimited file containing sequence alignment data, will be created.

For single-end data, type:

```
bowtie2 -x /pathToGenome/genome \
-U mydata.fastq -S mydata.sam --no-unal
```

For paired-end data, type:

```
bowtie2 -x /pathToGenome/genome \
-1 mydata_R1.fastq -2 mydata_R2.fastq \
-S mydata.sam --maxins 800 \
--no-mixed --no-discordant --no-unal
```

The option “no-unal” is used to ensure only the reads that can be properly aligned are taken into consideration for single-end samples, similar to the no-mixed, no-discordant options that are used to keep only the paired reads that are aligned in correct orientation and ignore the unpaired alignment. For more information on alignment option, please *see* Bowtie2 manual.

4. Alignment tools such as Bowtie create a BAM file, a binary version of the SAM file. Sort the BAM file and remove duplicates and indices. Read duplicates will most likely come from the PCR amplification during library preparation and are removed to avoid biases during analysis (*see* **Note 17**). The BAM file, together with the BAI file (BAM index), can be used to visualize the aligned IP and INPUT data Integrative Genomics Viewer [10] (IGV, http://software.broadinstitute.org/software/igv/).

```
samtools view -bS mydata.sam > mydata.bam
samtools sort mydata.bam mydata _sorted
java -Xmx4g -jar /picard-tools-1.103/MarkDuplicates.jar INPUT= mydata_sorted.bam OUTPUT mydata_rmdup_picard.bam METRICS_FILE=dup.txt VALIDATION_STRINGENCY=LENIENT REMOVE_DUPLICATES=true
samtools index mydata _rmdup_picard.bam
```

3.5.2 Analysis of TF ChIP-Seq Data: Peak Calling and Functional Analysis of the Peaks

1. Perform peak calling using MACS2 [11] for the ChIP-seq using the INPUT as a control (https://pypi.python.org/pypi/MACS2). The INPUT is important as this point and is used as a control to detect enriched peaks in the IP. MACS2 is optimized for sharp peaks obtained when doing ChIP-seq on transcription factors and is not recommended for ChIP of nucleosomes or chromatin marks (*see* Subheading 3.5.3 for the analysis of nucleosomes ChIP-seq). In bold are parameters that can be adjusted between experiments, depending on the fragment sizes.

```
macs2 callpeak -t myChIP_rmdup_picard.bam -c myINPUT_rmdup_-
picard.bam -g genomeSize --tsize sequensingSize\
-n myPeakFile --outdir myPeakFolder -B --SPMR --nomodel
--extsize fragmentZise
```

If doing peak calling on chromatin marks, it is recommended to use the parameter --broad-cutoff or to use other peak calling tools designed for broad peaks, such as SICER [12].

2. Create fold change or log ratio files of the IP versus the INPUT to visualize on IGV.

Fold change:

```
macs2 bdgcmp -t myPeakFile_treat_pileup.bdg -c myPeakFile_-
control_lambda.bdg -o myPeakFile_FE.bdg -m FE
bedGraphToBigWig myPeakFile_FE.bdg /genome/ChromInfo.txt my-
PeakFile_FE.bw
```

Log ratio:

```
macs2 bdgcmp -t myPeakFile_treat_pileup.bdg -c myPeakFile_-
control_lambda.bdg -o myPeakFile_logLR.bdg -m logLR -p 0.00001
bedGraphToBigWig myPeakFile_logLR.bdg /genome/ChromInfo.txt
myPeakFile_logLR.bw
```

3. Perform annotation of the peaks using PAVIS [13]. PAVIS is a program containing several utilities for the annotation and visualization of ChIP-seq peaks (http://manticore.niehs.nih.gov/pavis2/). PAVIS requires the peak files in BED format. A BED file is a TAB-delimited text file that defines the position of specific features, in this case the peaks identified by MACS2. To convert myPeakFile.narrowPeak to a BED file, type:

```
awk 'BEGIN{FS=OFS="\t"};{print $1,$2,$3}' myPeakFile.narrow-
Peak > myPeakFile.bed
```

To use PAVIS, go to http://manticore.niehs.nih.gov/pavis2/. In the "Species/Genome Assembly/Gene Set" drop-down menu, select "Arabidopsis thaliana Columbia TAIR10 (Thales cress)." Fill in the location of the peak file myPeakFile.bed, and leave the default parameters for the remaining options. Click on SUBMIT to run the tool. This list of closest downstream genes can be found under the link "The Full Annotation File."

4. Perform Gene Ontology (GO) enrichment of the genes obtained from PAVIS using AgriGo [14] (http://bioinfo.cau.edu.cn/agriGO/analysis.php), Gorilla [15, 16] (http://cbl-gorilla.cs.technion.ac.il/), or Gene Ontology Consortium [17, 18] (http://www.geneontology.org/).
5. Perform motif enrichment analysis for the peaks using MEME [19, 20]. The input for MEME should be a file in the FASTA format containing the sequences of the identified peaks. Other peak-finding tools merge overlapping areas of enrichment, and the resulting peaks tend to be much wider than the actual binding sites. It is important to subdivide the enriched areas by accurately partitioning enriched loci into a finer resolution set of individual binding sites. This will enhance the quality of the motif analysis. For this, redo peak calling with MACS2 using the --call-summits option.

```
macs2 callpeak -t myChIP_rmdup_picard.bam -c myINPUT_rmdup_-picard.bam -g genomeSize --tsize sequensingSize -n mySubPeak-File --outdir mySubPeakFolder -B --SPMR --nomodel --extsize fragmentZise -call-summits
```

Use bedtools to extract the sequences of the peak summits in a FASTA format:

```
bedtools getfasta -fi pathToGenome/genome.fa -bed mySubPeak-File.bed -fo mySubPeakFile.fa
```

To perform de novo motif discovery using MEME, open a web browser, go to the MEME website at http://meme-suite.org/, and choose the "MEME" tool. Fill in the necessary details: Upload the sub-peaks FASTA file mySubPeakFile.fa, indicate the number of expected motifs (1 per sequence), the width of desired motif (between 6 and 20), and the maximum number of motifs to find (3 by default). Start Search and follow the link "MEME html output" once the analysis is done. The TF motifs in MEME for *A. thaliana* originate from two studies using in vitro approaches [21, 22].

3.5.3 Analysis of MNase-Seq and ChIP-Seq for Chromatin Marks: Nucleosome Positioning and Profiles at Genes

1. Analyze nucleosome positioning using DANPOS [23]. Nucleosomes consist of histone octamers that are wrapped in 147 bp of DNA. Nucleosome positions can thus be inferred from MNase-seq and ChIP-seq of histones that show peaks with the widths close to 147 bp. To perform nucleosome positioning using DANPOS, type:

```
python danpos.py dpos mydata _rmdup_picard.bam
```

2. Analyze ChIP-seq profiles at transcription start sites (TSS) and transcription termination sites (TTS) of genes using deepTools [24] (http://deeptools.readthedocs.io/en/latest/). deepTools is a suite of tools for visualization, quality control, and normalization of next-generation sequencing data (*see* **Note 18**). deepTools can be installed as a stand-alone command line application or as a Galaxy tool. To analyze the ChIP-seq profile, the first step is to create a bigwig file from the BAM file that is going to be used for plotting, using bamCoverage:

```
bamCoverage -b mydata_rmdup_picard.bam -o mydata_rmdup_pi-
card.bw
```

To have more confidence on the profiles obtained, it is recommended to normalize the signal of the IP by its corresponding INPUT. To do so, use bamCompare:

```
bamCompare -b1 treatment_rmdup_picard.bam -b2 control_rmdup_-
picard.bam -o log2ratio.bw
```

Then create a matrix, which is required for plotting the ChIP profiles, with computeMatrix, using the mode to center profiles around TSS or TTS.

```
computeMatrix reference-point -S mydata_rmdup_picard.bw -R
<bed file> -a 1000 -b 1000 -out mydata_rmdup_picard.mat.gz
computeMatrix reference-point -S log2ratio.bw -R <bed file>
-a 1000 -b 1000 -out log2ratio.mat.gz
```

To plot profiles of ChIP-seq, use plotProfile:

```
plotProfile -m mydata_rmdup_picard.mat.gz -out Profile.png
plotProfile -m log2ratio.mat.gz -out Profile.png
```

4 Notes

1. Always collect the materials at the same time of the day to reduce circadian rhythm effects.
2. Wear gloves during the entire protocol.
3. Cross-linking should be done in a fume hood, as formaldehyde is toxic.

4. If working with transcription factors with low affinity to chromatin, it is possible to use other cross-linking reagents, as described by Tian and colleagues [25], for example.
5. The mentioned volumes of buffers for the chromatin extraction are for 1 g of powder. If using more powder, use more than 30 ml of buffer or split into multiple tubes.
6. The two layers of buffer 3 should be well separated before centrifugation. It is important to avoid any bubbles at this step as it can cause problems.
7. Sonication intensity, length of time, and number of cycles in this protocol were optimized with a Bioruptor® from Diagenode and are given as an example. Always optimize sonication the first time you perform this protocol and when working with new plant materials and/or sonicating equipment.
8. To remove supernatant more easily when working with beads, put a second 200 μl filter tip on top of the first 1000 nonfilter tip when pipetting.
9. Beads are well dried when cracks appear in the beads pellet. Beads should be well dried before elution as any ethanol remaining at this step can cause problems in the following steps.
10. Concentration of MNase enzyme and digestion times in this protocol are given as an example. Always optimize this part of the protocol the first time you perform this protocol and when working with new plant materials.
11. If working with transcription factors with low affinity with chromatin, you can reduce risks of losing the TF/DNA interaction by replacing the high-salt and LiCl buffers with the low-salt wash buffer during the washes.
12. It is important to transfer in a new tube after the last wash to reduce background signal.
13. This protocol was optimized using the TruSeq ChIP sample preparation kit from Illumina.
14. For TF ChIP-seq, single-end sequencing is normally sufficient for peak calling of TF binding events. It is recommended, however, to perform paired-end sequencing to define nucleosome positioning in the mono-nucleosome ChIP-seq.
15. Installation of analysis tools is not described. To know how to install them, follow the link to the manuals/descriptions of each tool in the protocol.
16. Code parameters are indicative and should be optimized each time.
17. In some cases, like for ChIP-seq of TF with few targets and sharp peaks, read duplicates can be informative, and you may

need to keep them. Always check the effect of removing duplicates on your results when performing ChIP on a new TF before deciding to remove or keep them.

18. plotHeatmap function in deepTools can also be used to generate heatmaps starting from the same matrix generated to plot the profile at TSS or TTS.

Acknowledgment

S.C. was supported by an EMBO long-term fellowship (ALTF 290-2013). V.C. lab is supported by the Thailand Research Fund (TRF) Grant for New Scholar (MRG6080235); Newton Advanced Fellowship through TRF (DBG60800003) and Royal Society (NA160153); the Faculty of Science, Mahidol University; and the Crown Property Bureau Foundation. The Wigge lab is supported by the Gatsby Charitable Foundation, the European Research Council, and the Biotechnology and Biological Sciences Research Council. The authors have no conflict of interest.

References

1. Bannister AJ, Kouzarides T (2011) Regulation of chromatin by histone modifications. Cell Res 21(3):381–395
2. Morohashi K, Xie Z, Grotewold E (2009) Gene-specific and genome-wide ChIP approaches to study plant transcriptional networks. Methods Mol Biol 553:3–12
3. Chow BY, Kay SA (2013) Global approaches for telling time: omics and the Arabidopsis circadian clock. Semin Cell Dev Biol 24 (5):383–392
4. Valouev A, Johnson DS, Sundquist A, Medina C, Anton E, Batzoglou S, Myers RM, Sidow A (2008) Genome-wide analysis of transcription factor binding sites based on ChIP-Seq data. Nat Methods 5(9):829–834
5. Ferrier T, Matus JT, Jin J, Riechmann JL (2011) Arabidopsis paves the way: genomic and network analyses in crops. Curr Opin Biotechnol 22(2):260–270
6. Solomon MJ, Varshavsky A (1985) Formaldehyde-mediated DNA-protein crosslinking: a probe for in vivo chromatin structures. Proc Natl Acad Sci U S A 82 (19):6470–6474
7. Ghavi-Helm Y, Zhao B, Furlong EE (2016) Chromatin immunoprecipitation for analyzing transcription factor binding and histone modifications in Drosophila. Methods Mol Biol 1478:263–277
8. Ballare C, Castellano G, Gaveglia L, Althammer S, Gonzalez-Vallinas J, Eyras E, Le Dily F, Zaurin R, Soronellas D, Vicent GP, Beato M (2013) Nucleosome-driven transcription factor binding and gene regulation. Mol Cell 49(1):67–79
9. Langmead B, Salzberg SL (2012) Fast gapped-read alignment with Bowtie 2. Nat Methods 9 (4):357–359
10. Robinson JT, Thorvaldsdottir H, Winckler W, Guttman M, Lander ES, Getz G, Mesirov JP (2011) Integrative genomics viewer. Nat Biotechnol 29(1):24–26
11. Zhang Y, Liu T, Meyer CA, Eeckhoute J, Johnson DS, Bernstein BE, Nusbaum C, Myers RM, Brown M, Li W, Liu XS (2008) Model-based analysis of ChIP-Seq (MACS). Genome Biol 9(9):R137
12. Zang C, Schones DE, Zeng C, Cui K, Zhao K, Peng W (2009) A clustering approach for identification of enriched domains from histone modification ChIP-Seq data. Bioinformatics 25(15):1952–1958
13. Huang W, Loganantharaj R, Schroeder B, Fargo D, Li L (2013) PAVIS: a tool for Peak Annotation and Visualization. Bioinformatics 29(23):3097–3099
14. Du Z, Zhou X, Ling Y, Zhang Z, Su Z (2010) agriGO: a GO analysis toolkit for the

agricultural community. Nucleic Acids Res 38 (Web Server issue):W64–W70

15. Eden E, Lipson D, Yogev S, Yakhini Z (2007) Discovering motifs in ranked lists of DNA sequences. PLoS Comput Biol 3(3):e39
16. Eden E, Navon R, Steinfeld I, Lipson D, Yakhini Z (2009) GOrilla: a tool for discovery and visualization of enriched GO terms in ranked gene lists. BMC Bioinformatics 10:48
17. Ashburner M, Ball CA, Blake JA, Botstein D, Butler H, Cherry JM, Davis AP, Dolinski K, Dwight SS, Eppig JT, Harris MA, Hill DP, Issel-Tarver L, Kasarskis A, Lewis S, Matese JC, Richardson JE, Ringwald M, Rubin GM, Sherlock G (2000) Gene ontology: tool for the unification of biology. The Gene Ontology Consortium. Nat Genet 25(1):25–29
18. Gene Ontology Consortium (2015) Gene Ontology Consortium: going forward. Nucleic Acids Res 43(Database issue):D1049–D1056
19. Bailey TL, Elkan C (1994) Fitting a mixture model by expectation maximization to discover motifs in biopolymers. Proc Int Conf Intell Syst Mol Biol 2:28–36
20. Bailey TL, Boden M, Buske FA, Frith M, Grant CE, Clementi L, Ren J, Li WW, Noble WS (2009) MEME SUITE: tools for motif discovery and searching. Nucleic Acids Res 37 (Web Server issue):W202–W208
21. Franco-Zorrilla JM, Lopez-Vidriero I, Carrasco JL, Godoy M, Vera P, Solano R (2014) DNA-binding specificities of plant transcription factors and their potential to define target genes. Proc Natl Acad Sci U S A 111 (6):2367–2372
22. O'Malley RC, Huang SS, Song L, Lewsey MG, Bartlett A, Nery JR, Galli M, Gallavotti A, Ecker JR (2016) Cistrome and epicistrome features shape the regulatory DNA landscape. Cell 165(5):1280–1292
23. Chen K, Xi Y, Pan X, Li Z, Kaestner K, Tyler J, Dent S, He X, Li W (2013) DANPOS: dynamic analysis of nucleosome position and occupancy by sequencing. Genome Res 23 (2):341–351
24. Ramírez F, Ryan DP, Grüning B, Bhardwaj V, Kilpert F, Richter AS, Heyne S, Dündar F, Manke T (2016) deepTools2: a next generation web server for deep-sequencing data analysis. Nucleic Acids Res 44(W1):W160–W165
25. Tian B, Yang J, Brasier AR (2012) Two-step cross-linking for analysis of protein-chromatin interactions. Methods Mol Biol 809:105–120

Chapter 19

μChIP-Seq for Genome-Wide Mapping of In Vivo TF-DNA Interactions in *Arabidopsis* Root Protoplasts

Alessia Para, Ying Li, and Gloria M. Coruzzi

Abstract

Chromatin immunoprecipitation (ChIP) is a widely used method to map the position of DNA-binding proteins such as histones and transcription factors (TFs) upon their interaction with particular regions of the genome. To examine the genomic distribution of a TF in specific cell types in response to a change in nitrogen concentration, we developed a micro-ChIP (μChIP) protocol that requires only ~5000 *Arabidopsis* cells transiently expressing the *Arabidopsis* TF *B*asic Leucine Zipper 1 (bZIP1) fused to the glucocorticoid receptor (GR) domain that mediates nuclear import in the presence of dexamethasone. The DNA fragments obtained from the immunoprecipitation of bZIP1-DNA complexes were analyzed by next-generation sequencing (ChIP-seq), which helped uncover genome-wide associations between a bZIP1 and its targets in plant cells upon fluctuations in nitrogen availability.

Key words Chromatin immunoprecipitation, Transcription factor, Plant protoplasts, Cross-linking, Library preparation, Next-generation sequencing

1 Introduction

The interaction between DNA and TFs is a fundamental aspect of the regulation of gene expression. Upon binding to specific *cis* motifs mainly found in regulatory regions such as promoters and 5′ UTRs, TFs can switch the expression of an array of genes on and off at a particular time and in a tissue-specific fashion, therefore coordinating the appropriate cellular response to the most diverse physiological, developmental, or environmental stimuli.

Chromatin immunoprecipitation (ChIP) is a powerful technique to uncover the points of physical contact between a TF and the genome, revealing when and where the interaction takes place in vivo. A critical requirement of ChIP is the ability to lock a TF to its binding site by chemical cross-linking to covalently stabilize the TF-DNA interactions (Fig. 1c). TF-DNA complexes can then be recovered and dissociated to isolate the once-bound DNA and

Daniela Ristova and Elke Barbez (eds.), *Root Development: Methods and Protocols*, Methods in Molecular Biology, vol. 1761, https://doi.org/10.1007/978-1-4939-7747-5_19,

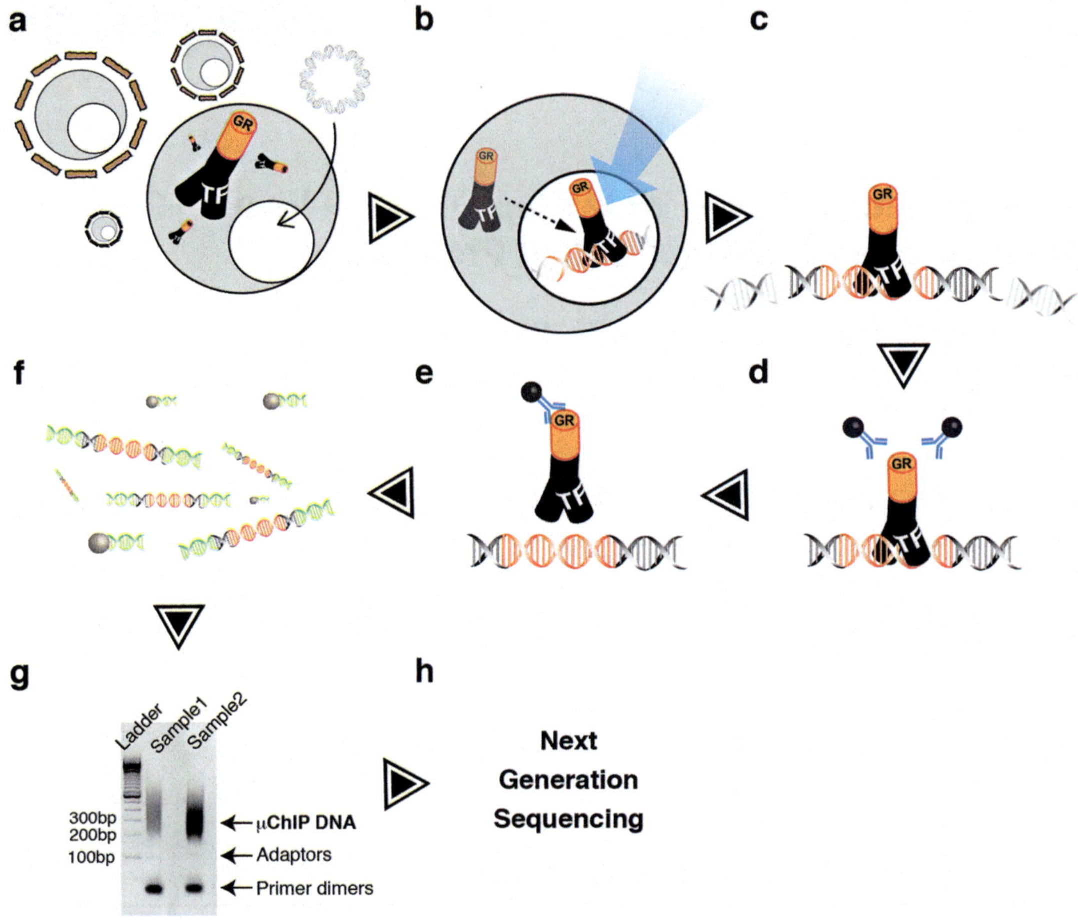

Fig. 1 Workflow for μChIP-seq in *Arabidopsis* root protoplasts. (**a**) The cell wall is removed from root cells, and protoplasts are transfected to achieve transient expression of a protein fusion between the TF and the GR domain to control nuclear translocation. (**b**) After triggering nuclear import (dotted line), the protoplasts are treated with cross-linking agent that penetrates into the nucleus (light blue arrow) and locks the TF onto the TF-binding motif (red). (**c**) Cell lysis brings the complexes into solution, and sonication sheers the DNA with the TF bound to it into shorter fragments. (**d**) TF-DNA complexes are captured with anti-GR Ab—conjugated magnetic beads. (**e**) Bound DNA is released from the complex, (**f**) ligated to adaptors (green), and amplified. Adapted dimers are removed by magnetic beads. (**g**) Size selection by agarose gel purification refines library size and removes library preparation artifacts before μChIP DNA is sent for sequencing (**h**)

determine its identity and quantity either by PCR (for single gene analysis) or by sequencing (for genome-wide mapping).

The identification of the binding sites of plant TFs by ChIP offers a particular challenge because of the presence of a rigid cell wall, which hinders the penetration of the cross-linking agent. Hence, a large amount of starting material is required to capture enough TF-DNA complexes for further analysis. This constraint, in addition to increasing the level of background noise from non-bound DNA, makes ChIP application cumbersome for small plant organs like young *Arabidopsis* roots. Moreover, by the time the cross-linking agent reaches the nucleus of intact plant cells, transient TF interactions are likely to have ceased. To improve cross-linking efficiency, we exploited the ability to make plant protoplasts by stripping plant cells of the cell wall (Fig. 1a) [1]. We expressed the *Arabidopsis* TF bZIP1 in protoplasts and triggered its nuclear localization using the dexamethasone-inducible system (Fig. 1b) [2]; next, we breifly applied the cross-linker to stabilize the TF-DNA iteractions and devised a μChIP protocol to isolate bZIP 1 bound DNA from a relatively small amount of cells (Fig. 1c–e) [3]. μChIP protocols have been previously established for mammalian cells by Dahl and coworkers [4–8], and we modified one for samples of ~5000 root protoplasts. We recovered bZIP1-target complexes using the GR domain as epitope tag and generated μChIP DNA sequences from a low-input DNA library through next-generation sequencing (Fig. 1d–g) [3]. The μChIP-seq dataset we obtained was combined with microarray gene expression data from the same cell sample to validate genome-wide bZIP1 primary targets (i.e., genes whose expression is directly controlled by the interaction of bZIP1 with their promoters) in the context of the gene network for nitrogen assimilation [3].

2 Materials

Stock solutions of each buffer can be prepared in advance using ultrapure water (resistivity of 18.2 MΩhm at 25 °C) and analytical grade reagents and stored at room temperature. The protease inhibitors must be added to the working solution just before use, due to their short half-life.

Because the procedure will yield a very small amount of DNA, the use of low DNA-binding microcentrifuge tubes, low protein-binding microcentrifuge tubes, and low retention tips is recommended.

2.1 Protein-DNA Cross-Linking

1. W5 buffer: 154 mM NaCl, 125 mM $CaCl_2$, 5 mM KCl, 2 mM MES (pH 5.7).
2. 36.5% formaldehyde.
3. 2 M glycine buffer: for 50 ml dissolve 7.507 g of glycine powder in 50 ml of water.

2.2 Cell Lysis and DNA Sheering

1. Lysis buffer: 50 mM Tris–HCl, pH 8.0, 10 mM EDTA, 1% (wt/vol) SDS, half protease inhibitor cocktail mini tablet, 1 mM PMSF. For 2 ml of lysis buffer, add 100 μl of 1 M Tris–HCl, pH 8.0, 40 μl of 0.5 M EDTA, and 100 μl of 20% (wt/vol) SDS in 1 ml of water. Just before use, dissolve half a tablet of protease inhibitor cocktail in 500 μl of water, and add

it to the buffer. Then add 20 μl of 100 mM phenylmethylsulfonyl fluoride (PMSF) (*see* **Note 1**). Bring the volume to 2 ml with water.

2. RIPA ChIP buffer: 10 mM Tris–HCl, pH 7.5, 140 mM NaCl, 1 mM EDTA, 0.5 mM EGTA, 1% (vol/vol) Triton X-100, 0.1% (wt/vol) SDS, 0.1% (wt/vol) Na deoxycholate, protease inhibitor cocktail mini tablet, 1 mM PMSF. For 5 ml, mix 50 μl of 1 M Tris–HCl, pH 7.5, 75 μl of 4 M NaCl, 10 μl of 0.5 M EDTA, 5 μl of 0.5 M EGTA, 500 μl of 10% (vol/vol) Triton X-100, 25 μl of 20% (wt/vol) SDS, and 50 μl of 10% (wt/vol) Na deoxycholate. Just before use, dissolve a tablet of protease inhibitor cocktail in the mixture, and then add 50 μl of 100 mM PMSF.
3. Bath-type sonicator.

2.3 Antibody-Bead Conjugation

1. Magnetic beads and magnetic concentrator.
2. Tube rotator.
3. Anti-GR antibody.
4. RIPA buffer: 10 mM Tris–HCl, pH 7.5, 140 mM NaCl, 1 mM EDTA, 0.5 mM EGTA, 1% (vol/vol) Triton X-100, and 0.1% (wt/vol) SDS. For 15 ml, dissolve 150 μl of 1 M Tris–HCl, pH 7.5, 524 μl of 4 M NaCl, 30 μl of 0.5 M EDTA, 15 μl of 0.5 M EGTA, 1.5 ml of 10% (vol/vol) Triton X-100, and 75 μl of 20% (wt/vol) SDS in 10 ml of water. Bring the volume to 15 ml with water.

2.4 Washes

1. RIPA buffer (*see* Subheading 2.3, **item 4**).
2. LiCl wash buffer: 0.25 M LiCl, 1% (wt/vol) Na deoxycholate, 10 mM Tris–HCl pH 8, 1% NP-40, 1 mM EDTA. For 20 ml, start with 10 ml of water; add 50 μl of 10 M LiCl, 2 ml of 10% (wt/vol) Na deoxycholate, 200 μl of 1 M Tris–HCl pH 8, 2 ml of 10% NP-40, and 40 μl 0.5 M EDTA; and bring the total volume to 20 ml with water.
3. TE buffer: 10 mM Tris–HCl pH 8.0 and 10 mM EDTA. For 10 ml solution, mix 100 μl of 1 M Tris–HCl (pH 8.0) and 20 μl of EDTA (0.5 M) with water up to 10 ml.
4. Magnetic concentrator.

2.5 Isolation of DNA (DNA Elution, Cross-Link Reversal, Proteinase K Digestion, and DNA Purification)

1. Elution buffer: 20 mM Tris–HCl, pH 7.5, 5 mM EDTA, and 50 mM NaCl. For 5 ml, add to 2 ml of water 100 μl of 1 M Tris–HCl, pH 7.5, 50 μl of 0.5 M EDTA, and 62.5 μl of 4 M NaCl, and bring the volume to 5 ml with water.
2. Complete elution buffer: 20 mM Tris–HCl, pH 7.5, 5 mM EDTA, 50 mM NaCl, 1% (wt/vol) SDS, and 50 mg/ml Proteinase K. For 5 ml, add to 2 ml of water 100 μl of 1 M

Tris–HCl, pH 7.5, 50 μl of 0.5 M EDTA, and 62.5 μl of 4 M NaCl. Just before use, add 250 μl of 20% (wt/vol) SDS and 12.5 μl of 20 mg/ml Proteinase K, and bring the volume to 5 ml with water.

3. Thermal shaker.
4. Proteinase K solution: dilute 20 mg/ml Proteinase K by adding 1 μl of 20 mg/ml Proteinase K to 9 μl of nuclease-free water.
5. Kit for enzymatic cleanup.
6. Alternative cleanup: phenol-chloroform-isoamylalcohol, 3 M NaAc (pH 7.0), 0.25% (wt/vol) acrylamide carrier, 96% ethanol.

2.6 End-Repair and dA-Tailing

1. T4 DNA ligase buffer with 10 mM ATP.
2. 10 mM dNTPs mix.
3. T4 DNA polymerase.
4. Klenow DNA polymerase and buffer.
5. T4 PNK.
6. PCR purification kit.
7. 1 mM dATP.
8. Klenow fragment (3′ to 5′ exo minus).

2.7 Adaptor Ligation and Cleanup

1. 2× DNA ligase buffer.
2. DNA ligase.
3. Adapter oligo mix (suitable for the sequencing platform of choice).
4. Magnetic beads for DNA size selection.
5. 80% ethanol.
6. 10 mM Tris–HCl, pH 8.0: for 100 ml dissolve 121 mg in 60 ml of water and adjust the pH to 8.0 by addition of HCl. Bring the final volume to 100 ml with water.

2.8 Enrichment of Adaptor-Modified μChIP DNA

1. 2× high-fidelity DNA polymerase and buffer.
2. PCR primer 1 (specific to the adaptor in Subheading 2.7).
3. PCR primer 2 (specific to the adaptor in Subheading 2.7).
4. Thermal cycler.
5. PCR purification kit.

2.9 Size Selection and Gel Purification

1. 2% SYBR gold gel: agarose, 1× TAE buffer, and SYBR gold. Add 1 g of agarose to 50 ml fresh TAE buffer, and warm it up until the agarose has dissolved. Cool down to 65 °C and add 5 μl of SYBR gold. For best resolution, cast a very thin gel.

2. Library loading buffer: 50 mM Tris–HCl pH 8.0, 40 mM EDTA, 40% (w/v) sucrose. Dissolve 2.5 ml of 1 M Tris–HCl (pH 8.0), 4 ml of 0.5 M EDTA, and 20 g of sucrose in 30 ml of water. Bring the volume to 50 ml with water.
3. 100 bp DNA ladder.
4. 1× TAE buffer: 40 mM Tris, 20 mM acetate, and 1 mM EDTA. Make a 50× TAE stock solution by dissolving 242 g of Tris free base and 18.61 g of EDTA in approximately 700 ml of water. Add 57.1 ml of glacial acetic acid and adjust the volume to 1 l.
5. Blue light transilluminator.
6. Disposable scalpels or blades.
7. Electrophoresis apparatus.
8. Gel purification kit.

3 Methods

3.1 Protein-DNA Cross-Linking

1. Collect about 5000 protoplasts expressing the TF of interest fused to the GR domain, and bring the volume to 500 μl with W5 buffer if necessary (*see* **Note 2**).
2. Add 13.5 μl of 36.5% formaldehyde (1% vol/vol final concentration), mix by swirling the tube, and incubate for 10 min at room temperature with gentle rotation.
3. Add 32 μl of 2 M glycine (125 mM final concentration) to quench the formaldehyde, and incubate for 5 min at room temperature with gentle rotation.
4. Pellet the cells by centrifuging at 2500 rpm (about 600 × *g*) for 2 min at 4 °C.
5. Add 500 μl of W5 buffer, mix gently, and pellet the cells by centrifuging at 2500 rpm for 2 min at 4 °C.
6. Carefully remove the supernatant. At this point, the pellet can be snap-freezed in liquid N2. Pellets of cross-linked cells can be stored at −80 °C for several months.

3.2 Antibody-Bead Conjugation

1. Prepare a slurry of protein A magnetic beads (if using IgGs). The stock bead suspension must be homogenous before pipetting. 20 μl of beads are used for each sample; thus for 4 μChIP reactions, pipette 100 μl of well-suspended protein A magnetic beads stock solution into a 1.5 ml microcentrifuge tube.
2. Place the tube in a magnetic concentrator, remove the buffer, add 300 μl RIPA buffer to the beads, and vortex briefly.
3. Repeat **step 2** twice.
4. After the third wash, vortex the magnetic beads in RIPA buffer, and place the tube on ice.

5. Add 10 µg of anti-GR antibody and rotate at 40 rpm in a tube rotator for 2 h at 4 °C.
6. Place the tube in the magnetic concentrator, remove the buffer, add 300 µl RIPA buffer, and vortex.
7. Place the tube in the magnetic concentrator, remove the buffer, and resuspend the beads in 90 µl RIPA buffer (*see* **Note 3**).

3.3 Cell Lysis and DNA Sheering

1. Add 120 µl of room temperature lysis buffer to the protoplasts, and vortex twice for 5 s.
2. Leave the tube on ice for 5 min, and resuspend the cells by vortexing (ensure that no liquid is trapped in the lid).
3. Using a bath-type sonicator, sheer the DNA by sonicating on ice for 30 s three times, with a 30-s pause on ice between each 30-s session (*see* **Note 4**).
4. Add 400 µl of RIPA ChIP buffer to the tube (which should contain about 140 µl of lysate), and mix by vortexing.
5. Centrifuge at 12,000 × *g* for 10 min at 4 °C, and carefully transfer the supernatant (chromatin) into a clean 1.5 ml micro-centrifuge tube chilled on ice (*see* **Note 5**).
6. Add 410 µl of RIPA ChIP buffer to the remaining pellet, mix by vortexing, and centrifuge at 12,000 × *g* for 10 min at 4 °C.
7. Remove the supernatant, leaving about 20 µl with the (invisible) pellet, and pool it with the first supernatant (*see* **Note 6**).
8. Bring the volume (about 920 µl) to 1 ml with RIPA ChIP buffer. Remove 100 µl of chromatin as 10% input and set aside. Discard the pellets.

3.4 Immuno-precipitation and Washes

1. Add 20 µl of the magnetic beads conjugated with the antibody to the chromatin suspension, and rotate the tube at 40 rpm on a tube rotator for 2 h at 4 °C (*see* **Note 7**).
2. Quickly spin down the tubes in a microcentrifuge, and capture the magnetic beads (now bound to the immune complexes between antibodies, DNA-binding protein, and DNA) by placing the tubes in the magnetic concentrator (pre-chilled).
3. Discard the supernatant and add 1 ml of ice-cold RIPA buffer.
4. Remove the tubes from the magnetic concentrator, resuspend the beads by gentle manual agitation, and rotate at 40 rpm on a tube rotator for 4 min at 4 °C.
5. Repeat the wash with 1 ml of RIPA buffer.
6. Wash the beads once with 1 ml of LiCl buffer (*see* **Note 8**).
7. Remove the buffer, add 1 ml of TE buffer, and rotate at 40 rpm on a tube rotator for 4 min at 4 °C.
8. Spin down the content of the tubes in a microcentrifuge, and transfer the content into a separate clean tube on ice.

9. Capture the complexes in the magnetic concentrator, and remove the TE buffer (*see* **Note 9**).

3.5 Isolation of DNA (DNA Elution, Cross-Link Reversal, Proteinase K Digestion, and DNA Purification)

1. Add 150 μl of complete elution buffer to each tube, and incubate for 2 h in a thermal shaker at 68 °C, 1300 rpm.
2. Meanwhile, prepare the input DNA by adding 200 μl elution buffer and 7.5 μl of 10× dilution (2 mg/ml) of the Proteinase K solution to input chromatin samples, vortex and incubate on a heating block for 2 h at 68 °C (*see* **Note 10**).
3. Remove the tubes from the thermal shaker, and centrifuge for 30 s in a microcentrifuge.
4. Capture the beads in the magnetic concentrator, collect the supernatant, and place it into a clean 1.5 ml low DNA-binding microcentrifuge tube.
5. Add 150 μl of elution buffer to the remaining ChIP material, and incubate on the thermal shaker for 5 min at 68 °C, 1300 rpm.
6. Remove the tube from the thermal shaker, capture the magnetic beads in the magnetic concentrator, collect the supernatant, and combine it with the first elution (300 μl total).
7. Use a kit for enzymatic cleanup to clean and concentrate the eluted ChIP material and the input DNA following the manufacturer instructions. Elute in 30 μl (*see* **Notes 11** and **12**).
8. Use 30 μl of ChIP DNA and 30 μl of input DNA for library preparation (*see* **Note 13**).

3.6 End-Repair and dA-Tailing

1. Mix 30 μl of ChIP DNA, 10 μl of T4 DNA ligase buffer with 10 mM ATP, 4 μl of dNTP (10 mM each), 5 μl of 3 U/ul T4 DNA polymerase, 2 μl of 1 U/μl Klenow DNA polymerase, 5 μl of 10 k U/μl T4 PNK with 44 μl of nuclease-free water.
2. Incubate for 30 min at 20 °C.
3. Purify the DNA with a PCR purification kit (*see* **Note 14**). The DNA can be stored at −20 °C.
4. To add 3′ A to the DNA fragments, mix 32 μl of the DNA from end-repair with 5 μl of 10× Klenow buffer, 10 μl of 1 mM dATP, and 3 μl of 5 U/μl Klenow 3′–5′ exo.
5. Incubate for 1 h at 37 °C.
6. Purify the DNA with a PCR purification kit (*see* **Note 14**). For best result, continue with the adaptor ligation immediately.

1. Prepare the adaptor ligation system by mixing 10 μl of DNA from the previous section, 11 μl of 2× ligase buffer, 0.5 μl or

3.7 Adaptor Ligation and Cleanup

1 µl of a 1:20 dilution of adaptor oligo mix, and 1 µl of DNA ligase (*see* **Notes 15** and **16**).

2. Incubate at room temperature for 10 min, and then bring the volume to 40 µl with 10 mM Tris–HCl, pH 8.0.
3. Fully resuspend the magnetic size selection beads by gently vortexing. Allow the beads to warm up to room temperature (about 15–30 min).
4. Add 40 µl of beads to each sample and pipette up and down to mix.
5. Incubate at room temperature for 15 min.
6. Place the tube in the magnetic concentrator for 2 min and remove the supernatant.
7. Add 200 µl of fresh 80% ethanol to the tube on the magnetic concentrator without disturbing the beads, and incubate for 30 s before removing the supernatant.
8. Repeat **step** 7.
9. Air dry the beads for 15 min.
10. Remove the tubes from the magnetic concentrator, and resuspend the beads in 42 µl of 10 mM Tris–HCl, pH 8.0, pipetting up and down to mix.
11. Incubate at room temperature for 2 min.
12. Put the tubes back in the magnetic concentrator for 2 min, and then transfer 40 µl of supernatant to a new tube (the supernatant now contains the DNA).
13. Repeat **steps 4–9**.
14. Remove the tubes from the magnetic concentrator. Resuspend the beads in 22.5 µl of 10 mM Tris–HCl, pH 8.0 by pipetting up and down.
15. Incubate at room temperature for 2 min, and place on a magnetic concentrator for 2 min.
16. Transfer 20 µl of supernatant to a new tube—the supernatant now contains the adaptor-modified DNA that can be stored at −20 °C.

3.8 Enrichment of Adaptor-Modified µChIP DNA

1. To enrich for adaptor-modified µChIP DNA, mix 20 µl of ligated DNA, 25 µl of 2× high-fidelity DNA polymerase, 1 µl of primer 1, and 1 µl of primer 2, and add water to 50 µl.
2. Amplify the DNA with the following cycle: 98 °C for 30 s, 18 cycles of 98 °C for 10 s, 65 °C for 30 s, 72 °C for 30 s, 72 °C for 5 min, and 4 °C ∞.
3. Purify with PCR purification kit, and elute the enriched DNA fragments in 12 µl of elution buffer.

3.9 Size Selection and Gel Purification

1. Mix 3 μl of library loading buffer with 11 μl of enriched DNA fragments, and load every other lane of a 2% SYBR gold gel (*see* **Note 17**). Include a 100 bp ladder.
2. Run the samples at 90–100 V for 45 min or until bands are visible and well separated. The size of the bands depends on the adaptor and primers used. From small to large size, the bands correspond to the PCR primer dimers, the adaptor self-ligation PCR products, and the enriched adaptor-modified μChIP DNA (a smear bigger than the other 2 bands) (Fig. 1g) (*see* **Note 18**).
3. On a blue light transilluminator, excise the band containing the enriched adaptor-modified μChIP DNA (use a new blade/scalpel for each sample to avoid cross-contamination).
4. Purify the DNA using a gel purification kit. Elute by loading 15 μl of elution buffer onto the column and incubating at room temperature for 5 min before spinning down to collect the eluate (about 13 μl).
5. Use a microfluidic-based platform or a fluorimeter to assess the quantity and integrity of DNA before sending it for sequencing.

4 Notes

1. PMSF is quickly inactivated in aqueous solutions with a half-life of ~35 min at pH 8.0. This short half-life means that aqueous solutions of PMSF can be safely discarded after they have been rendered alkaline (pH >8.6) and stored for several hours at room temperature.
2. For protoplast preparation from *Arabidopsis* roots, follow the protocol by Bargmann et al. [2]. If a fluorescent transformation marker is included in the vector for the transfection, FACS could be used to improve the signal/noise ratio.
3. This incubation step should be carried out during cross-linking, cell lysis, and chromatin preparation. If necessary, the incubation can be prolonged until all chromatin samples are ready for immunoprecipitation.
4. Sonication cycle and sonication time vary depending on the instrument. It is highly recommended to carry out a time course experiment (e.g., varying the time of "on" and "off" durations as well as the number of cycles) to determine the appropriate treatment to obtain the desired DNA size. Ideal chromatin fragments can range from 200 to >1000 bp. Because of the low amount of DNA, a nanofluidic device should be used to assess the extent of DNA fragmentation after a de-cross-linking step.

5. To avoid aspirating the sedimented material (which is practically invisible), leave about 50 µl of supernatant in the tube after aspiration. The sediment contains SDS-insoluble cellular debris, which may stick to the magnetic beads, resulting in unspecific background or, in the worst case, mask any specific enrichment caused by the subsequent immunoprecipitation.
6. Diluting the chromatin reduces the SDS concentration to 0.1% (wt/vol), which is suitable for immunoprecipitation with most antibodies. Chromatin can be stored at −80 °C for up to 1 year; chromatin from smaller cell numbers can be stored at −80 °C for at least 2 weeks. Prolonged storage has not been tested.
7. This step can be carried out overnight at 4 °C if necessary, but prolonged incubation may lead to enhanced background.
8. The LiCl wash improves the effective removal of non-specific chromatin interactions with the beads. Lithium is also soluble in buffers containing high amounts of SDS.
9. The tube surface is a source of unspecific binding of chromatin, so transferring the ChIP material in TE buffer to a clean tube reduces unspecific chromatin background and enhances specificity of the ChIP assay. Indeed, background becomes significant in µChIP due to low signal and due to an increase in surface-to-volume ratio with a reduction of sample volumes.
10. DNA elution from immune complexes, cross-link reversal, and protein digestion are carried out in a single 2-h step.
11. Alternatively, add 200 µl of elution buffer to the µChIP material, and input and extract DNA once (from µChIP samples and from input samples) with an equal volume of phenol-chloroform-isoamylalcohol. Centrifuge at 15,000 × *g* for 5 min to separate the phases, and transfer 460 µl of the aqueous (top) phase to a clean tube. Add 44 µl of 3 M NaAc (pH 7.0), 10 µl of 0.25% (wt/vol) acrylamide carrier, and 1 ml of 96% ethanol at −20 °C. Mix thoroughly, and incubate for at least 1 h at −80 °C (DNA can be stored at −80 °C for several h or days if more convenient). Thaw the tubes and centrifuge at 20,000 × *g* for 15 min at 4 °C. Remove the supernatant, add 1 ml of 70% ethanol at −20 °C, and vortex briefly to wash the DNA pellet. Centrifuge at 20,000 × *g* for 10 min at 4 °C. Repeat the wash. Remove the supernatant and dissolve the DNA in 30 µl of TE. DNA can be immediately used or stored at −20 °C for up to 1 week if necessary.
12. The volume of TE depends on the number of cells in the ChIP. Note that low DNA concentration leads to degradation of the DNA more rapidly than at high concentrations. Thus, we recommend to immediately use the DNA from ChIPs from less than 5000 cells. We have, however, stored DNA from

ChIPs from 50,000 cells at −20 °C for months without signs of degradation.

13. Because of the very low number of cells, the yield of μChIP DNA is below the detection level of most instruments commonly used for DNA quantification. However, amplification of the μChIP DNA during library preparation produces enough material for next-generation sequencing.
14. To minimize contamination, avoid using a vacuum manifold. Before eluting the DNA, add 32 μl of the elution buffer to the column, and let it stand for 5–10 min.
15. The adaptor ligation should be performed right after adding the 3′ A. The amount of adaptor to use depends on the amount of input DNA. Because of the low yield of μChIP, 0.5 μl of 1:20 dilution of adaptor is recommended.
16. Keep adaptor oligo mix in ice. Add adaptor to each tube, mix well, and then add DNA ligase to each tube separately. Avoid making a master mix of the adaptors and the ligase to prevent adaptor-adaptor ligation.
17. For DNA visualization, SYBR gold rather than ethidium bromide is recommended because the exposure to UV light can damage the μChIP DNA. In addition, and SYBR gold has higher sensitivity.
18. The intensity of each band can vary depending on the efficiency of the enrichment reaction, the beads selection, and the amount of DNA obtained after ChIP (Fig. 1g).

Acknowledgment

The work on μChIP [3] was supported by NIH R01-GM032877 to G.C.

References

1. Yoo SD, Cho YH, Sheen J (2007) Arabidopsis mesophyll protoplasts: a versatile cell system for transient gene expression analysis. Nat Protoc 2 (7):1565–1572. https://doi.org/10.1038/nprot.2007.199
2. Bargmann BOR, Marshall-Colon A, Efroni I et al (2013) TARGET: a transient transformation system for genome-wide transcription factor target discovery. Mol Plant 6(3):978–980. https://doi.org/10.1093/mp/sst010
3. Para A, Li Y, Marshall-Colon A et al (2014) Hit-and-run transcriptional control by bZIP1 mediates rapid nutrient signaling in Arabidopsis. Proc Natl Acad Sci U S A 111 (28):10371–10376. https://doi.org/10.1073/pnas.1404657111
4. Dahl JA, Collas P (2007) A quick and quantitative chromatin immunoprecipitation assay for small cell samples. Front Biosci 12:4925–4931
5. Dahl JA, Collas P (2008) μChIP—a rapid micro chromatin immunoprecipitation assay for small cell samples and biopsies. Nucleic Acids Res 36 (3):e15–e15. https://doi.org/10.1093/nar/gkm1158
6. Dahl JA, Collas P (2008) A rapid micro chromatin immunoprecipitation assay (microChIP). Nat Protoc 3(6):1032–1045. https://doi.org/10.1038/nprot.2008.68

7. Dahl JA, Collas P (2008) MicroChIP—a rapid micro chromatin immunoprecipitation assay for small cell samples and biopsies. Nucleic Acids Res 36(3):e15. https://doi.org/10.1093/nar/gkm1158
8. Dahl JA, Collas P (2009) MicroChIP: chromatin immunoprecipitation for small cell numbers. Methods Mol Biol 567:59–74. https://doi.org/10.1007/978-1-60327-414-2_4

Chapter 20

Proteome Analysis of *Arabidopsis* Roots

Natalia Nikonorova, Lam Dai Vu, Elisabeth Stes, Kris Gevaert, and Ive De Smet

Abstract

Plant health and development are directly depending on a plant's ability to react to a constantly changing environment. Sensing of water and nutrition levels and of the biotic environment is vital for a plant, making the root one of the key plant organs. Proteins are the key molecules that play numerous roles in a cell's everyday life. Quantitative proteome profiling of roots can provide a global overview on the molecular regulatory mechanisms and networks involved in plant growth and development and abiotic and biotic stress responses. Here, we provide a detailed proteomics workflow on *Arabidopsis thaliana* roots from plant growth up to proteomics data analysis.

Key words Root, *Arabidopsis thaliana*, Label-free proteomics, MaxQuant, Perseus

1 Introduction

Proper development and reaction to a constantly changing environment are undoubtedly essential for a plant's survival. Plants are continuously exposed to various types of biotic and abiotic stress factors. However, their sessile life forces them to develop strategies to overcome such stresses. The plant root system makes up a large portion of the plant biomass and is important for nutrient and water uptake, fixing the plant in the soil and sensing underground stimuli. The latter results in rapid, local responses and/or transmission of a signal to the aboveground part. The ease of growth, manipulation, and imaging made *Arabidopsis thaliana* roots a perfect model to study stress response, growth, and development. The past decades, "root science" made an enormous contribution to (plant) developmental biology, providing new insights into meristem maintenance and initiation and cell differentiation and identity [1–6].

Kris Gevaert and Ive De Smet contributed equally to this work.

Daniela Ristova and Elke Barbez (eds.), *Root Development: Methods and Protocols*, Methods in Molecular Biology, vol. 1761, https://doi.org/10.1007/978-1-4939-7747-5_20,

To date, plant science has accumulated a tremendous amount of evidence showing the role of individual genes and proteins in particular pathways. However, most of the current knowledge presents scattered snapshots of events occurring in a cell, and a global overview is missing. In the past decades, systems biology approaches have started to tackle this shortcoming, and transcriptional profiling was (and is) widely used to understand complex mechanisms involved in regulation of plant growth and environmental responses. Despite that transcriptomics has made a large contribution to our knowledge, this approach suffers from several disadvantages, since mRNA levels are very often just poor proxies for protein levels and protein activities [7].

Proteins are the central molecules that transform genomic information in physical action in cells. Multiple cellular tasks are realized by proteins, from basic metabolism maintenance to the fine regulation of signaling cascades. Knowledge of protein cellular localization, abundance, dynamic changes, and interactions has become crucial to our understanding of biological systems. In this respect, contemporary proteomics is a powerful method for unraveling the molecular regulatory networks involved in development and stress responses in various plant species [8]. Quantitative proteome data can be easily statistically analyzed and graphically represented in heat maps, protein profiles, and correlation scatterplots. In this context, recent advances in bioinformatics have enabled label-free protein quantification (LFQ) [9], which circumvents metabolic labeling of plant proteomes or post-metabolic labeling of proteins and peptides. Despite various algorithms, search engines and software are available for protein identification and data analysis; proteome-wide quantification with label-free approaches remains a challenge. In our workflow, we use the MaxQuant computational proteomics platform with the implemented Andromeda search engine and the MaxLFQ label-free quantification technology. MaxQuant is a freely available and user-friendly platform that proved itself as accurate, robust, and reliable tool in diverse biological projects [10–14]. Subsequent analysis of highly complex proteomics data is a challenge and requires some background in statistics and bioinformatics. In our workflow, however, we use the Perseus software package, a user-friendly, interactive workflow environment for shotgun proteomics data analyses [15]. Perseus contains a comprehensive range of statistical tools for data analysis, including multiple hypothesis testing, clustering, principal component analysis, and various options for data visualization that allows interpreting protein quantification and interaction.

Here, we provide a detailed proteomics workflow for *A. thaliana* roots including descriptions of plant growth conditions, protein extraction, protein identification, and statistical analysis of the proteomics data. However, a slightly modified setup can also be applied to other plants, such as wheat or maize [16, 17].

2 Materials

2.1 Plant Growth

1. *Arabidopsis thaliana* seeds.
2. 70% ethanol.
3. Sterilization solution: 15 ml of 12% sodium hypochlorite solution, 25 μl of Tween® 20, filled up to 50 ml with distilled water.
4. Sterile distilled water.
5. 1 l of sterile growth media containing 2.154 g Murashige and Skoog (MS) salt mixture, 0.1 g myoinositol, 0.5 g 2-(N-morpholino)ethanesulfonic acid (MES) monohydrate, 10 g sucrose, and 1% agar with the pH adjusted to 5.7.
6. 1.5 ml Eppendorf tubes.
7. Vortexer.
8. Microcentrifuge (optional).
9. pH indicator strips.
10. Pipette for 1000 and 20 μl.
11. Sterile tips for 1000 and 20 μl.
12. Sterile 0.1% agarose solution.
13. Sterile square Petri plates, 120 × 120.
14. Plant growth chamber.
15. Laminar flow cabinet.
16. Autoclave.
17. Sterile 50 ml Falcon tubes.
18. Breathable tape.

2.2 Protein Extraction

1. Homogenization buffer (pH 8) containing 30% sucrose, 250 mM Tris–HCl, 5 mM EDTA.
2. 500 mM dithiothreitol (DTT) (*see* **Note 1**).
3. Protease inhibitors (e.g., cOmplete™, EDTA-free).
4. Milli-Q water.
5. 80% acetone.
6. Methanol.
7. Chloroform.
8. 50 mM triethylammonium bicarbonate (TEAB).
9. 8 M urea in 50 mM TEAB (*see* **Note 2**).
10. Sonicator.

11. Centrifuge, refrigerated, with swinging buckets rotor (3220 × *g*).
12. Centrifuge with rotor for Eppendorf tubes (15,000 × *g*).
13. Vortexer.
14. Mortar and pestle.
15. Fume hood.
16. 15 and 50 ml Falcon tubes.
17. Scalpel.
18. Liquid nitrogen.
19. Ice.

2.3 Protein Reduction and S-Alkylation

1. 750 mM tris(2-carboxyethyl)phosphine hydrochloride (TCEP-HCl) in 50 mM TEAB.
2. 5 M NaOH.
3. 750 mM iodoacetamide in 50 mM TEAB (pH 8) (*see* **Note 3**).
4. Incubation oven for 30 °C and 37 °C.
5. Thermoblock.
6. pH indicator strips.

2.4 Protein Digestion

1. Endoproteinase LysC (Wako Chemicals) (*see* **Note 4**).
2. Trypsin (Promega sequencing grade) (*see* **Note 4**).
3. Trifluoroacetic acid (TFA).
4. NanoDrop spectrophotometer (*see* **Note 5**).

2.5 Desalting of the Peptide Mixtures Obtained Following Protein Digestion

1. Acetonitrile (100% and 80%).
2. Solvent A: 2% acetonitrile, 0.1% TFA.
3. Solvent B: 70% acetonitrile, 0.1% TFA.
4. C18 SPE columns (C18) solid-phase extraction cartridges: SampliQ C18 (*see* **Note 6**).
5. Vacuum manifold.

2.6 Data Analysis

1. MaxQuant (version 1.5.4.1) and Perseus software (*see* **Note 7**).
2. The latest version of protein sequence database in.fasta format (*see* **Note 8**).

3 Methods

The key steps for the protocol are indicated in Fig. 1.

3.1 Plant Growth

1. Distribute the seeds in 1.5 ml Eppendorf tubes.
2. In a laminar flow cabinet, add 1 ml of 70% ethanol to each tube, intensively shake, and leave for 2 min.

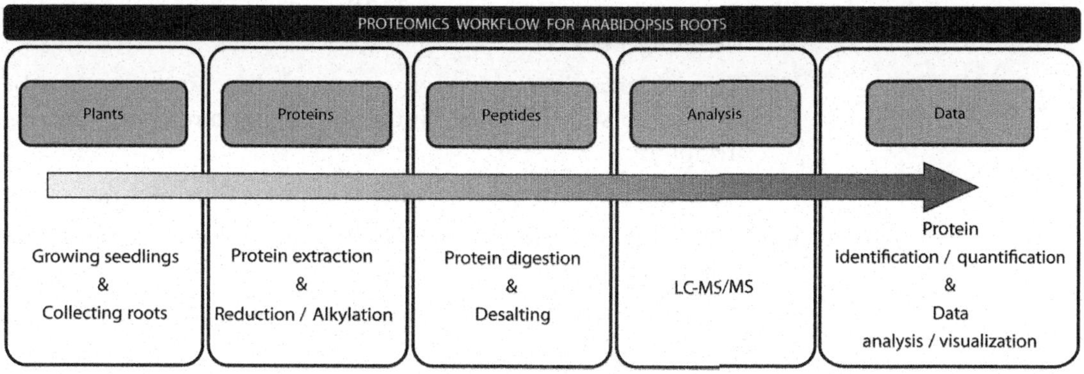

Fig. 1 Schematically represented proteomics workflow for *Arabidopsis thaliana* roots

3. Replace the ethanol with 1 ml of sterilization solution, intensively shake, and leave for 15 min (*see* **Note 9**).
4. Rinse the seeds three to five times with sterile distilled water until the pH is neutral (*see* **Note 9**).
5. After the last wash, replace water with 0.1% agarose.
6. Use a sterile Falcon tube, and pour approximately 50 ml of growth media per plate and leave for 30 min to solidify.
7. Resuspend the seeds by pipetting, and collect the necessary amount of seeds (*see* **Notes 10** and **11**).
8. Sow the seeds on the plate gently touching the surface of the solidified medium and releasing the drop of agarose with seeds.
9. Seal the plates with breathable tape, and store plates overnight at 4 °C in the dark (*see* **Note 12**).
10. Move plates to the growth chamber and place them vertically.

3.2 Protein Extraction

1. Cut roots with a scalpel and immediately freeze in liquid nitrogen (*see* **Note 13**).
2. Add 1 tablet of protease inhibitors and 200 μl of 500 mM DTT to 100 ml of homogenization buffer.
3. Quickly grind 0.1–1 g of plant material into a fine powder in liquid nitrogen, and transfer it to 15 ml Falcon tubes.
4. Add 5 ml of homogenization buffer.
5. Sonicate on ice (*see* **Note 14**).
6. Centrifuge for 15 min at 3220 × *g* and 4 °C (swinging bucket).
7. Collect supernatant in 50 ml centrifuge tube.
8. To each 5 ml of supernatant, add in the following order without mixing in between: 15 ml of methanol, 5 ml of chloroform, and 20 ml of distilled water. Shake vigorously.

9. Centrifuge for 10 min at 3220 × *g* at room temperature (swinging bucket).
10. Remove and discard upper aqueous phase from each tube without disturbing the interface containing the protein precipitate.
11. Add 20 ml of methanol in each tube. Mix thoroughly.
12. Centrifuge for 10 min at 3220 × *g* at room temperature (swinging bucket).
13. Discard supernatant and wash the pellet with 5 ml of 80% acetone.
14. Centrifuge for 10 min at 3220 × *g* at room temperature (swinging bucket).
15. Discard supernatant and dry 30 min at room temperature in open air.
16. Solubilize protein pellet in 500 μl of 8 M urea in 50 mM TEAB, and move to 1.5 ml Eppendorf tubes.

3.3 Reduction and S-Alkylation

1. Adjust 750 mM of TCEP-HCl to pH 8.0 with 5 M NaOH (add ~45 μl of 5 M NaOH to 100 μl of 750 mM of TCEP-HCl which gives ~570 mM TCEP-HCl) (*see* **Note 15**).
2. Add 14 μl of 570 mM TCEP-HCl, pH 8 and 20 μl of 750 mM iodoacetamide to 500 μl of solubilized protein (to a final concentration of 15 mM and 30 mM, respectively).
3. Place samples at 30 °C in the dark for 15 min (constantly shaking) (*see* **Note 16**).
4. Centrifuge for 5 min at 15,000 × *g* (rotor for Eppendorf tubes), and discard any precipitation or insoluble matters.

3.4 Protein Digestion

1. Measure the protein concentration with a NanoDrop spectrophotometer using 8 M urea in 50 mM TEAB as a blank.
2. Dissolve endoproteinase LysC in 50 mM TEAB to a final concentration of 1 μg/μl.
3. Take 1 mg of proteins and predigest with 10 μg of endoproteinase LysC per sample at 37 °C while constantly mixing for 2–4 h in the dark.
4. Dilute samples eight times with 50 mM TEAB.
5. Dissolve trypsin in 50 mM TEAB to a final concentration of 1 μg/μl.
6. Add 10 μg of trypsin per sample, and leave samples for digestion at 37 °C under constantly mixing and overnight.
7. Add TFA (final concentration of 0.2%) to stop the digestion (pH ≤3). Check pH with indicator strips, add more TFA if needed, and place samples on ice.

3.5 Desalting

1. Add acetonitrile (to a final concentration of 2%) and centrifuge at maximum speed at 4 °C for 10 min (*see* **Note 17**).
2. Equilibrate C18 SPE columns by applying 1 ml of 80% acetonitrile per 100 mg of column.
3. Apply vacuum (*see* **Note 18**).
4. Wash columns with 1 ml of Solvent A, apply vacuum, and discard wash elution.
5. Load samples on columns, apply vacuum, and discard wash elution.
6. Wash samples with 1 ml of Solvent A, apply vacuum, and discard wash elution. Repeat this step.
7. Apply 1 ml of Solvent B to elute the peptides.
8. Freeze samples in liquid nitrogen and dry in a speedvac.
9. Re-solubilize peptides in 30 μl of 2% (v/v) acetonitrile and 0.1% (v/v) TFA right before analysis.

3.6 LC-MS/MS

This step depends on the available facilities. An example of an LC-MS/MS protocol on an Ultimate 3000 RSLC nano LC (Thermo Fisher Scientific) in-line connected to a Q Exactive mass spectrometer (Thermo Fisher Scientific) can be found in our recent publication [17] (*see* **Note 19**).

3.7 Protein Identification and Quantification

1. Launch MaxQuant software (*see* **Note 20**).
2. Load.raw files.
3. Select the sample and give it a name by clicking "Set experiment" button.
4. Go to "Group-specific parameters" tab.
5. Click "Digestion" button and select "Trypsin/P" and set "Max. missed cleavages" to 1 or 2 (*see* **Note 21**).
6. Click "Modifications" button, and select variable modifications from the list (*see* **Note 22**).
7. Click "Label-free quantification" button and select "LFQ".
8. Click "Misc." button and set "Match type" as "Match from and to".
9. Go to "Global parameters" tab.
10. Click "Sequences" button, and load a protein database by clicking "Add file" button in "Fasta files" field. Select carbamidomethyl (C) as "Fixed modifications" from the list (*see* **Note 23**).
11. Click "Adv. Identification" button and check "Match between runs".

12. Click "Protein quantification" button, and specify Oxidation (M), Acetyl (Protein N-term) as "Modifications used in protein quantification".
13. Click "Folder locations" button and specify the locations (*see* **Note 24**).
14. Go to "Configuration" tab.
15. Click "Sequence databases button". Check if the desired protein database is in the list. If not proceed to the next step.
16. Click "Add" button and then "Select" button in "Fasta File name" field. Choose the fasta file of the protein database and click "Open". Specify "Source", "Taxonomy" and "Parse rules" (*see* **Note 25**).
17. Click "Modify table" and "Save changes" buttons.
18. Go to the "Performance" tab.
19. Specify "Number of threads" (*see* **Note 26**). Click "Start" button (*see* **Note 27**).
20. After the search is done, the result tables can be found in the "combined" folder which was specified previously.

3.8 Data Analysis, Visualization and Statistics

1. Launch the Perseus software (*see* **Note 28**).
2. Click "Generic matrix upload" icon on the "Load" panel (green arrow).
3. Click "Select" in the "File" field, and choose the "proteinGroups.txt" file in the "txt" folder of "combined" folder.
4. Move "LFQ intensity..." to the "Main columns" field and "Fasta headers" to the "Text" field. Click OK.
5. Click "Filter rows based on categorical column", choose "Only identified by site" in the "Column" field, "Remove matching rows" — in "Mode" and "Reduce matrix" — in "Filter mode". Repeat the procedure for "Reverse" and "Potential contaminant" in "Column" field (*see* **Note 29**).
6. Perform *log2* transformation of intensities by clicking "Transform" in "Basic" tab. Click OK (*see* **Note 30**).
7. Click "Annot. rows" tab and select "Categorical annotation rows". Group biological replicates by giving them the same name (*see* **Note 31**).
8. Optional. Click "Rearrange" tab and select "Rename columns" and remove "LFQ intensity" from sample names.
9. Click "Filter rows" tab and select "Filter rows based on valid values". Specify "Min. number of values" and "Mode". Click OK (*see* **Note 32**).
10. Click "Normalization" tab and select "Subtract".
11. Select "Columns" in "Matrix access" field. Click OK.

12. Click "Tests" tab, and select statistical test suitable for your experimental setup and truncation method (*see* **Note 33**).
13. Click "Filter rows based on categorical column", and select "ANOVA/t-test Significant" in the "Column" and "Keep matching rows" in "Mode" field.
14. Click "Normalization" tab and select "Z-score". Choose "Rows" in the "Matrix access" without grouping, check "Use median".
15. Click "Clustering/PCA" tab and select "Hierarchical clustering". Specify clustering parameters or use the default ones.
16. Click "Visualization" tab and select "Profile plot" (*see* **Note 34**).

4 Notes

1. Stock solution can be stored at −20°C for 6 months.
2. The use of urea can cause carbamylation at the N-termini of proteins and peptides and at the side-chain amino groups of lysine and arginine residues. Carbamylation interferes with protease digestion and affects mass spectrometry analysis by blocking peptide amino groups from isotopic/isobaric labeling and changing peptide charge states, retention times, and masses of proteins and peptides. Carbamylation occurs as a result of the reaction between protein amino groups and ammonium cyanate derived from urea after dissociation in aqueous solutions. To decrease the level of protein carbamylation, use freshly prepared urea solutions, and avoid heating.
3. Keep in the dark.
4. Was used according to the protocol provided by the manufacturer.
5. As an alternative, the protein concentration can be measured on a spectrophotometer using UV absorption at 280 nm, Coomassie (Bradford), Lowry, or bicinchoninic acid assays or commercial kits [18]. When applying another protein quantification method, check its compatibility with the buffer components, dilute your samples if needed, and build a calibration curve using a protein standard.
6. Was used according to the protocol provided by the manufacturer.
7. Freely available MaxQuant and Perseus software can be downloaded from http://www.coxdocs.org after a simple registration procedure. For newer versions the layout and search algorithm may (slightly) vary.

8. For example, use TAIR10 from https://www.arabidopsis.org/ (select Proteins in Download tab).
9. Here, seeds can be collected at the bottom of the tube following a short centrifugation step.
10. For pipetting seeds, one can use 1000 μl tips or 20 μl tips. Depending on the provider, some pipette tips are too narrow at the bottom, and seeds cannot enter the tip. This can be overcome by cutting 2–3 mm of tips with a sharp scalpel prior sterilization of tips in autoclave.
11. The number of seedlings per plate depends on the experimental setup. Avoid overloading as seedlings can undergo drought stress and starvation.
12. Can be longer, depends on the quality and age of seeds.
13. Our experience shows that 100 mg of root material is the minimum amount required.
14. Amplitude (we used 20% of possible amplitude), frequency, and duration (10 s in total, 2 s on/2 s off) settings depend on available facilities; please follow protocol from manufacturer. Avoid overheating of samples during sonication.
15. This pH adjustment is highly important, as at lower pH alkylation of cysteines and further digestion are inefficient.
16. Longer incubation time should be avoided as it promotes alkylation at amino acid groups (e.g., ε-amines of lysine side chains). The reaction needs to occur in the dark as iodoacetamide is light sensitive, leading to iodination of amino acids.
17. Discard any existing precipitate prior to loading samples on desalting columns.
18. Hereinafter, regulate vacuum pressure to have a flow of one drop per second.
19. Injection volume and/or concentration (which will also impact on Subheading 3.5, **step 9**) will depend on the available facilities.
20. For more details on all steps, see protocol from the developers [19].
21. If the digestion is perfect and no partial fragments are left, this parameter can be set at 0 (but this is unlikely to occur). If the digest mixture contains peptides with missed cleavage sites, one or two missed cleavage sites should be specified. Avoid specifying a higher number without reason as each additional level increases the number of calculated peptide masses to be matched against the experimental data.
22. Oxidation (M) and acetyl (protein N-term) are the most common occurring modifications.

23. The carbamidomethyl (C) modification was introduced to cysteine residues by the alkylation reaction with iodoacetamide.
24. Make sure you have sufficient space on the disk as the folders for data processing as well as the result output can be two times the size of the input.raw files.
25. For TAIR10 database "Identifier parse rule" is >([^]*) and "Description parse rule" is >.*\|(.*)\|.
26. Should never exceed the available number of (logical) cores on the computer.
27. Depending on the capacity of the computer and experiment/file size, the search can take several hours or days. You can also click "Partial processing" button. This will allow restarting the MaxQuant analysis from an intermediate step if the search crashed or a reanalysis of the data is needed. For more details on this option, see [19].
28. For more details, see protocol from the developers [15].
29. In this step, unwanted and incorrect protein identifications can be removed from subsequent analyses. Thus, positives in the "Reverse" column were identified against a protein database consisting of the reversed sequences; positives in "Potential contaminant" were identified against the database of the most common contaminants such as keratin and trypsin; positives in "Only identified by site" represent sites that were accepted by the algorithm but for which the whole protein couldn't pass the FDR threshold.
30. As hypothesis tests such as t-test and ANOVA deal with assumption of normality, often data needs to be transformed to meet assumptions of a statistical test or procedure. A log transformation achieves that assumption by reducing the skewness of the data and minimizing the impact of large values that can influence the outcomes.
31. You can specify several grouping factors that later can be used for statistical analysis, e.g., for samples of the same treatment or the same time point in a time course.
32. To increase the confidence of the analysis, at least 50% of the replicates in at least one group should have a valid value (e.g., 2 out of 3, 3 out of 4, and so on).
33. FDR correction is the most stringent truncation that will minimize false-positives in the list of significant hits.
34. When comparing only two conditions/samples, the volcano plot is a useful tool.

Acknowledgments

L.D.V. is the recipient of a VIB International Ph.D. program fellowship. E.S. was a postdoctoral fellow of the Research Foundation-Flanders. Kris Gevaert and Ive De Smet contributed equally to this work.

References

1. Pacifici E, Polverari L, Sabatini S (2015) Plant hormone cross-talk: the pivot of root growth. J Exp Bot 66:1113–1121
2. Stahl Y, Simon R (2012) Peptides and receptors controlling root development. Philos Trans R Soc Lond Ser B Biol Sci 367:1453–1460
3. Tian H, De Smet I, Ding Z (2014) Shaping a root system: regulating lateral versus primary root growth. Trends Plant Sci 19:426–431
4. Lavenus J, Goh T, Roberts I et al (2013) Lateral root development in arabidopsis: fifty shades of auxin. Trends Plant Sci 18:1360–1385
5. Sozzani R, Iyer-Pascuzzi A (2014) Postembryonic control of root meristem growth and development. Curr Opin Plant Biol 17:7–12
6. De Smet I, Beeckman T (2011) Asymmetric cell division in land plants and algae: the driving force for differentiation. Nat Rev Mol Cell Biol 12:273–273
7. Feder ME, Walser JC (2005) The biological limitations of transcriptomics in elucidating stress and stress responses. J Evol Biol 18:901–910
8. Jorrín-Novo JV, Pascual J, Sánchez-Lucas R et al (2015) Fourteen years of plant proteomics reflected in proteomics: moving from model species and 2DE-based approaches to orphan species and gel-free platforms. Proteomics 15:1089–1112
9. Cox J, Hein MY, C a L et al (2014) Accurate proteome-wide label-free quantification by delayed normalization and maximal peptide ratio extraction, termed MaxLFQ. Mol Cell Proteomics 13:2513–2526
10. Smaczniak C, Li N, Boeren S et al (2012) Proteomics-based identification of low-abundance signaling and regulatory protein complexes in native plant tissues. Nat Protoc 7:2144–2158
11. Zhang H, Zhou H, Berke L et al (2013) Quantitative phosphoproteomics after auxin-stimulated lateral root induction identifies an SNX1 protein phosphorylation site required for growth. Mol Cell Proteomics 12:1158–1169
12. Qian H, Lu H, Ding H et al (2015) Analyzing Arabidopsis Thaliana root proteome provides insights into the molecular bases of enantioselective imazethapyr toxicity. Sci Rep 5:11975
13. Fromm S, Senkler J, Eubel H et al (2016) Life without complex I: proteome analyses of an Arabidopsis mutant lacking the mitochondrial NADH dehydrogenase complex. J Exp Bot 67:3079–3093
14. Walton A, Stes E, Goeminne G et al (2016) The response of the root proteome to the synthetic strigolactone GR24 in Arabidopsis. Mol Cell Proteomics 15:2744–2755
15. Tyanova S, Temu T, Sinitcyn P et al (2016) The Perseus computational platform for comprehensive analysis of (prote)omics data. Nat Methods 13:731–740
16. LD V, Stes E, Van Bel M et al (2016) Up-to-date workflow for plant (Phospho)proteomics identifies differential drought-responsive phosphorylation events in maize leaves. J Proteome Res 15:4304–4317
17. LD V, Verstraeten I, Stes E et al (2017) Proteome profiling of wheat shoots from different cultivars. Front Plant Sci 8:1–11
18. Noble JE, Bailey MJAA (2009) Chapter 8 quantitation of protein. In: Methods Enzymol, 1st edn. Elsevier Inc., Amsterdam, pp 73–95
19. Tyanova S, Temu T, Cox J (2016) The MaxQuant computational platform for mass spectrometry-based shotgun proteomics. Nat Protoc 11:2301–2319

Chapter 21

Step-by-Step Construction of Gene Co-expression Networks from High-Throughput Arabidopsis RNA Sequencing Data

Orlando Contreras-López, Tomás C. Moyano, Daniela C. Soto, and Rodrigo A. Gutiérrez

Abstract

The rapid increase in the availability of transcriptomics data generated by RNA sequencing represents both a challenge and an opportunity for biologists without bioinformatics training. The challenge is handling, integrating, and interpreting these data sets. The opportunity is to use this information to generate testable hypothesis to understand molecular mechanisms controlling gene expression and biological processes (Fig. 1). A successful strategy to generate tractable hypotheses from transcriptomics data has been to build undirected network graphs based on patterns of gene co-expression. Many examples of new hypothesis derived from network analyses can be found in the literature, spanning different organisms including plants and specific fields such as root developmental biology.

In order to make the process of constructing a gene co-expression network more accessible to biologists, here we provide step-by-step instructions using published RNA-seq experimental data obtained from a public database. Similar strategies have been used in previous studies to advance root developmental biology. This guide includes basic instructions for the operation of widely used open source platforms such as Bio-Linux, R, and Cytoscape. Even though the data we used in this example was obtained from *Arabidopsis thaliana*, the workflow developed in this guide can be easily adapted to work with RNA-seq data from any organism.

Key words RNA-seq, Gene co-expression network, Differential gene expression, DESeq2, Cytoscape, Bioinformatics, Network generation, Correlation, Cytoscape, Bio-Linux, HISAT2, FastQC, Trimmomatic

1 Introduction

Global gene expression analyses, either by microarrays or RNA sequencing (RNA-seq), have been a useful tool to generate hypotheses that allow researchers to understand plant growth, development, or response to internal or external cues [1–3]. RNA-seq is today one of the most common tools used to analyze global gene

Orlando Contreras-López and Tomás C. Moyano contributed equally to this work.

Daniela Ristova and Elke Barbez (eds.), *Root Development: Methods and Protocols*, Methods in Molecular Biology, vol. 1761, https://doi.org/10.1007/978-1-4939-7747-5_21, © Springer Science+Business Media, LLC 2018

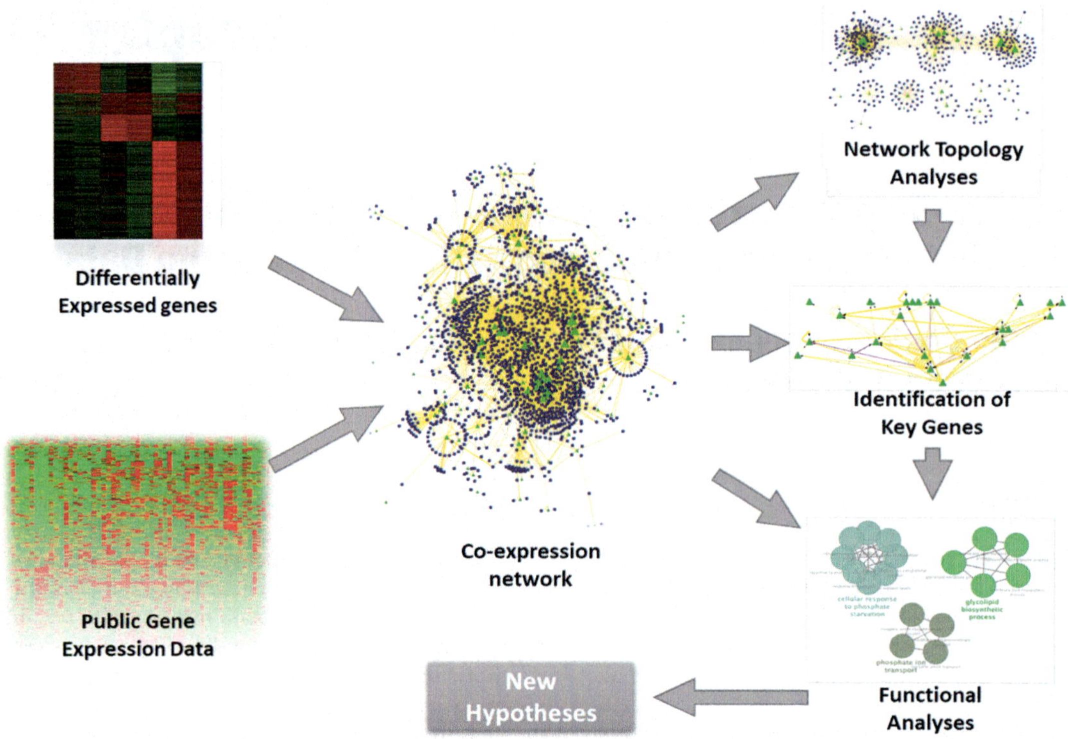

Fig. 1 A systems biology approach for transcriptome data analyses. Sets of differentially expressed genes (DEGs) of interest along with publicly available information (left) can be combined to construct co-expression gene networks (center). Gene network modeling and visualization are powerful tools from which biologists can derive testable hypotheses. For example, to infer groups of functionally related genes (upper right), identify highly connected genes that may be important to control expression of other genes (middle right), and describe the biological processes of these network modules (lower right). Finally new hypothesis generated in this exercise, such as new gene functions, should be tested experimentally

expression in any organism, using the number of reads assigned to each gene as an indicator of the expression level of the gene. The increased use of next-generation sequencing methodologies has been greatly catalyzed by continued reduction in costs [4]. In addition, RNA-seq presents several advantages over other technologies such as microarrays, including flexibility for a variety of applications from identification of differentially expressed genes (DEGs), gene discovery, noncoding RNA studies, characterization of mRNA processing intermediates, alternative splicing, RNA decay intermediates, and ribosomal profiling, among many others [5–7]. As a result, a fast-growing catalog of RNA-seq experimental data is now available from public data sources. For instance, from 2011 to 2016, there was a tenfold increase in the number of RNA-seq experiments performed solely in Arabidopsis found in ArrayExpress (http://www.ebi.ac.uk/arrayexpress/) [8] (Fig. 1).

Several approaches have been proposed to analyze RNA-seq data from a functional point of view. Differentially expressed genes

(DEGs) have been widely used to identify gene ontology (GO) enriched functions to highlight important biological processes that may be prominent and therefore relevant under given experimental conditions and derive useful biological insight from them [5, 6]. Another popular strategy is the use of undirected network graphs to represent gene co-expression data. Gene network analysis of transcriptomics data has proven to be an effective strategy for identification of candidate genes in plant biology [9]. Co-expression networks are typically represented as undirected network graphs where nodes represent genes (e.g., DEGs) and edges represent correlation of gene expression data. A co-expression network can be enriched with additional functional information from a variety of data sources, e.g., protein-protein interactions, protein-DNA interactions, transcription factor binding sites, and microRNA-target interactions [10, 11]. By integrating different sources of information, rich gene networks can be constructed that help biologists build testable hypothesis [12]. Because of the ubiquitous nature of gene expression data, it is by far the most accessible and easy to use data source to build gene networks. Despite or perhaps due to its simplicity, it is a great tool to select genes for functional studies as it has been shown repeatedly in recent years [13–17]. The underlying hypotheses here are, first, that covariance is an indirect measure of co-expression and, second, that relevant and functionally related processes or genes can be identified from their co-expressing characteristic [18].

Pioneering works showed that co-expression network analysis is a useful tool to identify gene function and coordination of gene expression in Arabidopsis [11, 19]. Using a network built from simple linear regression analysis, Wei and collaborators identified novel candidates, regulators, and control points for over 140 metabolic pathways providing new insight into the biological role of many genes which have no function annotated at the moment [19]. A more complex gene network model was built by integrating different data types (regulatory, metabolic, physical interaction) and used to uncover carbon (C) or nitrogen (N) functional relationships in Arabidopsis [11]. In this study, network analysis prompted the hypothesis that auxin signaling was implicated in Arabidopsis root responses to C or N metabolites [11]. Albeit this network model was qualitative and incomplete at this stage, this strategy proved very powerful to generate concrete testable hypothesis in this and a series of follow-up studies [14, 20–27].

Despite many online and user-friendly web resources to build gene networks exist, the options offered are not always easily customizable and rely on databases curated on the service provider side (e.g., VirtualPlant (10), CORNET [28], STRING [29], GeneMANIA [30], ATTED-II [31]). In order to democratize building gene networks and allow users to use arbitrary RNA-seq data sets, in here we provide step-by-step instructions, from downloading to integrating, analyzing, and visualizing RNA-seq data in a network context for users with basic bioinformatics skills. We describe a simple pipeline that can serve as a starting point to build gene regulatory networks for further analyses. The examples provided below use Arabidopsis root data, but the protocol should be easily applicable to any data set or organism for which similar data can be accessed.

2 Materials

1. Personal computer or server with access to the Internet. Computer requirements vary depending on the amount of data to be analyzed. In this guide, we use 16 GB RAM, 8 processor cores, and 1 TB of free disk space.
2. Bio-Linux (version Bio-Linux 8, *see* **Note 1**) is a Linux-based operative system that contains a series of preinstalled tools useful for data analyses. Bio-Linux is an open software platform for bioinformatics that can be installed on anything from a laptop to a server or run as a virtual machine even from USB device with no required installation [32]. It can be downloaded from http://environmentalomics.org/bio-Linux/.
3. R (version 3.2.0, *see* **Note 1**) is a free software designed for statistical computing and data visualization. The software is included in Bio-Linux, but different versions can be downloaded from http://www.r-project.org/.
4. Cytoscape (version 3.2.1, *see* **Note 1**) is a network visualization software with multiple applications for network analysis. It can be downloaded for free from http://www.cytoscape.org/.
5. Additional bioinformatics tools for RNA-seq processing (*see* **Note 1**): FastQC, version 0.11.5 (http://www.bioinformatics.babraham.ac.uk/projects/fastqc/); Trimmomatic, version 0.36 (http://www.usadellab.org/cms/?page=trimmomatic); HISAT2, version 2.0.5 (http://ccb.jhu.edu/software/tophat/index.shtml); DESeq2, version 1.16.1.t (http://bioconductor.org/packages/DESeq2/); EBSeq, version 1.17.0 (http://www.bioconductor.org/packages/EBSeq).
6. Gene expression information obtained from RNA-seq data. This data may be found on public databases or obtained in-house. Here we use RNA-seq information from *Arabidopsis thaliana* obtained from ArrayExpress detailed in the supporting material at http://virtualplant.bio.puc.cl/cgi-bin/Lab/tools.cgi.

3 Methods

The following sections describe a simple pipeline to generate a gene regulatory network starting from RNA-seq data (Fig. 2). The following steps will guide the user from obtaining data for selected RNA-seq experiments in FASTQ format to building a correlation network from which to infer biologically meaningful information.

This pipeline should be used primarily as a learning tool and starting point from which carefully evaluate and decide how to implement a pipeline in a research project (*see* **Note 2**). Supporting material for this guide is available at http://virtualplant.bio.puc.cl/cgi-bin/Lab/tools.cgi. Here, the user can find text files containing the experiment's URL and all command lines, including expected outputs at each step. The commands presented in this guide are for non-stranded single-end data (*see* Subheading 3.4 and **Note 3**).

3.1 Working with Data in Linux and R

Linux is an open source operative system. In Linux environment, it is necessary to recognize basic symbols and syntaxes. Remember Linux is case sensitive. Notice that syntax for Linux on this guide includes the use of the following symbols: "$" indicates start of command when it is located at the beginning of the line in the

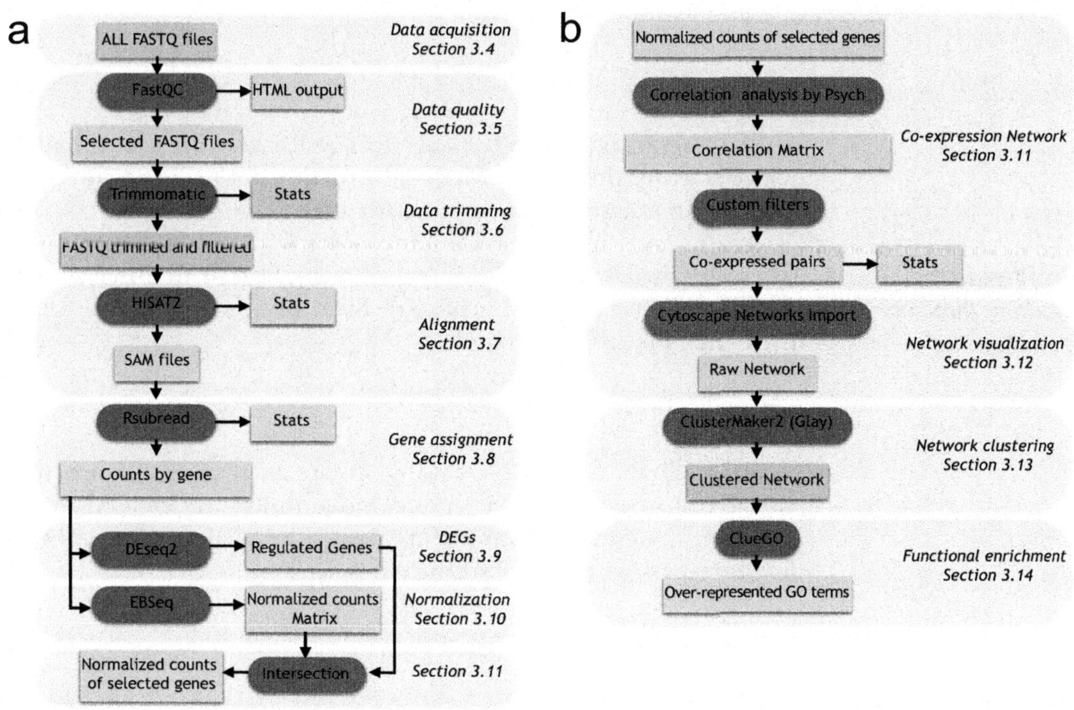

Fig. 2 Pipeline overview for data analysis used in this chapter. Light gray square boxes represent data files or data objects. Dark gray boxes represent tools or procedures. Black arrows correspond to direct steps. Light gray background shadows comprise sections in the chapter. (**a**) Data processing and (**b**) co-expression network generation

Linux console; monospace font represents the actual command. For example, we will use "`less`" command to show the content of the file "`myfile`". You can replace the name "myfile" with the name of any file of interest in your working directory:

```
$ less myfile
```

To exit this view, press the letter "q".

Commonly used commands in Linux can be found at Bio-Linux website (http://environmentalomics.org/bio-linux-training/). We strongly recommend the user to familiarize with Linux commands before starting this guide.

R is a very flexible programming platform with a large number of statistic and mathematical tools [33]. If the user wants to work in R using an interactive graphical interface, we recommend to download and install RStudio (http://www.rstudio.com/). Several tools for biologists can be found in R packages that are open source. One of the most complete and useful repositories for bioinformatics packages is Bioconductor (www.bioconductor.org). Bioconductor is an open source project that provides tools for understanding genomic and molecular biology data [34, 35]. Syntax for R on this guide are indicated with a greater than sign ">". For example, we will use the "sum" function in R to perform the arithmetic sum of 1 and 2:

```
> sum (1,2)
```

We also recommend the user to familiarize with R commands before starting this guide. More information about R can be found at http://cran.r-project.org/doc/manuals/R-intro.html.

3.2 Preparing a Working Directory Before Starting

To create a working directory, select a location in the user's computer hard drive or server running Bio-Linux, and create a folder using the "mkdir" command:

```
$ mkdir WorkDir
```

To facilitate further processing, create an alias of the directory path. To do this, the user should move inside the created folder with "`cd`" and use "`export`" to create an alias of the working directory path in the user's Linux session:

```
$ cd WorkDir
$ export WD=$(pwd)
```

3.3 Download and Install the Tools Used in This Chapter

The next step is to create a folder for all the tools to be used in this guide. To do so, create a directory called "`tools`" inside the "`WorkDir`" folder.

```
$ mkdir tools
$ cd tools
```

First, the user must install FastQC tool (*see* **Note 1**). We will use "`wget`" to download the program and unzip to decompress the downloaded file.

```
$ wget http://www.bioinformatics.babraham.ac.uk/projects/
fastqc/fastqc_v0.11.5.zip
```

Please note software versions change over time as improvements are implemented and released. This is true for FastQC and other tools used in this chapter. As a general rule, we recommend checking the developer's site and use the newest stable release, which can be found in the corresponding developer site (i.e., https://www.bioinformatics.babraham.ac.uk/for FastQC). Note also that different versions may update specific aspects of how the software runs. Please review the documentation for specific details of how to use each tool mentioned here.

```
$ unzip fastqc_v0.11.5.zip
```

To execute the FastQC tool, the user must give execution permissions to the folder with "`chmod`" command as the following:

```
$ chmod 777 FastQC/fastqc
```

Then, the user must install Trimmomatic, following similar steps as described above:

```
$ wget http://www.usadellab.org/cms/uploads/supplementary/
Trimmomatic/Trimmomatic-0.36.zip
$ unzip Trimmomatic-0.36.zip
```

Finally, download the HISAT2 align tool:

```
$ wget ftp://ftp.ccb.jhu.edu/pub/infphilo/hisat2/downloads/
hisat2-2.0.5-Linux_x86_64.zip
$ unzip hisat2-2.0.5-Linux_x86_64.zip
```

To return to the "`WorkDir`" folder, use:

```
$ cd..
```

3.4 Data Properties

Library generation for RNA-seq experiments based on the Illumina platform can be stranded or non-stranded. The former means that it is possible to assign a given read to the sense or the antisense genomic DNA strand. The data used in this chapter was obtained from libraries with no strand specificity. The user should make sure to select the proper strand specificity or adjust the commands indicated below if a different data set is used.

In addition to strand specificity, data can be either single-end (also known as single-read) or paired-end. As the names imply, single-end sequencing involves sequencing only one end for each molecule in the library generated, while paired-end sequences both ends of a fragment. Paired-end sequencing is typically more expensive than single-end. However, it is particularly useful when the reference genome is not well annotated, because it provides information about the two ends of the fragment and distance between them, increasing the number of reads that is unambiguously matched [36]. In the case of well-assembled and annotated genomes such as the Arabidopsis genome, it is common to find single-end data sets in public databases. In this example, non-stranded single-end data was used (*see* **Note 3**).

3.5 RNA-Seq Data Acquisition from Public Databases

RNA-seq data can be found in different publicly available repositories. The largest databases include Gene Expression Omnibus database from NCBI (http://www.ncbi.nlm.nih.gov/geo/), DNA Data Bank of Japan (DDBJ) (www.ddbj.nig.ac.jp/), and the Sequence Read Archive (SRA) from the European Nucleotide Archive (ENA). For easier data search and browsing, it is possible to access SRA using ArrayExpress portal web (http://www.ebi.ac.uk/arrayexpress/). RNA-seq data files can be downloaded using a standard web browser, FTP protocol, or implemented routines within software packages or applications. An updated list of molecular biology databases can be found at http://www.oxfordjournals.org/nar/database/a/ including those hosting RNA-seq experiments.

In this example, RNA-seq data from Illumina platform have been selected. Illumina is the most common RNA-seq platform today, with nearly 84% of all RNA-seq data generated using this technology [37]. RNA-seq data is commonly shared in files written in FASTQ format. This file format combines both sequence or base calling and the associated quality score per base for all reads passing the Illumina platform filtering. The usual extension of this type of file is "`.fastq`", and they can be automatically generated by the sequencer or BaseSpace, an Illumina cloud platform. FASTQ files can also be independently generated by bcl2fastq Conversion Software from Illumina. FASTQ files can also be obtained from other sequencing platforms and are compatible with the pipeline here described.

The complete list of RNA-seq experiments used in here was downloaded from ArrayExpress database (accession: E-MTAB-1668, E-MTAB-4124, E-MTAB-4298, E-MTAB-4391, E-MTAB-4396, and E-MTAB-4405). All the corresponding URLs and SRA identifiers are indicated in the file "fastq.url.txt" available for the user in the supporting material at http://virtualplant.bio.puc.cl/share/step-by-step/fastq-information.txt. This list of RNA-seq experiments considers profiles from root tissue of wild-type plants grown on different conditions or subjected to different treatments. This data is for example purposes only. For research applications, a careful selection of experiments should be performed to generate an appropriate data matrix.

To download the experimental data used in this guide, the user must have the file "`fastq.url.txt`" in the working directory "`WorkDir`". Use cat to read the file line by line and "`xargs`" to run "`wget`" in all files listed in "`fastq.url.txt`":

```
$ cat fastq.url.txt |xargs -n1 wget $1
```

3.6 Data Quality Assessment

The first step for data analysis of any genomic data set is to assess data quality (Fig. 2). Many are the factors that determine the quality of the data. There are different tools to obtain and filter reads based on quality. We recommend and use here FastQC (http://www.bioinformatics.babraham.ac.uk/projects/fastqc/). FastQC was developed and optimized for Illumina sequencing data. It accepts FASTQ files as input and outputs a summary of data quality in HTML format [38]. As a simple rule of thumb, the user should not use the sequencing data if the default "per base sequence quality" threshold is not passed. More details on data quality controls and the use of FastQC can be found at QC Fail website from the Babraham Bioinformatics Group (http://sequencing.qcfail.com/).

To run FastQC for every file, we will list with the Linux command "`ls`" all the FASTQ files present in the current folder and run the FastQC for each list element:

```
$ ls *.fastq.gz | xargs -n1 $WD/tools/FastQC/fastqc $1
```

To accelerate data processing, "`xargs`" command allows the user to select the number of threads used to process the data in parallel. This number should be selected based on the number of processors available in the user's computer or server to perform the analysis. For example, the above command line can be performed in parallel using two threads adding the option "`-P`":

```
$ ls *.fastq.gz | xargs -n1 -P2 $WD/tools/FastQC/fastqc $1
```

The FastQC output is html files which will be saved in the working directory and can be visualized with Chrome or any other browser.

3.7 Data Trimming

Read quality can be affected by read length and by the presence of adapter sequences. In Illumina sequencing, base quality decreases with read length. It is often desirable to trim those low-quality bases or sequences within reads that correspond to adapter or other sequences that are introduced during library preparation. Trimmomatic is a software tool that can identify portions of adapter sequences present in the read and trim them out [39]. Trimmomatic also takes advantage of Illumina quality score at each position to determine where the read should stop to ensure good-quality base calls, resulting in the retention of the high-quality 5′ portion and discarding poor-quality bases at the 3′ end of the read.

To run Trimmomatic, it is necessary to create a text file "script.trimmo.sh" containing the instructions to be executed (*see* **Note 4**). The user must make sure it is working in the "`WorkDir`" folder. Then use "`echo`" to display Trimmomatic instructions, and save them inside the "`script.trimmo.sh`" file using the following command line:

```
$ echo 'nohup java -jar $WD/tools/Trimmomatic-0.36/trimmo-
matic-0.36.jar SE $1 $1.trim.fil.gz LEADING:20 TRAILING:20
AVGQUAL:25 SLIDINGWINDOW:10:30 MINLEN:36 > $1.trim.nohup' >
script.trimmo.sh
```

We will next select all the FASTQ files and execute Trimmomatic on each one. "`--P2`" indicates the number of threads that will be performed in parallel (two in this case):

```
$ ls *.fastq.gz |xargs -n1 -P2 sh script.trimmo.sh
```

Trimmomatic output files are "`.fastq.fil.gz`". These contain sequences that passed the quality filter set in "script.trimmo.sh" file. The other files generated, with "`.trim.nohup`" extension, contain the number of reads and surviving sequence information.

3.8 Aligning Trimmed RNA-Seq Data to Reference Genome

Once the sequences have been quality checked and trimmed, the next step is to align the sequences to the reference genome (Fig. 2). For this example, we downloaded the *Arabidopsis thaliana* genome from Arabidopsis Information Portal (Araport) website. To do this, register in the website and go to Data > Downloads. First, from TAIR10_genome_release > assembly, download the file TAIR10_Chr.all.fasta.gz file. Then, go to Downloads > Araport11_Release_201606 and download Araport11_GFF3_genes_transposons.201606 ".gff.gz" and "gtf.gz" files. These files (GFF or GTF) contains a complete description of all annotated elements in the genome. These files will be used in Subheading 3.9. Please note that as new versions of the Arabidopsis genome are released, you may need to adjust this step to download the latest Arabidopsis genome and annotation.

The downloaded files must be decompressed with "gunzip" to continue working with them. The following commands do this:

```
$ gunzip TAIR10_Chr.all.fasta.gz
$ gunzip Araport11_GFF3_genes_transposons.201606.*.gz
```

For RNA-seq read alignment to the reference genome, we use HISAT2. Its alignment strategy that allows gaps gives speed, sensitivity, and accuracy across the alignment (*see* **Note 5**). This command generates the necessary files to perform the alignment with HISAT2:

```
$ $WD/tools/hisat2-2.0.5/hisat2-build TAIR10_Chr.all.fasta
TAIR10_Chr.all
```

Similar to what mentioned before (Subheading 3.7), create a simple text file named "script.align.sh" containing the instructions for the single-end read alignment:

```
$ echo 'nohup $WD/tools/hisat2-2.0.5/hisat2 -x TAIR10_Chr.all
-U $1 -S $1.sam > $1.align.stat.txt' > script.align.sh
```

This instruction creates a text file in SAM (Sequence Alignment/Map) format for each ".trim.fil.gz" file. SAM format is a generic alignment that describes the genome coordinates where the reads were aligned, the number of possible match sites in the genome, and other related information. The SAM files created by HISAT2 must be saved in the "WorkDir" directory. The following command executes the previous instructions for each trimmed file using two cores:

```
$ ls *.trim.fil.gz | xargs -n1 -P2 sh script.align.sh
```

The ".align.stat.txt" files contain the stats of the alignment. The alignment files with ".sam" extension are the SAM files.

3.9 Assignment of RNA-Seq Reads to Genes

Once all the SAM files from HISAT2 are generated, we have the coordinates where reads were aligned, but we still need to identify from which gene a read was generated. To do this, we will first create a file containing the gene identifiers for all genes of interest in the *Arabidopsis thaliana* genome. In this example, we will focus on protein-coding genes. The GFF file downloaded in Subheading 3.8will be used to create a file containing all annotated protein-coding genes. The "grep" command will allow the user to search in the GFF file the loci annotated as protein coding. Then the user has to save the found loci into a new file, named "`protein_coding`". All the described actions can be executed with the following commands:

```
$ grep "locus_type=protein_coding" Araport11_GFF3_genes_transposons.201606.gff |cut -f9 |cut -d';' -f1 |cut -d'=' -f2 > protein_coding
```

Then, the "`protein_coding`" identifiers file will be used below in this section to intersect with the mapped counts.

To assign the reads to a gene, we use R software [33]. In this guide, we will use the Rsubread package that facilitates the RNA--seq read data analyses, proportionating several metrics: quality assessment of sequence reads, read alignment, read summarization, and exon-exon junction, among others [40]. In the following instructions, lines that contain commands to execute in R will be indicated with a greater than sign ">".

Start R in the terminal or open Rstudio. Then, make sure that the working directory is "`WorkDir`". To install Rsubread package, press the following instructions:

```
> source("http://bioconductor.org/biocLite.R")
> biocLite("Rsubread")
> library(Rsubread)
```

Load the file containing the coordinates for all the genome elements (GFF or GFF3 file) accordingly to the reference genome; for this example we use the Araport11_GFF3_genes_transposons.201606.gff file from Araport downloaded in the previous Subheading 3.8 (*see* **Note 6**).

Then, it is necessary to run the feature counts for all SAM files. First, select the SAM files:

```
> sam.list <- dir(pattern=".sam$")
```

Since there are different types of library preparation and run configurations, the user should use the appropriate one (*see***Note 3**). In this example, we use the non-stranded (strandSpecific = 0) option.

The following command runs the program using two threads and storing the output in "`fc0`" object. The "`fc0`" will contain all the read counts assigned to every element found in the GFF file downloaded in Subheading 3.8 (*see* **Note 6**):

```
> fc0 <- featureCounts( sam.list, annot.ext= "Araport11_GFF3_genes_transposons.201606.gtf", isGTFAnnotationFile=T, allowMultiOverlap=T, isPairedEnd=F, nthreads=2, strandSpecific=0)
```

Then, save the counts and the associated statistics in tab-delimited files:

```
> write.table(fc0$counts, "fc0.counts.txt", sep="\t", col.
names=NA, quote=F)
> write.table(fc0$stat, "fc0.stat.txt", sep="\t", row.
names=F, quote=F)
```

Select protein-coding genes intersecting the "fc0$counts" and the "protein_coding" file generated at the beginning of this section:

```
> protein_coding <- as.matrix(read.table("protein_coding"))
> counts <- fc0$counts[protein_coding,]
```

Finally, we select protein-coding genes with at least one read and store them in "counts":

```
> counts <- counts[rowSums(counts)>0,]
```

The "counts" file contains the number of mapped reads to protein-coding loci for each of the RNA-seq experiment downloaded in Subheading 3.5.

3.10 Determination of Differentially Expressed Genes (DEGs) from RNA-Seq Data

We now turn our attention to differentially expressed genes in the subset of experiments in the list described in Subbheading 3.5. We decided to use the E-MTAB-4124 data set published by Ping Lan and colleagues in 2012 [41]. In this work, the authors investigate the acclimation to Pi starvation in *Arabidopsis*. We will analyze wild-type plants under the two different phosphate availability conditions: starvation and sufficiency. Wild-type plants were grown during 10 days on standard conditions and then transferred to fresh agar media without phosphate and grown for 3 days to starve the plants of phosphate. As control, wild-type plants were grown as described but transferred to fresh media and grown for 3 days keeping them under phosphate sufficiency conditions. After 3 days of treatment, total RNA extraction for RNA-seq was performed [41]. To determine differentially expressed genes under these contrasting conditions, we use the DESeq2 package from Bioconductor [42]. DESeq2 is one of the most popular packages for finding differentially regulated genes for a broad range of experimental designs [43]. The user can start in the same previous R session (Subheading 3.9) and do the following to install DESeq2:

```
> source("http://bioconductor.org/biocLite.R")
```

```
> biocLite("DESeq2")
> library(DESeq2)
```

DESeq2 requires the raw count data from the FASTQ files. The user can select the columns corresponding to the wild-type plants under the desired treatments, i.e., from columns 20–25 (named SRR420813, SRR420814, SRR420815, SRR420816, SRR420817, and SRR420818 in this experiment) in the object "`counts`" generated in Subheading 3.9.

```
> select <- counts[,20:25]
```

Next, the user must create an object describing the conditions (control or treatment) associated with each column in "`select`" object. In this case, the first three columns of "`select`" are assigned as "`control`" (or "`c`"), and the latter three are "`treatments`" (or "`t`"). The order of the assigned condition should match the column position within the data matrix as follows:

```
> condition <- c("c","c","c","t","t","t")
> coldata <- data.frame(row.names=colnames(select), condition)
```

The next step is to generate an object, called "`dds`", containing the counts and conditions to be compared as entry for DESeq2:

```
> dds <- DESeqDataSetFromMatrix( countData=select[rowSums(select)>0,], colData=coldata, design=~condition)
```

Run the DESeq2 function to start the statistical analysis. It retrieves a normalized expression level in logarithmic base 2 scale that will be stored in "`results`" object:

```
> results <- results(DESeq(dds))
```

Filter the data in "`results`" with the desired thresholds of fold change level and adjusted *p*-values. In this case fold change is log2FC >1 with adjusted *p*-values<0.01. Store the data in "`filter`" object:

```
> res <- na.exclude(as.data.frame(results))
> filter <- res[(abs(res$log2FoldChange)>1 & res$padj<0.01),]
```

The resulting table can be written as text file for further analyses:

```
> write.table(filter,"regulated.txt", quote=F,sep="\t", col.names = NA)
```

The "`regulated.txt`" table contains all regulated genes with more than two times differential expression level compared to the control and an adjusted *p*-value <0.01. In this example, we obtained 231 regulated genes.

3.11 Normalization of Gene Read Counts from RNA-Seq Data

In parallel to the analysis to obtain differentially expressed genes, read counts for each gene can be used to calculate correlation of gene expression (Fig. 2). In this example, read counts should be normalized to the number of reads that was effectively mapped. In this guide, we use median normalization of the data (*see* **Note 7**). This method can be applied using the EBSeq R package. EBSeq generates a normalized matrix using different methods including median normalization [44]. The normalization will be applied to all the mapped reads selected in Subheading 3.8 that were obtained from downloaded experiments in Subheading 3.5.

The following commands can perform the mentioned normalization. First install EBSeq:

```
> source("http://bioconductor.org/biocLite.R")
> biocLite("EBSeq")
> library(EBSeq)
```

Then normalize the data in "counts" file (Subheading 3.8) executing EBSeq, and store the normalized data in "`NormData`" object:

```
> NormData <- GetNormalizedMat(counts, MedianNorm(counts))
```

The resulting gene expression matrix "`NormData`" contains unique row identifiers and row counts obtained from different experiments from Subheading 3.5on each column. After normalization, it is recommended to delete very low count data or sum a unit to all data in order to avoid values equal to zero. Then it is useful to generate a logarithmic matrix of the data to standardize the variance [45]. All previous recommendations can be done by simply:

```
&gt; NormData.log &lt;- log2(NormData+1)
```

The "`NormData.log`" object stores normalized counts in logarithmic scale from all experiments downloaded in Subheading 3.5.

3.12 Calculating Correlation of Gene Expression for Every Pair of Genes

We are now getting close to the goal of this guide which is to generate a gene co-expression network. In this step, we will determine correlation of every pair of DEGs (Subheading 3.10) across all experimental conditions (Subheading 3.5). Please note that you should select a subset of genes before running the functions below. Calculating correlation for all possible pairs of genes may take a long time and may not be necessary in every case. In this case, we will select DEGs obtained by DESeq2 from the "`NormData.log`" table generated previously (Subheading 3.11) and extract their normalized counts, using the following command:

```
> Norm.interest <- NormData.log[rownames(filter),]
```

The "`Norm.interest`" object contains the normalized counts for DEGs. In order to build the gene co-expression network, the user needs to calculate correlation and correlation significance between each pair of DEGs in the whole experimental data set downloaded in Subheading 3.5.

There are several tools to calculate correlation; here we use the R package "`psych`" [46]. Psych is a general-purpose toolbox originally developed for psychometric analyses, but with useful tools for data analyses including correlation analysis. The user can download the package from the Comprehensive R Archive Network (CRAN) using the following commands:

```
> install.packages("https://cran.r-project.org/src/contrib/
Archive/psych/psych_1.7.5.tar.gz", repos=NULL, type="source")
> library("psych")
```

Then, set the instructions for the correlation calculation for every DEG pair. Between the available methods, select Pearson correlation to analyze the normalized data (*see***Note 8**). The correlation results will be stored in "`Norm.interest.corr`" object:

```
> Norm.interest.corr <- corr.test( t(Norm.interest), meth-
od="pearson", ci=F)
```

Among the many results of this function, there are two triangular matrices with the needed data. One matrix is symmetric and contains the correlation values. The other matrix contains the *p*-values in the upper part and the adjusted *p*-values in the lower part. To generate a table comprising the data organized properly to visualize the network, the user should execute the following commands:

```
> Norm.interest.corr$p[lower.tri( Norm.interest.corr$p,diag=-
TRUE)]=NA
> Pval.adj <- as.data.frame(as.table(Norm.interest.corr$p))
```

The "`Pval.adj`" object contains all *p*-values from the lower part of the matrix.

```
> Norm.interest.corr$r [lower.tri( Norm.interest.corr$r,di-
ag=TRUE)]=NA
> Correlation <- as.data.frame(as.table(Norm.interest.corr
$r))
```

The "`Correlation`" object contains all correlations from the matrix.

```
> Cor.table <- na.exclude(cbind( Correlation, Pval.adj))[,c
(1,2,3,6)]
```

The "`Cor.table`" object contains all correlations and *p*-values selected from the matrix. The following command adds the names to the columns in "`Cor.table`":

```
> colnames(Cor.table) <- c("gene1","gene2","cor","p.adj")
```

The generated "`Cor.table`" object can be filtered based on absolute correlation (0.9) and adjusted *p*-value (0.01) thresholds:

```
> Cor.table.filt <- Cor.table [(abs(Cor.table[,3])>0.9 & Cor.
table[,4] <0.01 ),]
```

The "`Cor.table.filt`" object contains the filtered correlations and *p*-values.

For convenience, adjusted *p*-values can be converted to logarithmic base 10. To avoid problems with very low *p*-values, which are reported as 0 by the "`corr.test`" function, we will replace zero values with the smallest positive floating-point number interpreted as different from zero in your R configuration.

```
> p.adj <- Cor.table.filt[,4]
> p.adj[p.adj==0] <- as.numeric(unlist(format(.Machine)))[1]
> Cor.table.filt <- cbind(Cor.table.filt, log10(p.adj))
> write.table(Cor.table.filt, "Cor.table.filter.txt",
sep="\t", row.names=F, quote=F)
```

At this point, we generated the "`Cor.table.filter.txt`" file containing the statistically significant correlations across the whole data set (Subheading 3.5) for every pair of differentially expressed genes (Subheading 3.10). The "`Cor.table.filter.txt`" file will be used in Subheading 3.13 to visualize the network.

3.13 Network Visualization

Co-expression networks help associate genes that are involved in similar biological functions. The analysis and visualization of gene networks is a key and powerful step to identify relationships and discover important elements in the network. Analysis of gene networks also offers us the opportunity to formulate hypotheses about key genes and implicated biological functions (Fig. 1). Below, we describe some simple steps to visualize a gene co-expression network.

Before we generate a network view, we will calculate a few useful network statistics. The user can later graphically represent network statistics on the same network. For instance, in Cytoscape (a popular software platform to view and analyze networks), statistics can be calculated and added as attributes to the nodes or edges. The number of connections of a node in a network is known as "degree," and it is a useful statistic to identify relevant nodes in biological networks, which are typically highly connected nodes or hubs. Another important node statistic is the number of times that a path passes through the node, which represents the influence this node exerts over different network modules and their potential interactions in the network known as "betweenness centrality" [47]. To calculate network statistics in this guide, we used "igraph" (http://igraph.org) [48]. This R package can process simple as well as large graphs or networks. In addition, it offers functions for generating graph visualization and complete network statistics. The basic statistics of the network, degree and betweenness, can be calculated using "`igraph`" R package. Download and install "`igraph`" with the following commands:

```
> install.packages("igraph")
> library(igraph)
```

Select the columns 1 and 2 in the "`Cor.table.filt`" object (Subheading 3.12). These columns contain all the DEG pairs with high correlation. Store them in "`g`" object, which is the network:

```
> g <- graph.data.frame( Cor.table.filt[,1:2], directed=-
FALSE)
```

Calculate the degree and betweenness for each node in "`g`" object, and store the results in the "`degree`" and "`betweenness`" objects, respectively:

```
> degree <- degree(g)
> betweenness <- betweenness(g)
> Node_nw_st <- data.frame( degree, betweenness)
```

The "`Node_nw_st`" object contains all the calculated statistics for each node in "`g`" object. Below we integrate the two statistics into a single table that can be used to rank the nodes in the network. Typically, nodes with highest degree or betweenness centrality are chosen for further analysis. The statistics of interest for every gene and the average rank (degree, betweenness centrality) can be combined into a single table using the following commands:

```
> Rank_stat <- rowMeans(cbind(rank(Node_nw_st[,1]), rank(Node_nw_st[,2])))
> Node_nw_st <- cbind(Node_nw_st, Rank_stat)
> write.table(Node_nw_st,file="Node_nw_st.txt", sep="\t", col.names = NA, quote=F)
```

The "`Node_nw_st.txt`" file contains all the calculated statistics for each node. These statistics will be used to complement the visualization of the "`Cor.table.filter.txt`" network created in Subheading 3.12.

We will use Cytoscape to visualize the network and corresponding statistics. Cytoscape is an open source software platform [49, 50]. Cytoscape can be downloaded from http://www.cytoscape.org/ and requires JAVA™ JRE or JDK.

First, we launch Cytoscape, and then we import the network table "`Cor.table.filter.txt`" (Subheading 3.12). This can be done by selecting File > Import > Network > File. After selecting the file, the user should indicate where relevant information is stored in the file. In our case, "source interaction" is in the first column which is labeled "gene1," and "target interaction" is in the second column which is labeled "gene2." These columns contain the gene ID information that Cytoscape will use to identify the nodes in each interacting pair in the network (Fig. 3). It is useful to keep the information contained in the table shown in the "Preview" window as an edge (e.g., correlation value, adjusted *p*-values); the user should click the corresponding column header to activate it. After the network has been displayed (Fig. 4a), and for better visualization, nodes can be arranged using different layouts. One of the most common displays is the "organic layout" which can be found under the Layout > yFiles Layout > Organic menu within Cytoscape. This layout will display nodes based in repulsive forces between the nodes and attractive forces induced by edges [51], facilitating identification of highly connected nodes (Fig. 4b).

Since both degree and betweenness centrality are measures of the function of a node in network connectivity, we will use the combined ranking in the node attribute file "`Node_nw_st.txt`" generated early in this section (Subheading 3.13) to map the size of the nodes to represent node importance; the file is located in the "`WorkDir`" folder.

To load the node information calculated before, the user should go through File > Import > Table > File and select "Node_nw_st.txt". In the pop-up window, the drop-down list "Network Collection," select the imported network, in this case "Cor.table.filter.txt". Import the table as "Node table columns" in the "Import Data as:" drop-down list, and be sure all columns are

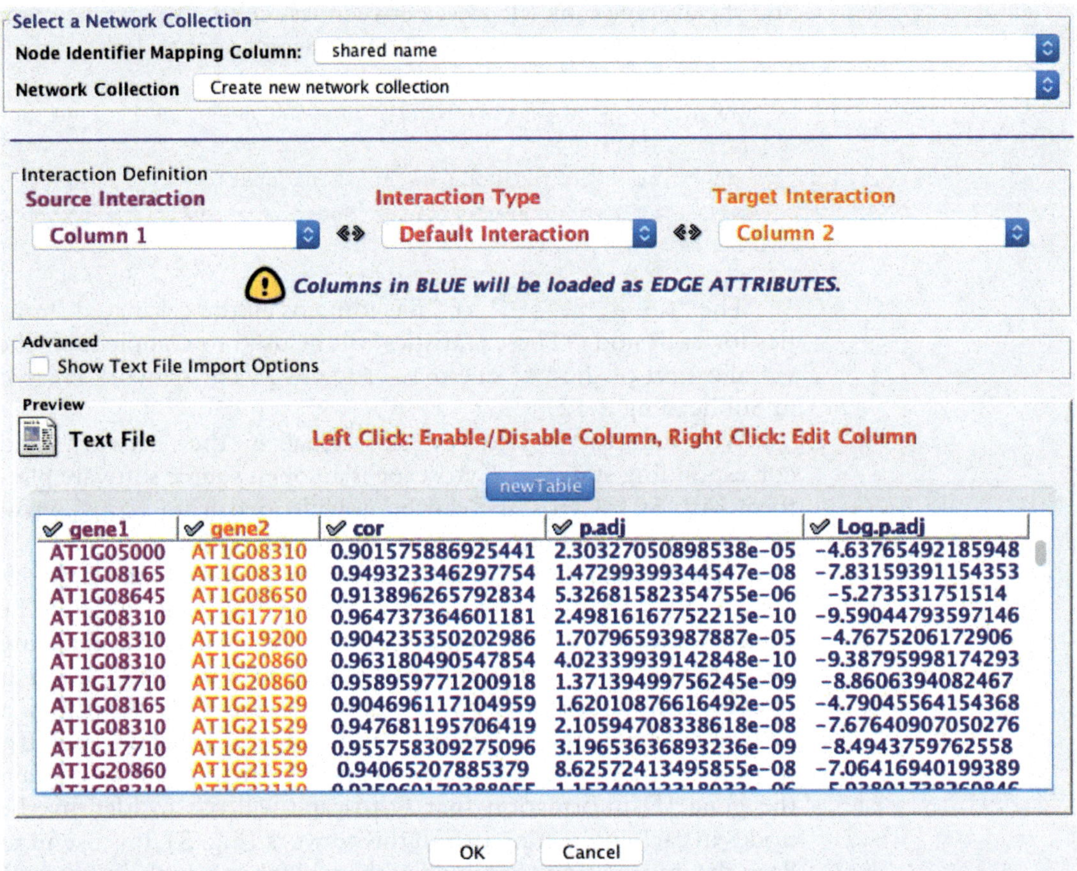

Fig. 3 Detail of the Cytoscapegraphical user interfaceused to import a network file. Screenshot of the "Import Network From Table" dialog table. Select columns for source and target interaction genes (Columns 1 and 2, order does not matter for undirected network). Double-click columns will check columns and will be loaded as edge attributes

checked. Node size can be set accordingly by selecting the "Style" tab under the "Control Panel" box. The "Style" tab also allows the user to set graphic properties on edges and the whole network in their corresponding sections in the lower part of the box.

Under the "Node" section in the "Style" tab of the "Control Panel" box, set the size of the node by checking "Lock node width and height," and then select "Size." On the new menu displayed, click on "Column" to show a new menu in which select "Rank_-stat" as the attribute to determine the size. "Mapping type" should be in "Continuous Mapping" option to distribute the sizes continuously along the previously generated combined ranking for each node (Fig. 4c). The size variation can be adjusted in the graph that appears in the "Current Mapping" section.

Edge appearance can also be customized in a similar way to nodes. In the same "Style" tab in the "Control Panel," we select the

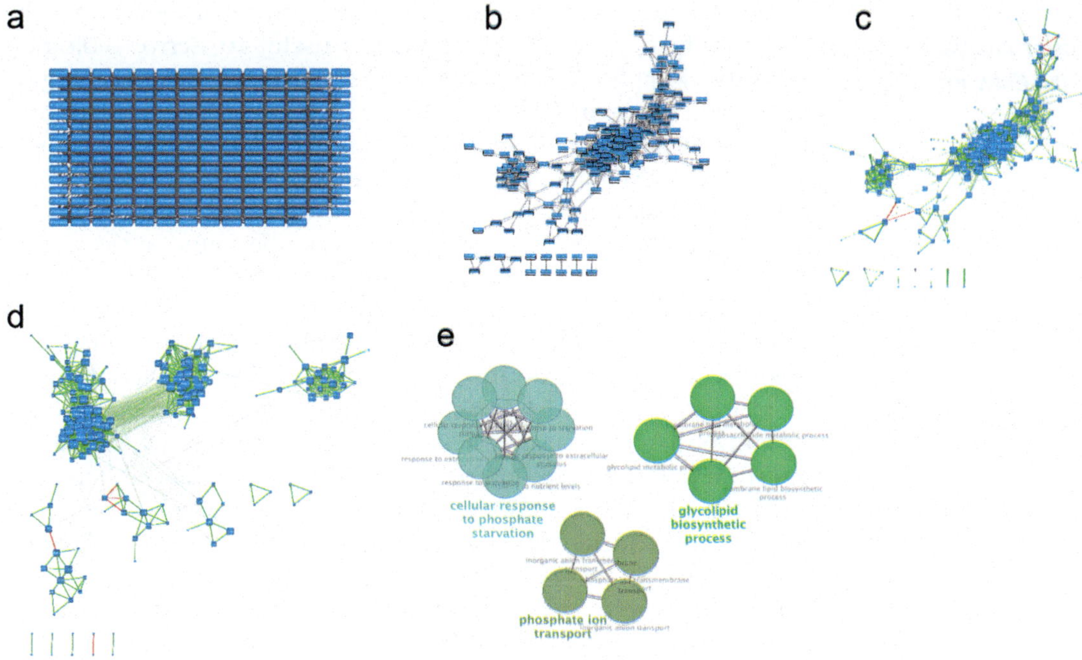

Fig. 4 Network visualization and analysis. Screenshot of the step-by-step networks obtained by the analysis pipeline. (**a**) Network view when data is first imported into Cytoscape. (**b**) Organic layout of the network. (**c**) Network with edge color and width associated to the correlation and adjusted *p*-values, respectively. (**d**) Community cluster algorithm output network. (**e**) ClueGO analysis output for the first network cluster

"Edge" tab in the lower part of the section. Click the option "Edge color to arrows." In the option "Column," select "cor" as the mapping attribute. Then in "Mapping Type" select "Continuous Mapping" option. Clicking the graph is possible to adjust the colors and intensities. The user must set the minimum and maximum values every time, in this case −1 and 1, respectively. In this example, we select red for negative correlations and green for positive correlations. Similar steps can be used to adjust the width of the edge using the "Log.p.adj" associated to the pair in the option width. This representation shows a global vision of the behavior of the genes in our data set (Fig. 4d).

In the "Table Panel" the user can see the attributes of each node and edges. These tables can be ordered by any of the attributes clicking on column headers. The most important genes of the network can be identified easily by ordering the node table using the "Rank_stat" column or by its size. The best ranked and the most connected gene (hub) of the whole network is the *MGD3*gene (AT2G11810.1). Interestingly, *MGD3*codes for an enzyme in galactolipid metabolism that is involved in phosphate starvation responses.

3.14 Performing Cluster Analysis of the Network

Analysis of network topology or the structure that determines the way in which genes are connected is useful to derive biological insights. For instance, subnetworks of genes that are highly connected (network cluster) in a co-expression network are usually involved in similar biological functions (Fig. 1). To find groups of genes that may be acting in a coordinated manner, the user can perform a cluster analysis of the network.

Numerous network clustering apps can be found in the Cytoscape application store (http:// apps.cytoscape.org/). For the purpose of this example, download, install, and use "clusterMaker" plug-in [52]. clusterMaker offers many options to perform cluster analysis that can be addressed in its manual. We use "community clustering" (GLay) with the default options, because it provides an optimized layout for large networks and a structured visualization for more efficient exploration and analysis of biological networks [53]. To display the clustering results as a new network, we select "Create new clustered network," and we check "Assume edges are undirected." In this case, directionality cannot be assumed, because relationships were established based on correlation of gene expression. To keep the gene-gene connection, we select "Restore intercluster edges after layout." This analysis generates a new network which contains the same nodes than the previous but arranged according to the connectivity of the nodes in five different clusters (Fig. 4e).

3.15 Biological Function Enrichment

To investigate if a group of genes are functionally related, the user can perform a gene ontology (GO) enrichment analysis and determine the biological functions associated to them. This analysis can be performed over the whole network or in a selected subset or cluster of interest. To perform the analysis, download and install from the Cytoscape application store the ClueGO application [54]. ClueGO allows to determine which GO categories are statistically overrepresented in a given set of genes from the network.

Select the nodes comprised in the larger cluster identified from the network, and then go to "Apps" > "ClueGO." It will appear a new tab in the control panel, which contains all ClueGO adjustable parameters. First, the user must select the nodes of interest in the network. Then, in the left panel, go to "Load Market List(s)" section and select the species (in this case, *Arabidopsis thaliana*). Then, press the "Network" button and a new option "Load Attribute" will appear. From the drop-down list, select "shared name." Below, the user can find a box in which a list containing the gene identifiers should be loaded. To do this, press the folder icon at the right side of the blank box. After this, the box will show the gene identifiers selected from the network.

In ClueGO the user has different options to perform an overrepresentation analysis that can be set in the "Clue GO Settings" section. In here, we selected "GO Biological Process" with evidence code "All." Then select the option "Show only Pathways with p..." available below, and keep default value of 0.05. We proceed with the default options, but the user can customize the analysis, following the ClueGO documentation (http://www.ici.upmc.fr/cluego/ClueGODocumentation.pdf). Finally, to run the analysis, press "Start."

The first output of ClueGO includes a new window with basic information of the analysis. Then a functionally grouped annotation network is generated with the results. These networks group gene ontology terms by similarity of their associated genes. The size of the nodes reflects the statistical significance of each term. The name of each group is given by the most significant term of the group [54]. Depending of the experimental setting, the results can show different overrepresented biological processes. In this example, we found that the larger cluster is enriched in several biological processes, from which the most representative are "cellular response to phosphate starvation," "inorganic anion transport," "glycolipid biosynthetic process," and "xyloglucan metabolic process." This enrichment analysis suggests that the highly connected genes from the largest cluster in the network are involved in the regulation of complementary processes triggered by phosphate starvation. We also perform enrichment analysis on the other clusters from the network. The second larger cluster is enriched in "cellular response to phosphate starvation," while the third and last cluster does not show enrichment of any process in this analysis. These results help highlight groups of genes with coherent biological functions that are presumably acting in coordination in response to the experimental factor.

3.16 Final Remarks

We presented a simple bioinformatics pipeline that allows users to identify (a) a list of differentially expressed genes (DEGs) from RNA-seq data, (b) a network with highly correlated genes across different experiments, (c) a ranking with relevant genes in the network, (d) clusters of highly connected genes in the co-expression network, and (e) biological processes enriched in these clusters (Fig. 1). In this particular example, we were able to identify genes and biological processes involved in the plant response to phosphate starvation. The top five ranked genes from the co-expression network are *MGDC*, encoding a monogalactosyl diacylglycerol synthase; *SQD2*, encoding a UDP-sulfoquinovose; *AT1G08310*, encoding an alpha/beta-hydrolases; *AT4G23000*, encoding a metallo-phosphoesterase; and *PFA-DSP*, encoding an atypical phosphatase. Three of these genes (MGDC, SQD2, and PFA-DSP) have already been implicated in the plant response to phosphate starvation. The other two top ranked genes have not been described yet in the context of phosphate metabolism, offering interesting hypotheses about the role of these genes in plant response to phosphate starvation. *MGDC* and *SQD2* are both top ranked in the cluster 1 and 2, respectively. In order to evaluate the functional relevance of the identified clusters, we used overrepresentation of gene ontology terms. When we examined the biological function overrepresented in cluster 1, we found "cellular response to phosphate starvation." Other biological functions related with the response to phosphate starvation were also found (Fig. 4e), such as "phosphate ion transport." All these results are in agreement with the experimental conditions assayed by Lan and colleagues [41].

This case study shows the effectiveness of this simple strategy to analyze RNA-seq data in order to predict important new genes and their biological functions in response to environmental cues. The user should note that the data, methods, and parameters used are intended to serve as guide only and by no means should be taken as the only or best way to proceed with RNA-seq data analyses in every case. What tools and criteria to use to analyze a given experiment should be determined on a case-by-case basis. Notwithstanding, we hope this chapter represents a starting point for the uninitiated to carry out data analysis of RNA-seq data.

4 Notes

1. The user should be aware that the program versions may change in the future resulting in different available options or software incompatibility.
2. The user should notice that data, methods, and parameters used at each step are intended as a simple first guideline for RNA-seq data analyses. Each software tool utilized has parameters that can be optimized, and new tools appear regularly in the literature. The user should also be aware that changes in methods and parameters can impact the final results. A rigorous data analysis should consider evaluating multiple tools and options to identify the best performing pipeline according to the experimental design and questions.
3. In here, we use non-stranded single-end data, but the user can find a detailed example of stranded paired-end reads in the supporting material available online. We provide to the user a fictitious example with commands needed to process stranded paired-end data in the file "stepbystep_sPE.txt" available at http://virtualplant.bio.puc.cl/cgi-bin/Lab/tools.cgi. The SLIDINGWINDOW:10:30 calculates the average quality in ten base windows, cutting the reads if the average quality of them is below 30. As a rule of thumb, the size of the window should be set around 10% of read length. For example, in reads of 50 bases length, the usual window is set to 4 or 5.

4. Additionally, Trimmomatic offers the ILLUMINACLIP option to trim remaining Illumina adapters from the reads. More information is available at software documentation.
5. There are numerous tools for RNA-seq alignment [18]. One of the most popular choices is TopHat2 [55]. Nowadays, TopHat developers recommend HISAT2 which provides similar functionality in a more efficient way [56].
6. The user can optimize this step using a GTF file instead of a GFF. The GTF file is a less detailed version of the GFF file. Using the GTF will speed up the calculus time and memory usage. From Araport website download Araport11_GFF3_-genes_transposons.201606.gtf.gz.
7. There is no consensus about what is the best normalization method adjustable to a broad range of data. The most used normalization methods are quantile, lowess, and median [45].
8. There are several correlation methods besides Pearson's. For detailed information see the package documentation (http://CRAN.R-project.org/package=psych).

Acknowledgments

Research in our group is funded by Fondo de Desarrollo de Areas Prioritarias (FONDAP) Center for Genome Regulation (15090007), MIISSB Iniciativa Científica Milenio-MINECON, Fondo Nacional de Desarrollo Científico y Tecnológico (FONDE-CYT) 1141097, and EvoNet (DE-SC0014377).

References

1. Usadel B, Fernie AR (2013) The plant transcriptome—from integrating observations to models. Front Plant Sci 4:1–3
2. Moustafa K, Cross JM (2016) Genetic approaches to study plant responses to environmental stresses: an overview. Biology (Basel) 5:1–18
3. Malik VS (2016) RNA sequencing as a tool for understanding biological complexity of abiotic stress in plants. J Plant Biochem Biotechnol 25:1–2
4. Wetterstrand, KA (2016). DNA sequencing costs: data from the NHGRI large-scale genome sequencing program. www.genome.gov/sequencingcostsdata, Accessed 4 Sep 2016
5. Martin LBB, Fei Z, Giovannoni JJ, Rose JKC (2013) Catalyzing plant science research with RNA-seq. Front Plant Sci 4:66
6. Weber APM (2015) Discovering new biology through RNA-Seq. Plant Physiol 169 (3):1524–1531. 01081.2015
7. Hrdlickova R, Toloue M, Tian B (2017) RNA-Seq methods for transcriptome analysis. Wiley Interdiscip Rev RNA 8:e1364
8. Kolesnikov N, Hastings E, Keays M, Melnichuk O, Tang YA, Williams E et al (2015) ArrayExpress update-simplifying data submissions. Nucleic Acids Res 43: D1113–D1116
9. Serin EAR, Nijveen H, Hilhorst HWM, Ligterink W (2016) Learning from co-expression networks: possibilities and challenges. Front Plant Sci 7:444

10. Katari MS, Nowicki SD, Aceituno FF, Nero D, Kelfer J, Thompson LP et al (2010) VirtualPlant: a software platform to support systems biology research. Plant Physiol 152:500–515
11. Gutiérrez RA, Lejay LV, Dean A, Chiaromonte F, Shasha DE, Coruzzi GM (2007) Qualitative network models and genome-wide expression data define carbon/nitrogen-responsive molecular machines in Arabidopsis. Genome Biol 8:R7
12. Yang C, Wei H (2015) Designing microarray and RNA-Seq experiments for greater systems biology discovery in modern plant genomics. Mol Plant 8:196–206
13. Bassel GW, Gaudinier A, Brady SM, Hennig L, Rhee SY, De Smet I (2012) Systems analysis of plant functional, transcriptional, physical interaction, and metabolic networks. Plant Cell 24:3859–3875
14. Canales J, Moyano TC, Villarroel E, Gutiérrez RA (2014) Systems analysis of transcriptome data provides new hypotheses about Arabidopsis root response to nitrate treatments. Front Plant Sci 5:22
15. Long TA, Brady SM, Benfey PN (2008) Systems approaches to identifying gene regulatory networks in plants. Annu Rev Cell Dev Biol 24:81–103
16. Rasmussen S, Barah P, Suarez-Rodriguez MC, Bressendorff S, Friis P, Costantino P et al (2013) Transcriptome responses to combinations of stresses in Arabidopsis. Plant Physiol 161:1783–1794
17. Ruffel S, Krouk G, Coruzzi GM (2010) A systems view of responses to nutritional cues in Arabidopsis: toward a paradigm shift for predictive network modeling. Plant Physiol 152:445–452
18. Conesa A, Madrigal P, Tarazona S, Gomez-Cabrero D, Cervera A, McPherson A et al (2016) A survey of best practices for RNA-seq data analysis. Genome Biol 17:13
19. Wei H, Persson S, Mehta T, Srinivasasainagendra V, Chen L, Page GP et al (2006) Transcriptional coordination of the metabolic network in Arabidopsis. Plant Physiol 142:762–774
20. Alvarez JM, Riveras E, Vidal EA, Gras DE, Contreras-López O, Tamayo KP et al (2014) Systems approach identifies TGA1 and TGA4 transcription factors as important regulatory components of the nitrate response of Arabidopsis Thaliana roots. Plant J 80:1–13
21. Gutierrez RA, Stokes TL, Thum K, Xu X, Obertello M, Katari MS et al (2008) Systems approach identifies an organic nitrogen-responsive gene network that is regulated by the master clock control gene CCA1. Proc Natl Acad Sci 105:4939–4944
22. Gutiérrez RA, Gifford ML, Poultney C, Wang R, Shasha DE, Coruzzi GM et al (2007) Insights into the genomic nitrate response using genetics and the Sungear software system. J Exp Bot 58:2359–2367
23. Krouk G, Mirowski P, LeCun Y, Shasha DE, Coruzzi GM (2010) Predictive network modeling of the high-resolution dynamic plant transcriptome in response to nitrate. Genome Biol 11:R123
24. Nero D, Krouk G, Tranchina D, Coruzzi GM (2009) A system biology approach highlights a hormonal enhancer effect on regulation of genes in a nitrate responsive 'biomodule'. BMC Syst Biol 3:59
25. Vidal EA, Araus V, Lu C, Parry G, Green PJ, Coruzzi GM et al (2010) Nitrate-responsive miR393/AFB3 regulatory module controls root system architecture in Arabidopsis Thaliana. Proc Natl Acad Sci U S A 107:4477–4482
26. Vidal EA, Moyano TC, Krouk G, Katari MS, Tanurdzic M, McCombie WR et al (2013) Integrated RNA-seq and sRNA-seq analysis identifies novel nitrate-responsive genes in Arabidopsis Thaliana roots. BMC Genomics 14:701
27. Araus V, Vidal EA, Puelma T, Alamos S, Mieulet D, Guiderdoni E et al (2016) Members of BTB gene family of scaffold proteins suppress nitrate uptake and nitrogen use efficiency. Plant Physiol 171:1523–1532
28. De Bodt S, Carvajal D, Hollunder J, Van den Cruyce J, Movahedi S, Inze D (2010) CORNET: a user-friendly tool for data mining and integration. Plant Physiol 152:1167–1179
29. Szklarczyk D, Franceschini A, Wyder S, Forslund K, Heller D, Huerta-Cepas J et al (2015) STRING v10: protein-protein interaction networks, integrated over the tree of life. Nucleic Acids Res 43:D447–D452
30. Zuberi K, Franz M, Rodriguez H, Montojo J, Lopes CT, Bader GD et al (2013) GeneMANIA prediction server 2013 update. Nucleic Acids Res 41:W115–W122
31. Aoki Y, Okamura Y, Tadaka S, Kinoshita K, Obayashi T (2016) ATTED-II in 2016: a plant coexpression database towards lineage-specific coexpression. Plant Cell Physiol 57:e5
32. Field D, Tiwari B, Booth T, Houten S, Swan D, Bertrand N et al (2006) Open software for biologists: from famine to feast. Nat Biotechnol 24:801–803
33. R Core Team (2015) R: a language and environment for statistical computing. R a lang.

environ. stat. comput. R Core Team, Vienna, Austria

34. Gentleman R, Carey V, Bates D, Bolstad B, Dettling M, Dudoit S et al (2004) Bioconductor: open software development for computational biology and bioinformatics. Genome Biol 5:R80
35. Huber W, Carey VJ, Gentleman R, Anders S, Carlson M, Carvalho BS et al (2015) Orchestrating high-throughput genomic analysis with bioconductor. Nat Methods 12:115–121
36. Wilhelm BT, Landry J-R (2009) RNA-Seq-quantitative measurement of expression through massively parallel RNA-sequencing. Methods 48:249–257
37. Kodama Y, Shumway M, Leinonen R (2012) The sequence read archive: explosive growth of sequencing data. Nucleic Acids Res 40:2011–2013
38. Andrews, S (2010). FastQC: a quality control tool for high throughput sequence data. Available online at:http://www.bioinformatics.babraham.ac.uk/projects/fastqc
39. Bolger AM, Lohse M, Usadel B (2014) Trimmomatic: a flexible trimmer for Illumina sequence data. Bioinformatics 30:2114–2120
40. Liao Y, Smyth GK, Shi W (2013) The subread aligner: fast, accurate and scalable read mapping by seed-and-vote. Nucleic Acids Res 41(10):e108
41. Lan P, Li W, Schmidt W (2012) Complementary proteome and transcriptome profiling in phosphate-deficient Arabidopsis roots reveals multiple levels of gene regulation. Mol Cell Proteomics 11:1156–1166
42. Love MI, Huber W, Anders S (2014) Moderated estimation of fold change and dispersion for RNA-seq data with DESeq2. Genome Biol 15:550
43. Schurch NJ, Schofield P, Gierliński M, Cole C, Sherstnev A, Singh V et al (2016) How many biological replicates are needed in an RNA-seq experiment and which differential expression tool should you use? RNA 22:839–851
44. Leng N, Dawson JA, Thomson JA, Ruotti V, Rissman AI, Smits BMG et al (2013) EBSeq: an empirical Bayes hierarchical model for inference in RNA-seq experiments. Bioinformatics 29:1035–1043
45. Morgun A, Yambartsev A, Thomas L, Shulzhenko N, Ramsey S, Dong X (2015) Reverse enGENEering of regulatory networks from big data: a roadmap for biologists. Bioinform Biol Insights 9:61–74
46. Revelle, W (2017) psych: Procedures for Personality and Psychological Research, Northwestern University, Evanston, Illinois, USA. Available at: https://CRAN.R-project.org/package=psychVersion=1.7.8.
47. Yoon J, Blumer A, Lee K (2006) An algorithm for modularity analysis of directed and weighted biological networks based on edge-betweenness centrality. Bioinformatics 22:3106–3108
48. Csardi G, Nepusz T (2006) The igraph software package for complex network research, InterJournal, Complex Systems 1695. Available at: http://igraph.org
49. Kohl M, Wiese S, Warscheid B (2011) Cytoscape: software for visualization and analysis of biological networks. Methods Mol Biol 696:291–303
50. Shannon P, Markiel A, Ozier O, Baliga NS, Wang JT, Ramage D et al (2003) Cytoscape: a software environment for integrated models of biomolecular interaction networks. Genome Res 13:2498–2504
51. Cline M, Smoot M, Cerami E, Kuchinsky A, Landys N, Workman C et al (2007) Integration of biological networks and gene expression data using Cytoscape. Nat Protoc 2:2366–2382
52. Morris JH, Apeltsin L, Newman AM, Baumbach J, Wittkop T, Su G et al (2011) clusterMaker: a multi-algorithm clustering plugin for Cytoscape. BMC Bioinformatics 12:436
53. Su G, Kuchinsky A, Morris JH, States DJ, Meng F (2010) GLay: community structure analysis of biological networks. Bioinformatics 26:3135–3137
54. Bindea G, Mlecnik B, Hackl H, Charoentong P, Tosolini M, Kirilovsky A et al (2009) ClueGO: a Cytoscape plug-in to decipher functionally grouped gene ontology and pathway annotation networks. Bioinformatics 25:1091–1093
55. Kim D, Pertea G, Trapnell C, Pimentel H, Kelley R, Salzberg SL (2013) TopHat2: accurate alignment of transcriptomes in the presence of insertions, deletions and gene fusions. Genome Biol 14:R36
56. Kim D, Langmead B, Salzberg SL (2015) HISAT: a fast spliced aligner with low memory requirements. Nat Methods 12:357–360

Check for updates

Chapter 22

GWA-Portal: Genome-Wide Association Studies Made Easy

Ümit Seren

Abstract

Genome-wide association studies (GWAS) are an effective method for investigating the genetics of natural phenotypic variation in many different model organisms.

Here we present GWA-Portal, an interactive web application that enables researchers to upload their phenotypes and easily carry out GWAS directly in the browser. We will present all the steps needed—from uploading the phenotype to interpreting the results—using a published root phenotype.

Key words Genome-wide association studies, GWAS, 1001 genomes, Natural variation, Manhattan plots, Population genetics, Web application, Portal

1 Introduction

In nature, organisms and specifically plants exhibit a diverse range of phenotypic variability. Understanding how genetic variation translates into phenotypic variation has been a prime focus of population geneticists, and we have made substantial progress in the *Arabidopsis* community [1, 2].

Since the first genome-wide association studies (GWAS) in plants [1, 3], many research groups have generated and mapped high-quality phenotypic data ranging from disease resistance phenotypes to high-dimension molecular phenotypes—e.g., RNA expression, DNA methylation, metabolites. Recently, a detailed map of genetic variation was released by the 1001 Genomes Consortium for the *Arabidopsis* community [4]. With this high-quality genotype and phenotype data on hand, we can conduct GWAS in *Arabidopsis thaliana* for a worldwide sample of 1135 *A. thaliana* accessions. In the past, carrying out GWAS was not always straightforward: researchers had to get hold of the genotype data, convert it into the right format, download a software package and install it

Electronic supplementary material: The online version of this chapter https://doi.org/10.1007/978-1-4939-7747-5_22 contains supplementary material, which is available to authorized users.

Daniela Ristova and Elke Barbez (eds.), *Root Development: Methods and Protocols*, Methods in Molecular Biology, vol. 1761, https://doi.org/10.1007/978-1-4939-7747-5_22, © Springer Science+Business Media, LLC 2018

with all its dependencies, and then run the GWA analysis from the command line. We developed GWA-Portal, a user-friendly and interactive web application that allows researchers to easily carry out GWAS by simply uploading their phenotype of interest and clicking through an interactive front end. Additionally, GWA-Portal can enrich the samples in the uploaded phenotypes with additional genotypic and accession-related information to help the user uncover interesting patterns in the data. GWA-Portal is the successor of the GWAPP [5], a web application for running GWAS "on the fly."

We will showcase all the steps needed using a published cellular trait related to root development in *A. thaliana* [6]. In the first step, I will show how the user can upload a phenotype and create a GWAS study; and further we will demonstrate how to interpret the GWAS results using interactive Manhattan plots and visualizations; and finally, we will show how to examine specific associations in depth.

2 Materials

GWA-Portal can be accessed using the URL https://gwas.gmi.oeaw.ac.at with a modern browser (Firefox, Chrome, or Safari). To conduct the steps described in Subheading 3, an account on GWA-Portal has to be created.

1. Open the URL https://gwas.gmi.oeaw.ac.at in your browser (Chrome, Safari, or Firefox).
2. Click on the "Login" menu item in the top navigation bar.
3. Choose one of two registration methods:
   (a) Sign in using an existing third-party account (Google, GitHub, or Facebook).
   (b) Sign up with username and password.
4. Make sure that you are logged in ("Login" menu item should be replaced with "My Account" menu item).

For the showcase we will be using the meristem zone length trait [6], which can be downloaded here: https://goo.gl/gKEIKe (*see* **Note 1** for how to use your own traits).

3 Methods

3.1 Upload Phenotype and Create GWAS Using a Wizard (Video tutorial)

1. From the landing page, click on the "Create" button in the "New GWAS analysis" card. This will open a 6-step wizard.
2. In the first step, a new study needs to be created by clicking on the "Create new study" card. In the pop-up fill out the study

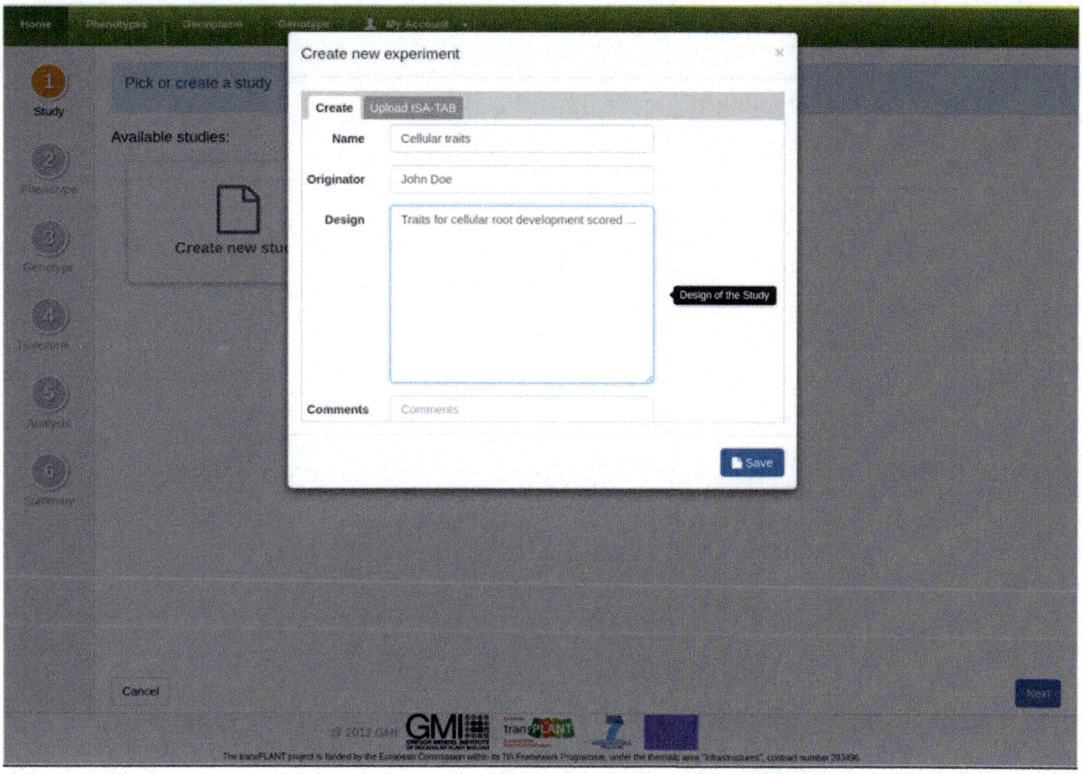

Fig. 1 Create a new study. Studies provide a logical grouping for one or more phenotypes. The user needs to specify a name and an originator

and the originator fields (*see* **Note 2**), and click "Save" which will save the study and automatically select it (Fig. 1). Click the "Next" button on the lower right side of the screen.

3. Next, the phenotype has to be uploaded as a comma-separated values (csv) file (*see* **Note 3**). Click on the "Upload phenotype card," and either click on the "Add files" button and choose the previously downloaded trait file or drag and drop the file onto the rectangle area. The trait file will be validated. If there were no errors, press on the "Start Upload" button. Once the upload is finished, GWA-Portal will parse the trait file and display the results. Click on the phenotype card labeled "MZ_length" and more information about the selected phenotype will be displayed (it is possible to upload multiple phenotypes at the same time). By default, a table with all the phenotyped samples and corresponding phenotypic values are displayed. The user can also view additional information such as the histogram of the phenotypic distribution, the geographic distribution of the samples, a map with the locations of the samples, and the phenotype-explorer visualization that can overlay phenotypic values of the samples with geographic information (Fig. 2).

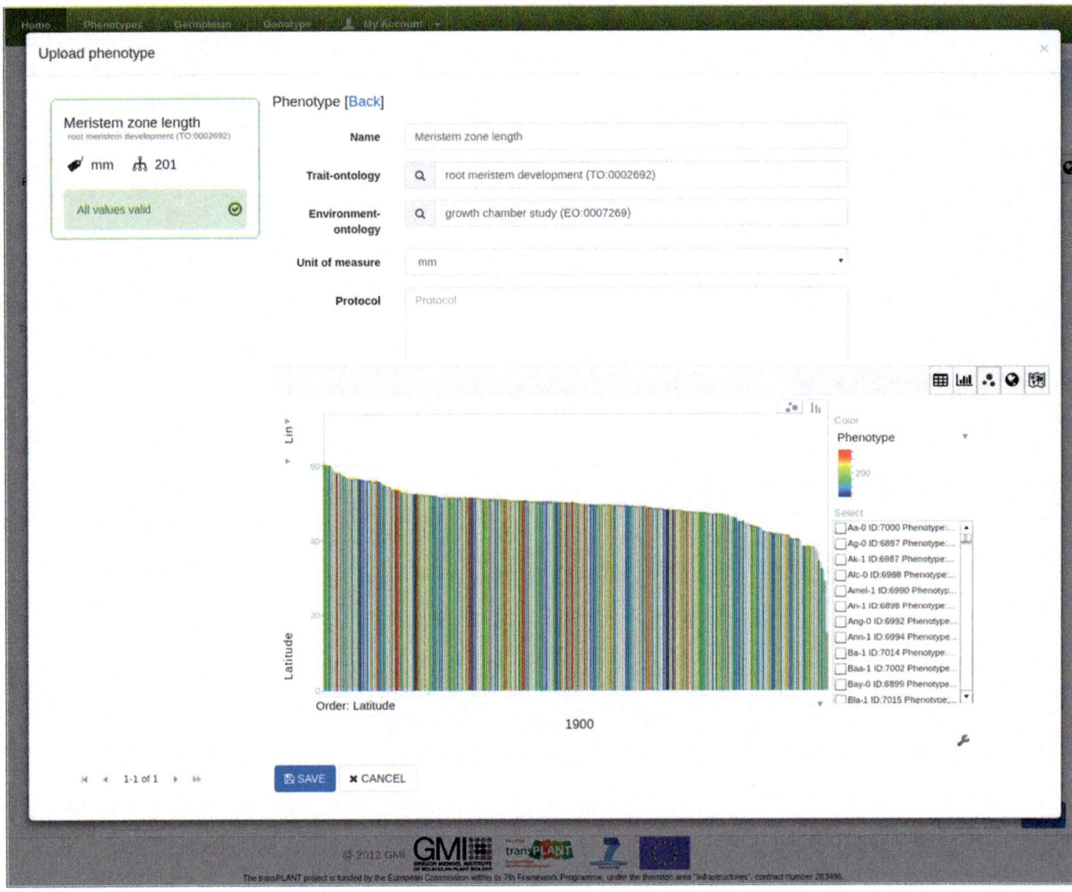

Fig. 2 Upload phenotype. On the left side, the phenotype card shows the phenotypes that were uploaded. Clicking on one of them will show a form with information about the phenotype. Users can edit this information. Additionally, the user can select from a range of different visualizations from the icon action bar. Here the phenotype-explorer visualization is shown

4. By default, the filename will be used as the name for the trait. Replace the name with a more meaningful name such as "meristem zone length." In the Trait-ontology field, search for "root meristem development," and choose the corresponding ontology term from the search result. Repeat the same step for Environment-ontology field, this time searching for and selecting the "growth chamber" ontology term. "In the unit of measure," choose "mm." Finally click on the "Save" button to actually save the uploaded trait. The saved trait will be automatically selected. Click "Next" to proceed.

5. Next, we need to choose a genotype dataset for our GWAS analysis. Currently, GWA-Portal supports four different genotype datasets (*see* Table 1): the 250k SNP dataset [7], the Swedish genome dataset [8], the 1001 genomes dataset [4], and the imputed full sequence dataset (*see* **Note 4**). They are

Table 1
Various genotype datasets that can be used in GWA-Portal for GWAS

Genotype dataset	# Samples	# SNPs
250k SNP	1386	214.051
Swedish genome	259	~ 6 million
1001 genomes	1135	~ 10 million
Imputed full sequence	2029	~ 10 million

displayed as cards on the left side of the view. Additionally, the overlap between the samples in our selected trait and the samples of the corresponding genotype is displayed as a number and percentage inside the card. The font color (red, yellow, and green) aids the user to choose the right genotype dataset with the highest overlap. The missing samples for a genotype dataset with less than 100% overlap can be displayed as a table by clicking on a card (Fig. 3). Originally, the GWAS for the meristem zone length trait was conducted using the 250k SNP dataset [7]. Hence, we will choose that dataset and click on "Next."

6. In the fourth step, we need to choose a transformation. GWA-Portal currently supports three different transformations: logarithmic (LOG), square root (SQRT), and Box-Cox [9]. To help the user choose the right transformation, the histogram of the transformation together with the pseudo-heritability and the Shapiro-Wilk score is displayed. By default, the "No transformation" card is selected, which was the same transformation that was also used in the published GWAS study [6]. In order to reproduce the results, we will use the "No transformation" option (*see* **Note 5**) for our analysis and click the "Next" button to move to the next step.
7. In the fifth step, we will choose one of three different GWA methods (Fig. 4): a nonparametric Wilcoxon rank sum test (KW) [10], a simple linear regression (LM), or an accelerated linear mixed model (AMM). The KW method is the fastest but also the simplest. The LM is slightly slower, and the AMM method is the slowest one but is the only method that does account for population structure (*see* **Note 6**). AMM first performs a genome-wide scan using the approximate inference [11, 12] and then updates the smallest 200 *P* values using an exact mixed model inference [13]. For each GWA method, the estimated runtime, which depends on the chosen genotype, the number of phenotyped samples, and the selected method, is displayed. We choose the AMM method and make sure that the "Create HPC Job" toggle button is set to "ON" (we leave the

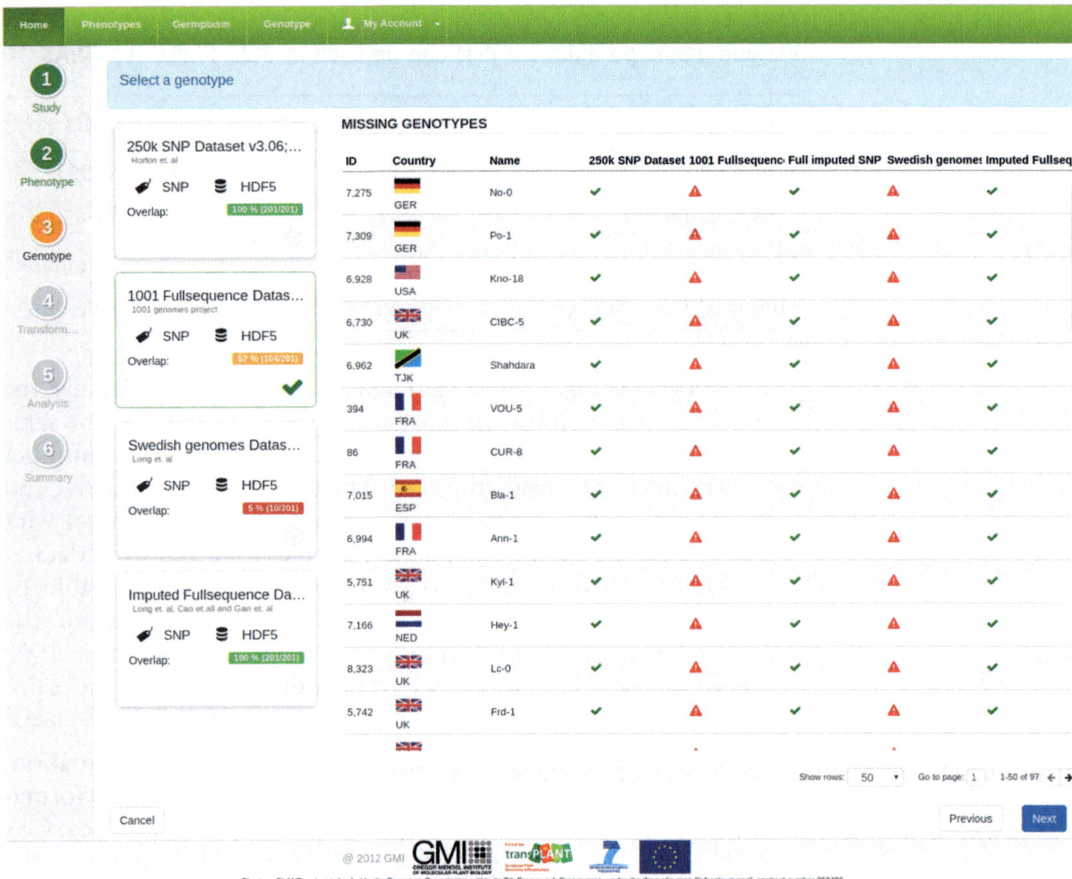

Fig. 3 Select a genotype. Each genotype card on the left side represents one of four available genotype datasets. Because the genotyped samples do not necessarily overlap with the phenotyped samples, the overlap is displayed as a number and percentage. A color code (green, yellow, and red) aides the user in selecting the genotype dataset with the highest overlap. The missing genotypes table is displayed when the user selects a genotype dataset with less than 100% overlap

"Run enrichment analysis" option "OFF" for now, *see* **Note 7**). We need to fill out the name of the analysis. We choose a descriptive name such as "Meristem_Zone_Length_250k_raw_amm" (*see* **Note 8**) and then click on the "Next" button.

8. In the last step, we see an overview of all options that we chose in the previous five steps. If we want to choose a different option in one of the previous steps, we can still go back using the "Previous" button or cancel the entire wizard by clicking on "Cancel." We click on the "Finish" button. GWA-Portal will create the GWAS in the database and create a job on our high-performance cluster (HPC), and we will be automatically navigated to the detailed page of the GWAS analysis, where we can observe the status of the analysis via a donut chart (Fig. 5).

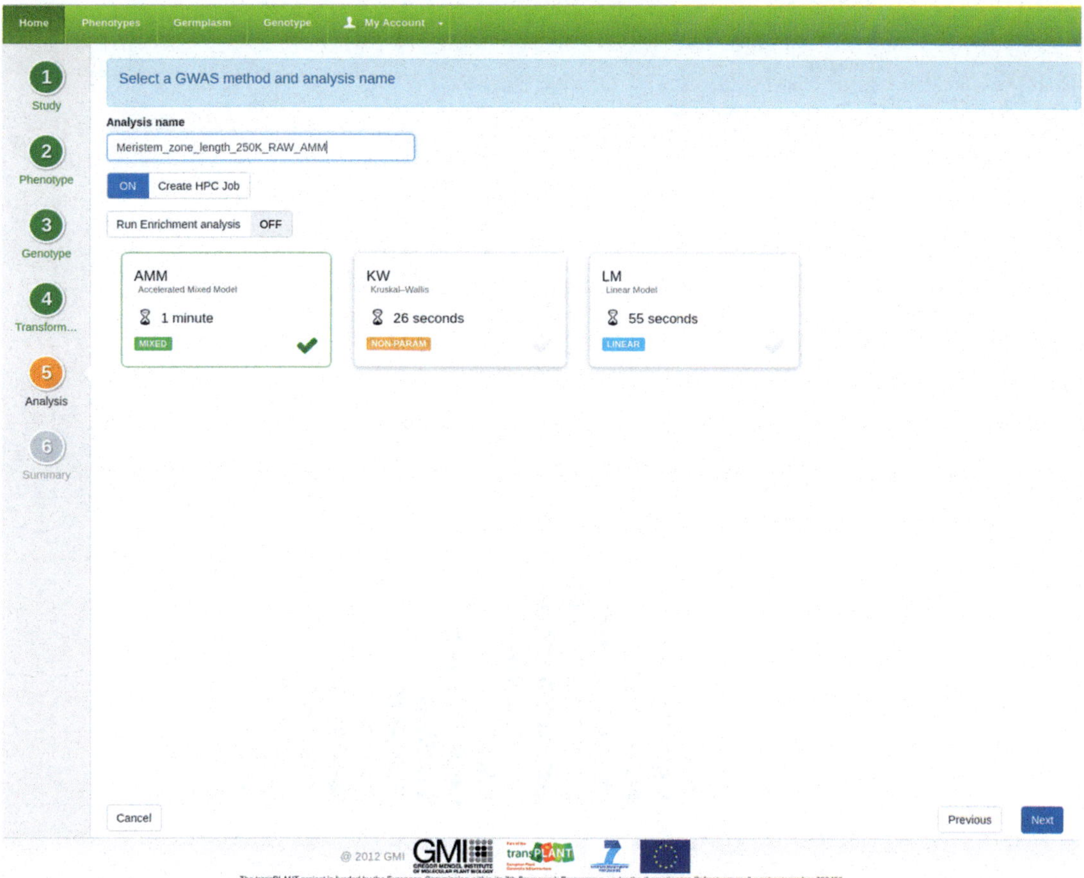

Fig. 4 Select an analysis. In the last step, the user has to provide a name for the GWAS analysis, choose whether to automatically run the analysis, optionally run enrichment analysis, and choose an appropriate GWAS method. Additionally, the estimated runtime, which depends on sample size and chosen genotype dataset, is displayed for each GWAS method

The status of the analysis will transition from "Waiting" to "Queued," then to "Running," and finally to "Finished." Additionally, we will get notifications in our "My Account" panel for each state transition.

3.2 Interactive Manhattan Plots

1. Once the GWAS analysis is finished, an additional tab named "Plot (AMM)" will be displayed next to the "Overview" tab.
2. When the user clicks on the "Plot (AMM)" tab, five vertically stacked interactive Manhattan plots will be displayed, one for each chromosome (Fig. 6). Each marker in the Manhattan plot represents a single nucleotide polymorphism (SNP) along the chromosomal position (x-axis). The y-axis represents the score of the association. The higher the score, the more significant the association (mathematically, the score is the -log10 of the *P* value). Assuming arbitrary dependence between SNPs, the

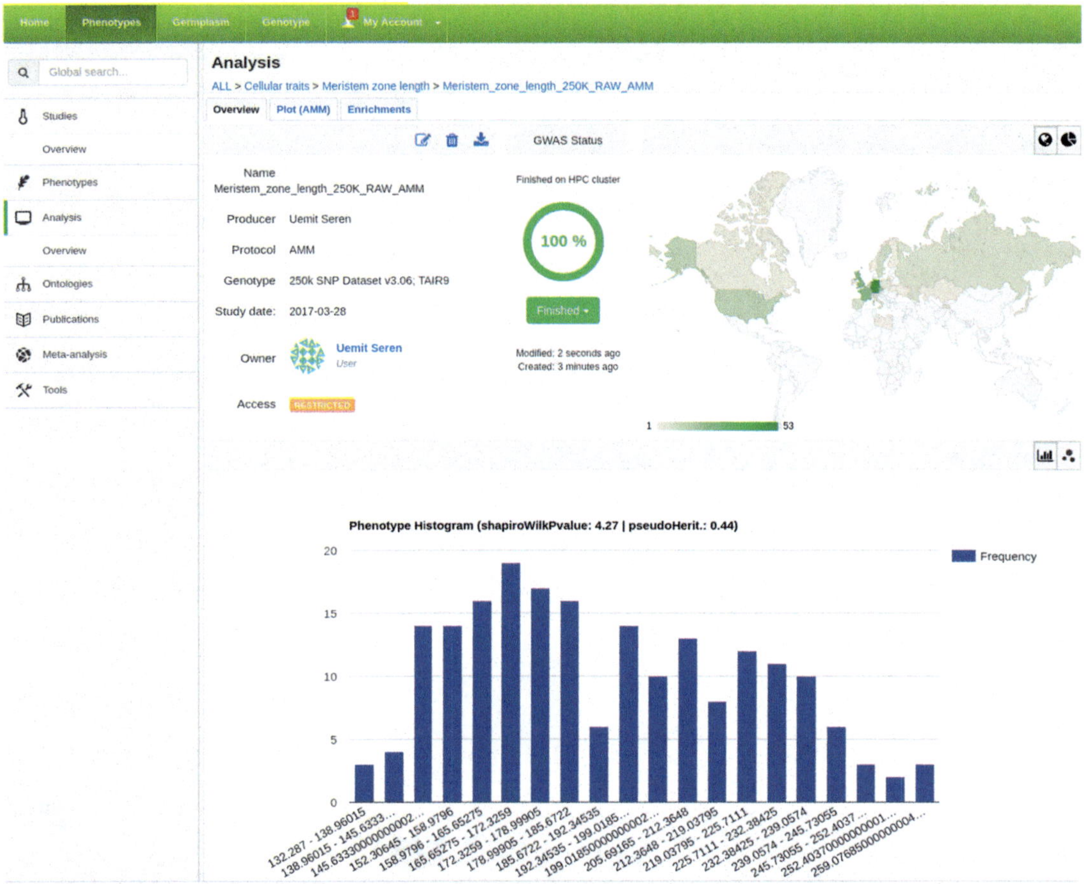

Fig. 5 Detailed view of the GWAS analysis. Apart from general information, the histogram of the phenotype is displayed in the lower panel as well as the geographic distribution of the phenotyped samples on the upper right side. The status of the GWAS analysis is displayed as a donut chart. Whenever the status of the analysis changes (i.e., Running → Finished), the user will get a notification in the "My Account" tab. An icon action bar provides access to commonly used functions, and once the analysis is finished, the user can display the Manhattan plots by clicking on the "Plots (AMM)" tab.

5% false discovery rate (FDR) threshold is plotted as a dashed horizontal line. Associations above the threshold are considered as significant. SNPs in the Manhattan plot are rendered in one of three different shapes. The shape encodes the annotation of the SNP: triangles for non-synonymous, rectangles for synonymous, and circles for exonic SNPs and SNPs with no information.

3. The user can hover the mouse over a certain association/SNP in the Manhattan plot, and additional information such as the genomic position, the SNP annotation, the score, the minor allele frequency (MAF), and minor allele count (MAC) will be displayed. Furthermore, the user can change the display settings of the Manhattan plots by clicking on the gear icon, which is located above the chromosome name. This will open a

Fig. 6 The Manhattan plots. For each chromosome, a Manhattan plot is displayed. The user can change filtering and display settings from a pop-up window by clicking on the gear icon. Moving the mouse over a SNP will display additional information such as position, MAC and SNP annotation on the upper right side. Clicking on a SNP will display context menu with additional actions. Users can also search for candidate gene lists or genes using the TAIR gene id to display them on the Manhattan plot. Publication quality versions of the QQ and Manhattan plots are also available for download

pop-up where the user can change the MAC or MAF filtering (by default SNPs with a minor allele count of at least 15 are displayed, *see* **Note 9**), only display certain SNPs (i.e., genic or non-synonymous), color the SNPs in the Manhattan plot based on their MAF, and display additional statistics tracks such as gene density.

4. The user can display the QQ plots and download static versions of the Manhattan plots by clicking on the corresponding buttons in the top right area of the view. Additionally, a search box allows the user to search for candidate gene lists (*see* **Note 10**) or single genes using the "TAIR gene id" and display them as vertical lines at their genomic positions in the Manhattan plot.

Fig. 7 Zoom in view of Manhattan plots. Once the user zooms into a specific region using a "click, hold, and drag" mouse gestures, an additional gene annotation track is displayed underneath the Manhattan plot. Moving the mouse over a SNP will draw a blue vertical line in the gene annotation track and thus show if the SNP is located in a specific part of the gene (UTR, intron, etc.). Similarly, moving the mouse over a gene will show a pop-up with the gene description. Genes that were found using the search function will be highlighted in red

5. The GWA analysis of the example root trait *meristem zone length* shows a promising association on chromosome 1 around position 22244990. To find out more about the genes that are underlying this peak, we can zoom into this region. This can be done with the mouse using a "click, hold, and drag mouse" gesture that defines the area for the zoom action. If the zoom level is below a specific threshold (~1.5 Mb), a gene annotation view is displayed (Fig. 7). Moving the mouse over a point in the Manhattan plot will also display a vertical line in the gene annotation view. Zooming in further below 150 kb, a more detailed gene annotation view containing gene features (e.g., the coding sequence region and the untranslated regions) will be shown. Resetting the zoom level can be done by double-clicking

anywhere in the Manhattan plot. Moving the mouse over a specific gene in the gene annotation track will display a pop-up with additional functional description for the gene. Clicking on the gene will direct the user to the *Arabidopsis* Information Resource (TAIR) website for the gene, containing more detailed information. In our GWAS example, the most significantly (P value $= 3.46 \times 10^{-7}$) associated SNP is present in the previously uncharacterized F-box gene (At1g60370). Meijon et al. functionally validated the role of that F-box gene in the meristem zone length trait.

6. Sometimes GWAS peaks can be quite broad and complex. For those cases checking the Linkage disequilibrium (LD) structure around the peak can be useful. Although our peak does not look too complex, we check LD by choosing a SNP as a focal point; we choose the most significant associated SNP at position 22244990 by clicking on it in the Manhattan plot and choosing the option "Calculate exact LD in this region." This will display a LD triangle plot below the gene annotation panel and color code the SNPs in the Manhattan plot (Fig. 8). The LD triangle plot displays pairwise r^2 values of at most 500 SNPs around the selected SNP. Hovering over SNP in the Manhattan plot will color code all neighboring SNPs according to their r^2 value. At the same time, all pairwise r^2 values in the triangle plot will be highlighted. Similarly, when a specific r^2 value in the triangle plot is selected, the corresponding pair of SNPs in the Manhattan plot and the triangle plot is highlighted with corresponding color coding.

3.3 Examine Specific Associations in Depth

1. We want to know more about the most significantly associated SNP in our GWAS, so we choose the option "Detailed SNP information" from the option menu that is displayed upon clicking on the SNP. This will show detailed information such as function, effect and codon change about the chosen SNP, as well as the minor allele frequency and count in our phenotyped population. Additionally, a table with information about those samples together with their phenotypic values is displayed. This way the user can easily check if a sample has a specific combination of phenotypic value and allele. The user can sort the table by phenotypic value or allele. For a broader overview that uncovers interesting patterns in the data, we can click on the icon with the three circles.

2. This will show the phenotype-explorer visualization that overlays sample information, such as geographic coordinates and country of origin, with phenotypic values and alleles for the selected SNP of each sample. By default, it will display a scatter plot where each circle represents a phenotyped sample. On the x-axis the longitude and on the y-axis the latitude of the

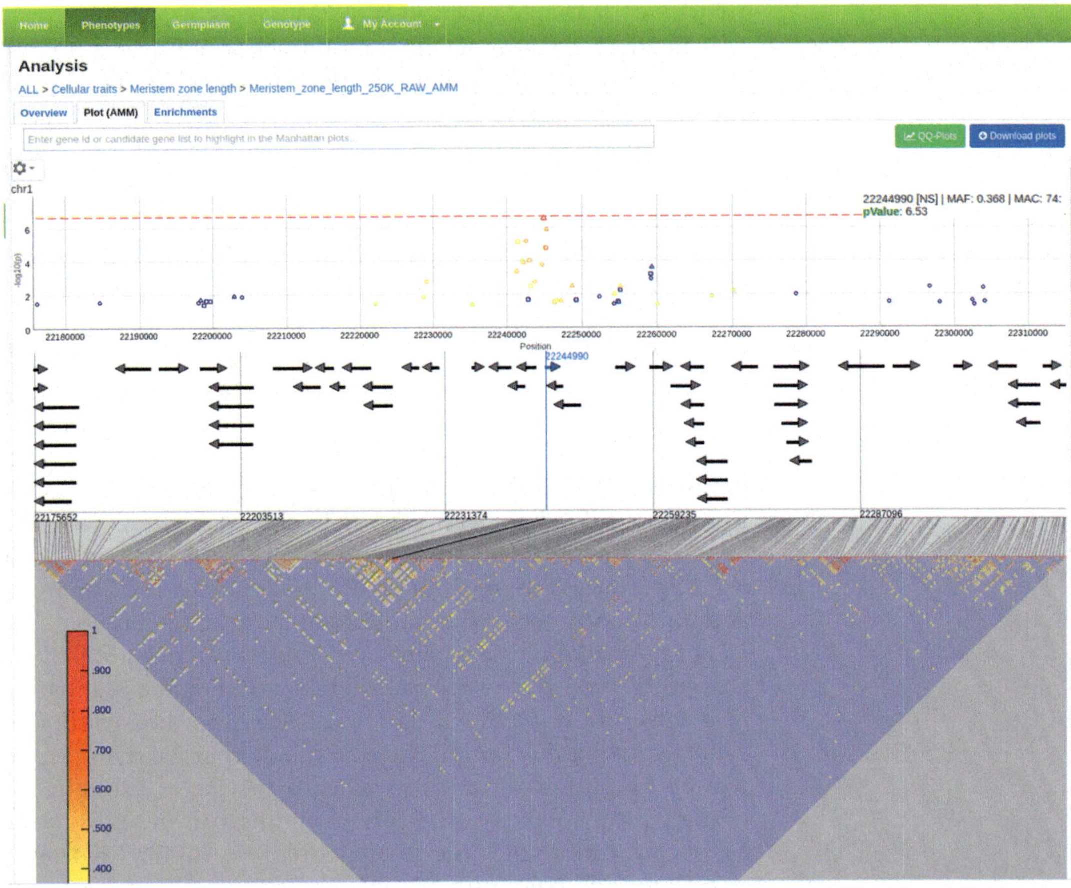

Fig. 8 Linkage disequilibrium (LD) plots. The user can choose a SNP as a focal point and let GWA-Portal calculate linkage disequilibrium structure in a certain region around the focal point. This will display an additional LD triangle plot underneath the gene annotation track. By moving the mouse over SNPs in the Manhattan plot, neighboring markers will be color coded based on their r^2 value ranging from blue (low r^2) to red (high r^2). Additionally, the corresponding rows in the LD triangle plot are highlighted. Alternatively, moving the mouse over a certain datapoint in the LD triangle plot will highlight the corresponding pair of SNPs in the Manhattan plot

phenotyped sample are displayed. We can choose to color code the circles based on the allele they carry (blue, A, and green, G). Switching the phenotype-explorer display from a scatterplot to a bar plot allows us to display the geographic distribution of the phenotypic values or alleles (Fig. 9). We can check different patterns by choosing different configurations for the x-axis and y-axis (i.e., latitude, longitude, and phenotype).

3. Finally we can check what kind of effect the SNP has on our phenotype by clicking on the bar chart icon (Fig. 10). This will display a strip chart and a candlestick chart that shows the effect of the reference and alternative allele of the chosen SNP on the trait values. By default, it will show the allelic effect for all

Fig. 9 Association detail view. The user can display detailed information for a specific association by choosing the corresponding option from the SNP context menu. The SNP detail view will display general information—e.g., effect, function, codon change, and the minor allele count in the phenotyped population. In the main panel, the user can choose from three different visualizations using an icon action bar. Here a phenotype-explorer is displayed. Each bar represents a phenotyped sample, and the color of the bar represents the allele (reference or alternative) that the sample carries. The x- and y-axis can be ordered by latitude, longitude, or phenotype

phenotyped samples. As GWA-Portal provides the information where the samples came from, we can easily display the effect for a specific subset. For example, by choosing the United Kingdom from the drop-down list, we can display the effect for samples that come from the United Kingdom.

4 Notes

1. GWA-Portal supports two phenotype formats: ISA-Table [14] and a simple CSV file format. Both formats support uploading multiple phenotypes for multiple replicates and maintaining the correct relationship between replicate and phenotype value.

Fig. 10 Allele effect view. The phenotypic effect of the two alleles is displayed using a candlestick chart and a strip chart. By default, it is displayed for all the samples. Users can choose to restrict it to subsets from certain countries using a drop-down list

2. Studies represent an experiment, which provides a logical grouping for one or more phenotypes. The Atwell study (https://gwas.gmi.oeaw.ac.at/#/study/1/overview) is an example of that.
3. For uploading phenotypes using the CSV file format, the first column must contain the accession id of your phenotyped sample. The corresponding values for one or more phenotypes must be stored in columns 2 to N. Missing values are represented as empty cells. For replicates, just repeat the accession id and the corresponding phenotypic values multiple times.
4. The "imputed full sequence" genotype dataset is a combined dataset of the 250K SNP dataset and the 1001 genomes dataset using imputation (*see* Table 1).
5. Some GWAS models expect the samples to come from a population that has a normal distribution. Shapiro-Wilk is a

statistical test used to assess whether your data are normally distributed. Then if the *P* value (Shapiro-Wilk *P* value) associated with the test statistics is lower than αα, where αα is a statistical significance level, you can reject the null hypothesis and conclude that your data are not normally distributed. Conversely, when the *P* value associated with the test statistics is greater than αα, you fail to reject the null hypothesis and conclude that your data are normally distributed.

6. Confounding due to population structure can represent a problem for GWAS if the phenotype in question is correlated with the structure. It can lead to inflation of *P* values and consequently to false positives [15–17]. Linear mixed models [11–13] such as AMM can help to account for population structure.
7. Usually the user wants the toggle switch "Create HPC job" to be set to ON. This will make sure that the GWAS analysis will be automatically run on our HPC as soon as the GWAS wizard is finished (it is also possible to run the analysis at a later point of time, if the user forgets to set it to ON). Sometimes the user might want to use a different GWAS method outside of GWA-Portal. For those cases, the user can decide to keep the switch to OFF, still create the GWAS analysis, and later upload the *P* value file that was generated outside of GWA-Portal. Setting the "Run enrichment analysis" to ON will make sure that enrichment analyses against all available candidate gene list is automatically conducted after the GWAS analysis.
8. Users are free to choose any naming schema for the GWAS analysis. The authors' recommendation is to use a naming schema which contains all the relevant information about the phenotype, the chosen genotype, the transformation, and the GWAS method (i.e., "PHENOTYPE_GENOTYPE_TRANSFORMATION_METHOD," "Meristem_Zone_Length_250-K_NO_AMM"). This naming scheme helps with discoverability when doing a full text search.
9. Rare SNPs in GWAS are usually not very meaningful. As a result a default MAC filter of 15 is defined. This depends of course on the number of samples. Thus, the user can set the filtering on a case by case basis.
10. GWA-Portal provides a list of published candidate gene lists [1] that can be used for enrichment analysis and filtering or highlighting of genes. Additionally users can create their own candidate gene lists (this function can be found under the menu item "Meta-analysis"), share them with collaborators, and use them the same way as the published ones.

Acknowledgment

GWA-Portal was developed in the course of the transPLANT project, which was funded by the European Commission within its 7th Framework Programme, under the thematic area "Infrastructures," contract number 283496.

References

1. Atwell S, Huang YS, Vilhjalmsson BJ, Willems G, Horton M, Li Y, Meng D, Platt A, Tarone AM, TT H, Jiang R, Muliyati NW, Zhang X, Amer MA, Baxter I, Brachi B, Chory J, Dean C, Debieu M, de Meaux J, Ecker JR, Faure N, Kniskern JM, Jones JD, Michael T, Nemri A, Roux F, Salt DE, Tang C, Todesco M, Traw MB, Weigel D, Marjoram P, Borevitz JO, Bergelson J, Nordborg M (2010) Genome-wide association study of 107 phenotypes in Arabidopsis Thaliana inbred lines. Nature 465(7298):627–631. https://doi.org/10.1038/nature08800
2. Brachi B, Morris GP, Borevitz JO (2011) Genome-wide association studies in plants: the missing heritability is in the field. Genome Biol 12(10):232. https://doi.org/10.1186/gb-2011-12-10-232
3. Todesco M, Balasubramanian S, TT H, Traw MB, Horton M, Epple P, Kuhns C, Sureshkumar S, Schwartz C, Lanz C, Laitinen RA, Huang Y, Chory J, Lipka V, Borevitz JO, Dangl JL, Bergelson J, Nordborg M, Weigel D (2010) Natural allelic variation underlying a major fitness trade-off in Arabidopsis Thaliana. Nature 465(7298):632–636. https://doi.org/10.1038/nature09083
4. Genomes Consortium. Electronic address mngoaa, Genomes C (2016) 1,135 genomes reveal the global pattern of polymorphism in Arabidopsis Thaliana. Cell 166(2):481–491. https://doi.org/10.1016/j.cell.2016.05.063
5. Seren U, Vilhjalmsson BJ, Horton MW, Meng D, Forai P, Huang YS, Long Q, Segura V, Nordborg M (2012) GWAPP: a web application for genome-wide association mapping in Arabidopsis. Plant Cell 24(12):4793–4805. https://doi.org/10.1105/tpc.112.108068
6. Meijon M, Satbhai SB, Tsuchimatsu T, Busch W (2014) Genome-wide association study using cellular traits identifies a new regulator of root development in Arabidopsis. Nat Genet 46(1):77–81. https://doi.org/10.1038/ng.2824
7. Horton MW, Hancock AM, Huang YS, Toomajian C, Atwell S, Auton A, Muliyati NW, Platt A, Sperone FG, Vilhjalmsson BJ, Nordborg M, Borevitz JO, Bergelson J (2012) Genome-wide patterns of genetic variation in worldwide Arabidopsis Thaliana accessions from the RegMap panel. Nat Genet 44(2):212–216. https://doi.org/10.1038/ng.1042
8. Long Q, Rabanal FA, Meng D, Huber CD, Farlow A, Platzer A, Zhang Q, Vilhjalmsson BJ, Korte A, Nizhynska V, Voronin V, Korte P, Sedman L, Mandakova T, Lysak MA, Seren U, Hellmann I, Nordborg M (2013) Massive genomic variation and strong selection in Arabidopsis Thaliana lines from Sweden. Nat Genet 45(8):884–890. https://doi.org/10.1038/ng.2678
9. Box GEP, Cox DR (1964) An analysis of transformations. J R Stat Soc Series 26(2):211–252
10. Wilcoxon F (1946) Individual comparisons of grouped data by ranking methods. J Econ Entomol 39:269
11. Kang HM, Sul JH, Service SK, Zaitlen NA, Kong SY, Freimer NB, Sabatti C, Eskin E (2010) Variance component model to account for sample structure in genome-wide association studies. Nat Genet 42(4):348–354. https://doi.org/10.1038/ng.548
12. Zhang Z, Ersoz E, Lai CQ, Todhunter RJ, Tiwari HK, Gore MA, Bradbury PJ, Yu J, Arnett DK, Ordovas JM, Buckler ES (2010) Mixed linear model approach adapted for genome-wide association studies. Nat Genet 42(4):355–360. https://doi.org/10.1038/ng.546
13. Kang HM, Zaitlen NA, Wade CM, Kirby A, Heckerman D, Daly MJ, Eskin E (2008) Efficient control of population structure in model organism association mapping. Genetics 178(3):1709–1723. https://doi.org/10.1534/genetics.107.080101
14. Cwiek-Kupczynska H, Altmann T, Arend D, Arnaud E, Chen D, Cornut G, Fiorani F, Frohmberg W, Junker A, Klukas C, Lange M, Mazurek C, Nafissi A, Neveu P, van Oeveren J, Pommier C, Poorter H, Rocca-Serra P, Sansone SA, Scholz U, van Schriek M, Seren U, Usadel B, Weise S, Kersey P, Krajewski P

(2016) Measures for interoperability of phenotypic data: minimum information requirements and formatting. Plant Methods 12:44. https://doi.org/10.1186/s13007-016-0144-4

15. Korte A, Farlow A (2013) The advantages and limitations of trait analysis with GWAS: a review. Plant Methods 9:29. https://doi.org/10.1186/1746-4811-9-29

16. Vilhjalmsson BJ, Nordborg M (2013) The nature of confounding in genome-wide association studies. Nat Rev Genet 14(1):1–2. https://doi.org/10.1038/nrg3382

17. Platt A, Vilhjalmsson BJ, Nordborg M (2010) Conditions under which genome-wide association studies will be positively misleading. Genetics 186(3):1045–1052. https://doi.org/10.1534/genetics.110.121665

Index

A

B

C

D

Daniela Ristova and Elke Barbez (eds.), *Root Development: Methods and Protocols*, Methods in Molecular Biology, vol. 1761, https://doi.org/10.1007/978-1-4939-7747-5, © Springer Science+Business Media, LLC 2018

E

F

G

H

I

K

L

M

S

T

W

Z

Printed by Printforce, the Netherlands